T0335626

THE PARTITION METHOD FOR A POWER SERIES EXPANSION

THE PARTITION METHOD FOR A POWER SERIES EXPANSION

Theory and Applications

Victor Kowalenko

Department of Mathematics and Statistics
The University of Melbourne

AMSTERDAM • BOSTON • HEIDELBERG • LONDON
NEW YORK • OXFORD • PARIS • SAN DIEGO
SAN FRANCISCO • SINGAPORE • SYDNEY • TOKYO

Academic Press is an imprint of Elsevier

ACADEMIC
PRESS

Academic Press is an imprint of Elsevier
125 London Wall, London EC2Y 5AS, United Kingdom
525 B Street, Suite 1800, San Diego, CA 92101-4495, United States
50 Hampshire Street, 5th Floor, Cambridge, MA 02139, United States
The Boulevard, Langford Lane, Kidlington, Oxford OX5 1GB, United Kingdom

Notices

Knowledge and best practice in this field are constantly changing. As new research and experience
broaden our understanding, changes in research methods, professional practices, or medical
treatment may become necessary.

Practitioners and researchers must always rely on their own experience and knowledge in evaluating
and using any information, methods, compounds, or experiments described herein. In using such
information or methods they should be mindful of their own safety and the safety of others,
including parties for whom they have a professional responsibility.

To the fullest extent of the law, neither the Publisher nor the authors, contributors, or editors,
assume any liability for any injury and/or damage to persons or property as a matter of products
liability, negligence or otherwise, or from any use or operation of any methods, products,
instructions, or ideas contained in the material herein.

Library of Congress Cataloging-in-Publication Data
A catalog record for this book is available from the Library of Congress

British Library Cataloguing-in-Publication Data
A catalogue record for this book is available from the British Library

ISBN: 978-0-12-804466-7

For information on all Academic Press publications
visit our website at https://www.elsevier.com

  Working together
to grow libraries in
developing countries

www.elsevier.com • www.bookaid.org

Publisher: Nikki Levy
Acquisition Editor: Glyn Jones
Editorial Project Manager: Charlotte Kent
Production Project Manager: Omer Mukthar
Designer: Greg Harris

Typeset by VTeX

CONTENTS

PREFACE

The partition method for a power series expansion was first used by the author in the early 1990's in the solution of an asymptotic problem in the response theory of the magnetized charged Bose gas, one of the two fundamental systems in condensed matter physics. Soon after, it was realized that the method could be applied to numerous problems, where standard methods, in particular Taylor/Maclaurin series, failed to produce a power series representation for a mathematical function. Power series expansions are important for understanding the behaviour and isolating properties of the original function. They also provide a means of obtaining extremely accurate values of the functions they represent. Even in instances where standard techniques can be applied, the partition method for a power series expansion yields the same power series except from a different perspective. This not only can be illuminating, but it can also result in new mathematical properties for the original function. Therefore, since the first application, the method has been applied to various intractable problems such as cosecant, secant and the logarithm of $(1 + z)$, all taken to arbitrary powers and all three elliptic integrals. All these applications together with the interesting mathematics discovered in the process have appeared in separate publications. Thus, the first aim of this book has been to combine these separate works and extend them in one reference.

Another aim of this book is to provide the general theoretical framework behind the method since the previous applications only sketched out the method as it applied to the specific problem under investigation. Consequently, this book remedies the present situation by presenting a very general theory that is capable of going beyond the previously mentioned applications. In particular, the later chapters introduce further applications and developments resulting from the method that have not been previously published.

In the partition method for a power series expansion the coefficient of each power in the resulting series is determined by summing all the contributions due to all the integer partitions summing to the power under consideration. In calculating the contribution from each partition each part or element is assigned a specific value based on a partial expansion of the original function. In addition, one needs to know the number of occurrences or the multiplicity of each part in the partition. Hence the method requires the composition of each partition in order to calculate the coefficients in the resulting power series.

Whilst it is possible to write down or calculate by hand the first few orders of the resulting power series, it becomes impractical to do so for higher orders typically greater than the tenth order, where there are already 42 partitions required to determine the coefficient. At this stage it is better to develop a computational approach for determining the coefficients. Therefore, this book also presents the programming methodology behind the method. The methodology begins with the problem of generating partitions in the appropriate form so that a code based on the general theory can be created. By appropriate form it is meant that the partitions are expressed in the multiplicity representation rather than the standard lexicographic form, which most existing codes adopt. Consequently, a code based on representing the partitions in the form of a special non-binary tree known as a partition tree is presented. The bivariate recursive central partition or BRCP algorithm has the major advantage that it can be adapted with minor modification to handle a host of problems in the theory of partitions, which often require completely different programs.

With the aid of the BRCP algorithm, the general theory is developed into a programming methodology whereby a C/C++ code expresses the coefficients of the resulting power series in a general symbolic form. Then the output from the code can be transported to different computing platforms where mathematical packages such as Mathematica can be used to calculate the coefficients for specific problems either in rational or algebraic form, neither of which can be handled effectively in standard C/C++ with its floating point arithmetic. By importing the general symbolic forms for each power into Mathematica, one can exploit the package's integer arithmetic routines to express the coefficients in integer form, thereby avoiding (1) round-off errors due to scientific notation, and (2) under(over)flow due to under(over)sized numbers. On the other hand, one can use the package's symbolic routines to express the coefficients in algebraic form, e.g., as polynomials, for problems involving other variables and parameters unrelated to the variable in the power series.

Although the method has been applied in a number of interesting and challenging problems in this book, a sure sign of its power, it is still in its infancy with many more exciting developments waiting to be uncovered. For example, the material in the final chapters represents a cursory introduction into modifying the method to infinite products, which abound in applied mathematics and theoretical physics, particularly statistical mechanics. Therefore, it is hoped that this book will not only interest, but will also encourage the reader to apply the method in new and unforeseen applications.

I am indebted to Dr Glyn Jones and his colleagues at Elsevier for encouraging and supporting the production of this book, especially as it presents a totally novel approach from enumerative combinatorics to the subject of power series expansions. I also wish to thank Professor A.J. Guttmann for extending the facilities of the ARC Centre of Excellence for Mathematics and Statistics of Complex Systems, where this work was carried out. Finally, I thank Dr. D.X. Balaic, Australian X-Ray Capillary Optics Pty Ltd, for producing the first programs incorporating the BRCP algorithm. Although not used here, they were helpful in developing the programming methodology for the method and enabling the codes to be developed to tackle various problems in the theory of partitions mentioned in this book. Many of these programs, which represent a by-product of the BRCP algorithm, are listed in Chapter 7 and the second appendix for the reader, who may not necessarily be interested in power series expansions, but may have an interest in the properties and classes of partitions, a discipline in its own right.

Victor Kowalenko
Melbourne, Australia
August, 2016

CHAPTER 1

Introduction

The partition method for a power series expansion was first introduced in the derivation of an asymptotic expansion for the particular Kummer or confluent hypergeometric function that arises in the response theory of the charged Bose gas [1]. In mathematical terms, this means that it was employed during the derivation of a large $|\alpha|$-expansion for the confluent hypergeometric function, $_1F_1(\alpha, \alpha + 1; z)$, which was unusual because it was the parameters and not the variable that affected the final asymptotic result. Consequently, it was possible to determine for the first time ever the physical properties of this fundamental quantum system of condensed matter physics in small or weak magnetic fields [2]. Although the charged Bose gas was considered initially, it was soon realized that the new expansion could be applied to the magnetized degenerate electron gas because the Bose Kummer function is a variant of the incomplete gamma function, upon which the response theory of this alternative, and far more important, system in condensed matter physics is based. Moreover, in an interesting twist the weak magnetic field expansion for the Bose Kummer function was later used in Ref. [1] to derive a large or strong magnetic field expansion, thereby allowing the strong magnetic field behaviour of the system to be presented in Ref. [2].

Because the partition method for a power series expansion had been applied effectively to an important special function of mathematical physics, a natural question was whether it could be applied to other special functions, particularly those where the standard techniques, e.g. Taylor/Maclaurin series, for obtaining power series expansions broke down. In response to this question, over the next decade or so, the partition method for a power series expansion was developed further so that it could be applied to a host of mathematical functions culminating in a series of papers, Refs. [3–5]. Included in these studies were not only basic functions such as $1/\ln(1+z)$, $\sec z$, $z \csc z$ and arbitrary powers of them, but also more esoteric ones such as the Legendre-Jacobi elliptic integrals, $F(\psi, k)$, $E(\psi, k)$ and $\Pi(\psi, n, k)$, and generalizations of them. Moreover, as a result of this activity another application was found where the method was used in the evaluation of trigonometric sums with inverse powers of cosine or sine [6,7]. These,

The Partition Method for a Power Series Expansion.
DOI: http://dx.doi.org/10.1016/B978-0-12-804466-7.00001-2

however, will not be discussed here, even though they have produced some fascinating number-theoretical results.

Before proceeding any further, a good question to ask at this stage is: What do we mean by a power series expansion? In this book we shall use the following definition from Ref. [8], which states that

A power series in a variable z is an infinite sum of the form $\sum_{i=0}^{\infty} a_i\, z^i$, where a_i are integers, real numbers, complex numbers, or any other quantities of a given type.

This broad definition becomes confusing when we ask ourselves what a generating function is. According to Ref. [9], a generating function $f(z)$ is defined as a formal power series of the form $f(z) = \sum_{n=0}^{\infty} a_n\, z^n$, where the coefficients a_n give a sequence $\{a_0, a_1, a_3, \dots\}$. Obviously, from these definitions there is much overlap between the definitions and hence confusion arises. For example, one usually refers to $1 + z + z^2/2! + z^3/3! + \dots$ as the power series expansion for $\exp(z)$, not as its generating function, but the coefficients form a sequence given by $a_n = 1/n!$. Moreover, it is the aim of this book to derive expressions where the coefficients are represented by a sequence, no matter how complicated it may be. We shall find that the derivatives in Taylor/Maclaurin series can even be expressed as sums over partitions. Nevertheless, we shall refer to the results as power series expansions, but in cases, where the generating function terminology is more familiar, we shall apply the latter expression. This will occur particularly in the later chapters where the sequences for the coefficients involve special functions such as the partition–number function denoted here by $p(k)$. So as a rule, we shall regard generating functions as power series expansions whose coefficients involve special functions. It should also be mentioned here that in the literature the partition–number function is often called the partition function, but we shall refrain from doing so here because it means something quite different to those working in statistical mechanics.

As stated in the introduction to Ref. [5], the partition method for a power series expansion is not only capable of yielding power series where standard techniques, e.g., Taylor/Maclaurin series or the Hardy-Littlewood circle method [10], break down, but also in those cases where a standard technique can be applied. In these cases the power series expansions are ultimately identical to one another, but give rise to a totally different perspective. As we shall see, the cross-fertilization of both approaches is frequently responsible for the derivation of new mathematical results and properties. A particularly fascinating property of the partition method for a power series expansion is that the discrete mathematics of integer partitions results in power series expansions for continuous functions. Thus, discrete mathematics turns out to be the base of continuous mathematics.

Broadly speaking, the partition method for a power series expansion is composed of two major steps. For first step one requires the composition of each integer partition summing to the order k of each term in the resulting power series expansion. By composition it is meant that not only each part or element of a partition is known, but also their number of occurrences or multiplicities, commonly denoted by λ_i for each part i in a partition. Because the first step involves the generation of (integer) partitions, it can also be of importance to the general theory of partitions, especially when it results in alternative algorithms to the standard methods of generating partitions as discussed in Refs. [11–13]. Despite this, however, the more important step is the second one since it is responsible for producing the power series expansions for mathematical functions. In this step each partition summing to k provides a specific contribution to the coefficient of the k-th power in the power series expansion. As will be seen later in this chapter, this contribution is obtained by: (1) assigning a value to each part in the partition, and (2) multiplying these values by a multinomial factor composed of the factorial of the total number of parts or the length of the partition, viz. $N_k!$, divided by each $\lambda_i!$. Occasionally, there is a third step, but this will be addressed at the appropriate place.

For the lowest order terms in the resulting power series expansion it is possible to calculate the coefficients by hand, but after these have been determined, it becomes increasingly onerous to calculate the coefficients. This is due to the exponential increase in the number of partitions summing to higher powers, which is known as combinatorial explosion. For example, the number of partitions that sum to 10 is 42. Therefore, one needs to sum 42 distinct contributions in order to calculate the tenth order term in the power series expansion. At this stage it is far more expedient to calculate the coefficients by computer programs, which in turn means developing a programming methodology for the method. Such a methodology requires not only being able to generate the partitions at each order, but also calculating their contributions in symbolic form stemming from a general theory for the method. However, before a general theory can be created, first we need to examine some simple examples in this chapter to understand how the method can be applied to problems. Then in the following chapter we shall study more advanced examples, which will enable the theory to become as general as possible.

1.1 COSECANT EXPANSION

The first application of the partition method for a power series expansion after the derivation of the asymptotic power series expansion for the Bose Kummer function was the cosecant power series expansion. According to No. 1.411(11) of Gradshteyn and Ryzhik [14], the power series expansion for cosecant is given by

$$\csc x = \frac{1}{x} + \sum_{k=1}^{\infty} \frac{2(2^{2k-1} - 1)}{(2k)!} \, |B_{2k}| \, x^{2k-1} \, , \tag{1.1}$$

for $x^2 < \pi^2$. In this result, which has been taken originally from Ref. [15], B_k represent the Bernoulli numbers. These famous numbers are defined by the following generating function:

$$\sum_{k=0}^{\infty} B_k \frac{t^k}{k!} \equiv \frac{t}{e^t - 1} \, . \tag{1.2}$$

The condition below (1.1) implies that x must be real, but as we shall see, x can be complex. In addition, note the appearance of the equivalence symbol instead of an equals sign in (1.2). This is because the lhs can become divergent, i.e., it equals infinity for $|t| \geq 2\pi$, whereas the rhs is always finite. Therefore, the statement is now an equivalence statement or equivalence, for short. The equivalence symbol means that the rhs represents the (finite) regularized value of the series on the lhs when the infinity is removed in the remainder for $|t| > 2\pi$. For those values of t, where the lhs is convergent, viz. $|t| < 2\pi$, one can replace the equivalence symbol by an equals sign. Since the process of regularization will be applied throughout this book, it is suggested that the reader, who is unaware of this concept, consult Appendix A for a more detailed exposition.

Now we are in a position to derive the power series expansion for csc x via the partition method for a power series expansion with the following theorem.

Theorem 1.1. There exists a power series expansion for cosecant, which can be written as

$$\csc x = \frac{1}{\sin x} \equiv \sum_{k=0}^{\infty} c_k \, x^{2k-1} \, , \tag{1.3}$$

where the cosecant numbers, c_k, are special rational numbers with $c_0 = 1$, $c_1 = 1/6$, $c_2 = 7/360$, etc. More generally, they are given by

$$c_k = (-1)^k \sum_{\substack{\lambda_1,\lambda_2,\lambda_3,\ldots,\lambda_k=0 \\ \sum_{i=1}^{k} i\lambda_i = k}}^{k,\lfloor k/2 \rfloor, \lfloor k/3 \rfloor, \ldots, 1} (-1)^{N_k} N_k! \prod_{i=1}^{k} \left(\frac{1}{(2i+1)!} \right)^{\lambda_i} \frac{1}{\lambda_i!}, \qquad (1.4)$$

where λ_i represents the number of occurrences or multiplicity of the part or element i in each partition that sums to k and N_k represents the length or the total number of parts in a partition summing to k, i.e., $N_k = \sum_{i=1}^{k} \lambda_i$. Moreover, the expansion represents a small $|x|$-expansion with a finite disk of absolute convergence, whose radius is given at least by $|x^2| < 6$.

Remark. As in the case of the generating function for the Bernoulli numbers, viz. (1.2), the equivalence symbol has been introduced to indicate that the power series on the rhs of (1.3) can become divergent when $|x^2| \geq 6$. Later, we shall see that the power series expansion is absolutely convergent for $|x| < \pi$, which implies that it only needs to be regularized whenever $|x| \geq \pi$, not less than π as stated in the remark to Theorem 1 of Ref. [5].

Proof. By introducing the convergent power series expansion for sine, e.g., see Ref. [16], we can write $\csc x$ as

$$\csc x = \frac{1}{x(1 - x^2/3! + x^4/5! - x^6/7! + \cdots)}. \qquad (1.5)$$

As explained in Appendix A, the geometric series, viz. $\sum_{k=0}^{\infty} z^k$, yields a limit value of $1/(1-z)$ for $\Re z < 1$, which means in turn that the series does not require regularization for these values of z. According to the literature, e.g. Ref. [17], the geometric series is only equal to $1/(1-z)$ when $|z| < 1$. This, however, represents the condition for absolute convergence. In actual fact, it emerges from the more detailed analysis in Refs. [18] and [19] and summarized in Appendix A that outside the unit circle of absolute convergence, but to the left of the barrier at $\Re z < 1$, the series is conditionally convergent. On the other hand, for $\Re z > 1$, the series is divergent. Then it must be regularized by removing the infinity in the remainder to yield a finite value, much like the Hadamard finite part of a divergent integral [20–22]. Once the series is regularized, one is left with the same limit value of $1/(1-z)$ as in the domain of convergence. Therefore, for all values of z, the regularized value of the geometric series can be expressed as

$$\sum_{k=0}^{\infty} z^k \equiv \frac{1}{1-z}, \qquad \forall\, z. \qquad (1.6)$$

Whenever we are only concerned with the values of z for which the geometric series is convergent, which will arise in various problems or theorems in this book, we shall replace the equivalence symbol by an equals sign. Now if we introduce (1.6) into (1.5), then we obtain

$$\csc x \equiv \frac{1}{x} \sum_{k=0}^{\infty} \left(\frac{x^2}{3!} - \frac{x^4}{5!} + \frac{x^6}{7!} - \frac{x^8}{9!} + \frac{x^{10}}{11!} - \cdots \right)^k . \tag{1.7}$$

Note that the introduction of an equivalence statement into (1.5) has resulted in another equivalence (statement). Since the above equivalence is a function of even powers, it can be expressed as

$$\csc x \equiv \frac{1}{x} \sum_{k=0}^{\infty} c_k \, x^{2k} . \tag{1.8}$$

The series on the right-hand side (rhs) of (1.8) is absolutely convergent provided $|x^2/3! - x^4/5! + x^6/7! - \cdots| < 1$. From the inequality, No. 3.7.29 in Ref. [23], this means that

$$\left| \left| \frac{x^2}{3!} \right| - \left| \frac{x^4}{5!} - \frac{x^6}{7!} + \frac{x^8}{9!} - \cdots \right| \right| \le \left| \frac{x^2}{3!} - \frac{x^4}{5!} + \frac{x^6}{7!} - \frac{x^8}{9!} + \cdots \right| < 1 . \tag{1.9}$$

Therefore, from the left-hand and right-hand sides or members of (1.9), we find that $|x^2/3!| < 1 + |x^4/5! - x^6/7! + \cdots|$, which means, in turn, that the power series expansion for $\csc x$ is at least absolutely convergent for $|x^2/3!| < 1$ or $|x^2| < 6$ as stated in the theorem. For these values of x, the equivalence symbol can be replaced by an equals sign. Soon, we shall see that this means that the innermost singularity of the expansion in (1.8) occurs at $x^2 = 6$ by the above method, although $\csc x$ does not possess singularities at these values of x.

Let us turn our attention to the cosecant numbers or the c_k in (1.3). From (1.7), it is obvious that $c_0 = 1$, while c_1 is easily evaluated from the lowest order term in the $k = 1$ term, which yields a value of $1/3!$. The coefficient c_2 consists of two contributions, one from the $k = 1$ term and the other from the $k = 2$ term in (1.7). The $k = 1$ contribution to c_2 is represented by the coefficient of x^4, viz. $-1/5!$, while the $k = 2$ contribution comes from squaring $x^2/3!$, which in turn yields $1/3!^2$. Hence we find that $c_2 = 1/3!^2 - 1/5! = 7/360$.

So far, everything has been relatively simple except perhaps for the concept of regularization. Nevertheless, we can see that the evaluation of the higher orders of the cosecant numbers is becoming more complex. Consequently, two algorithms for evaluating them will be presented. Although

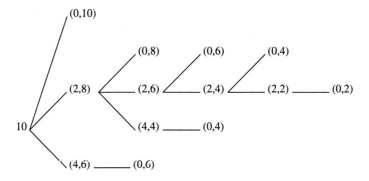

Figure 1.1 The final tree diagram for the cosecant number, c_5.

the second of these methods is better from an optimization point of view, the first needs to be described as the second evolves from it.

For the first algorithm we still require the partition method for a power series expansion, which as stated earlier, was first introduced by the author in Refs. [1] and [2]. This method begins with constructing tree diagrams for each c_k. One begins by drawing branch lines to all pairs of numbers that can be summed to the order, $2k$, where the first number in the tuple is an even integer less than or equal to k as depicted in Fig. 1.1. For example, in the figure, which represents the final tree diagram for c_5, there are branch lines to $(0, 10)$, $(2, 8)$, and $(4, 6)$ emanating from the seed number 10. Whenever a zero appears in the first element or member of a tuple, the path stops, as can be seen by $(0, 10)$. We shall refer to those tuples with zero as their first member as terminating tuples because there are no other branches emanating from them. For the remaining pairs or tuples, one draws branch lines to all pairs with even integers that sum to the second number under the same prescription that the first member of each new tuple is less than or equal to half the second member of the previous tuple and is, of course, even. Therefore, for $(2, 8)$ in the figure there are paths branching out to $(0, 8)$, $(2, 6)$ and $(4, 4)$. The path to $(4, 4)$, however, does not possess a branch to $(2, 2)$ since Fig. 1.1 represents the final version of the tree for c_5. Its omission will be explained shortly. This recursive approach is applied repeatedly until all paths emanating from the seed number terminate with a tuple containing a zero. Hence the number of tuples with a zero in them represents the total number of paths, which in turn, as we shall see, gives the total number of partitions summing to 10 with only even parts.

It is evident that all the first members of the tuples plus the second member in the final or terminating tuple for each path constitute a partition summing to $2k$, but where all the parts are only even numbers. For example, in Fig. 1.1 the path consisting of $(2, 8)$, $(2, 6)$, $(2, 4)$ and $(0, 4)$ represents the partition, $\{2, 2, 2, 4\}$, while the path consisting of $(4, 6)$ and $(0, 6)$ represents the partition, $\{4, 6\}$. Therefore, the tree diagram is composed of all the even partitions that sum to $2k$. In the case of Fig. 1.1 this is 7, which is identical to the same number of terminating tuples. Unfortunately, the above prescription is not all that is required to produce the final version of the tree diagram for each cosecant number. Duplicated paths involving permutations of the same partition can arise and must be removed. That is, each partition must appear only once in the diagram. For example, if the prescription of the previous paragraph is applied to $(4, 4)$ in the figure, then there would also be the branches $(2, 2)$ and $(0, 2)$ emanating from $(4, 4)$ yielding the path of $(2, 8)$, $(4, 4)$, $(2, 2)$ and $(0, 2)$. This is merely a permutation of the partition $\{2, 2, 2, 4\}$, which is represented by the path $(2, 8)$, $(2, 6)$, $(2, 4)$, and $(0, 4)$ in Fig. 1.1. Therefore, a search algorithm seems to be required to ensure that each integer partition only appears once in the diagram. In other words, we only require distinctly different partitions irrespective of the order in which the parts appear. This means that we are not interested in compositions [24]. When the search process is carried out, one ends up with the final diagram as depicted in Fig. 1.1. It should be noted that as we become more familiar with the partition method for a power series expansion, we shall be able to dispense with these tree diagrams, which we shall call partition trees.

Once all the partitions have been determined, we can turn to the next step in the evaluation of the cosecant numbers. This involves the evaluation of the contribution made by each partition, which is dependent upon the number of parts or elements in each partition. If l_i represents the distinct parts of a partition appearing in the partition tree for c_k, then we let $n_1, \ldots, n_l$ represent the number of occurrences or frequency each part appears in a partition. For example, in the partition $\{2, 2, 2, 4\}$ there are three twos and one four. Hence $l_1 = 2$ and $l_2 = 4$. On the other hand, since l_1 refers to the twos in the partition, $n_1 = 3$, whereas $n_2 = 1$ since it represents the number of fours. Furthermore, if we let N and L represent respectively the total number of parts and the number of distinct parts in a partition, then we have $\sum_{i=1}^{L} n_i = N$ and $\sum_{i=1}^{L} n_i l_i = 2k$. Moreover, if $l_i \bmod 4 = 0$, then each l_i will contribute a value of $-1/(l_i + 1)!$. On the other hand, when $l_i \bmod 4 \neq 0$, each l_i will contribute a value of $1/(l_i + 1)!$. Finally, the total contribution for each partition is multiplied by the multinomial factor of $N!/n_1! \cdots n_L!$. Thus, the contribution to the c_k by an arbitrary partition is given by

$$C\Big([l_1, n_1], [l_2, n_2], \ldots, [l_L, n_L]\Big) = (-1)^{n_1(l_1/2+1)} \cdots (-1)^{n_L(l_L/2+1)} \frac{N!}{n_1! \cdots n_L!}$$

$$\times \frac{1}{((l_1+1)!)^{n_1} \cdots ((l_L+1)!)^{n_L}} . \quad (1.10)$$

Alternatively, the above result can be expressed in a more succinct form as

$$C\Big([l_1, n_1], [l_2, n_2], \ldots, [l_L, n_L]\Big) = N! \prod_{i=1}^{L} \frac{(-1)^{n_i(l_i/2+1)}}{n_i! ((l_i+1)!)^{n_i}} . \quad (1.11)$$

For the partition whose path is $(2, 8)$, $(2, 6)$, $(2, 4)$ and $(0, 4)$ in Fig. 1.1, the contribution to c_5 is $-4!/(3! \cdot 1!) \cdot (3!)^3 \cdot (5)!$ according to (1.10). Therefore, by summing over all seven partitions appearing in the partition tree via the same coding approach, we obtain the final value for c_5.

From Fig. 1.1 we see that the longest path or the partition with the most parts will be the partition composed only of twos. For each c_k this path will possess k twos. According to the coding approach described above, this partition contributes a value of $(1/3!)^k$ to each c_k. The second longest path will contain one four and all twos. Since this path possesses $k-2$ twos and one 4, it contributes a value of $-1/(3!)^{k-2} \cdot 5!$ multiplied by the multinomial factor of $k-1$ obtained from $(k-1)!/(k-2)!$. From then on, the path lengths become shorter until there is only one branch corresponding to the single part partition $\{2k\}$, which contributes a value of $(-1)^{k+1}/(2k+1)!$ to each c_k.

From the preceding discussion we see that the number of contributions to each cosecant number is finite and that each contribution does not involve any irrational numbers. Therefore, the cosecant numbers are rational. Values for cosecant numbers up to $k = 12$ are displayed in the second column of Table 1.1. Except for the highest order values, viz. $k = 10$ to $k = 12$, these values have been calculated by determining the partitions via partition trees. For $k > 9$, this method becomes awkward. At this stage a computer program is required. The first step in such a program involves determining or processing all the unique partitions for each k, while the second step involves evaluating the contributions due to each partition and summing them to yield the coefficients of the resulting power expansions. It is obvious from (1.10) that the second step cannot be made more efficient. It is, however, possible to improve the first step since much time can be expended in searching for duplicated or redundant partitions. Therefore, the second algorithm aims to dispense with the requirement of having to

Table 1.1 Cosecant, secant and reciprocal logarithm numbers up to $k = 12$

k	c_k	d_k	A_k
0	1	1	1
1	1/6	1/2	1/2
2	7/360	5/24	−1/12
3	$31/3 \cdot 7!$	61/720	1/24
4	$127/15 \cdot 8!$	1385/8!	−19/720
5	$511/66 \cdot 9!$	50521/10!	3/160
6	$1414477/105 \cdot 13!$	2702765/12!	−863/60480
7	$8191/6 \cdot 13!$	199360981/14!	275/24192
8	$237036478/30 \cdot 17!$	19391512145/16!	−33953/3628800
9	$5749691557/21 \cdot 19!$	2404879675441/18!	8183/1036800
10	$915462773587/165 \cdot 20!$	370371188237525/20!	−3250433/479001600
11	$1792042792463/69 \cdot 22!$	69348874393137901/22!	4671/788480
12	$1982765468311237/1365 \cdot 24!$	15514534163557086905/24!	826464439/201180672000

search for redundant partitions. As we develop the method further, we shall see that we do not even need the partitions, just the numbers of distinct parts in each partition and their numbers of occurrences, which are more commonly known as multiplicities.

To observe how the generation of partitions can be avoided in the calculation of the cosecant numbers, we adopt an alternative approach where all even parts ranging from 2 to $2k$ possess multiplicities λ_1 to λ_k. That is, the multiplicity λ_i corresponds to the number of occurrences of the part $2i$ in each partition. The main difference between the previous approach resulting in (1.10) and this new approach is that the multiplicities always have a specific value even though they may be frequently equal to zero. In addition, λ_i will always lie between 0 and $\lfloor k/i \rfloor$, where $\lfloor x \rfloor$ is the floor function representing the greatest integer less than or equal to x, because $\lfloor k/i \rfloor$ is the maximum value that λ_i can equal. Moreover, the specific partitions that provide a contribution to c_k are constrained by the fact that

$$2 \sum_{i=1}^{k} i\lambda_i = 2\lambda_1 + 4\lambda_2 + 6\lambda_3 + \cdots + 2k\lambda_k = 2k. \tag{1.12}$$

From this constraint we observe that λ_k will always equal zero for all partitions except the partition, $\{2k\}$, in which case it will equal unity with all the other λ_i equal to zero. As a consequence, this approach incorporates much redundancy as many of the λ_i are equal to zero in the partitions. Moreover, we can divide throughout by a factor of two, which means that we are now considering all partitions summing to k rather than the even partitions summing to $2k$. Consequently, the cosecant numbers in (1.10)

Table 1.2 Multiplicities of partitions summing to 5

Partition	λ_1	λ_2	λ_3	λ_4	λ_5
{5}					1
{4, 1}	1			1	
{3, 2}		1	1		
{3, 1, 1}	2		1		
{2, 2, 1}	1	2			
{2, 1, 1, 1}	3	1			
{1, 1, 1, 1, 1}	5				

can be written as

$$c_k = \sum_{\substack{\lambda_1,\lambda_2,\lambda_3,\ldots,\lambda_k=0 \\ \lambda_1+2\lambda_2+3\lambda_3+\cdots+k\lambda_k=k}}^{k,\lfloor k/2\rfloor,\lfloor k/3\rfloor,\ldots,1} N_k! \prod_{i=1}^{k}\left(\frac{(-1)^{(i+1)}}{(2i+1)!}\right)^{\lambda_i}\frac{1}{\lambda_i!}. \tag{1.13}$$

Since the length, $N_k = \sum_{i=1}^{k}\lambda_i$, and $\sum_{i=1}^{k} i\lambda_i = k$, we finally arrive at (1.4). This formulation of the cosecant numbers involves constraints that avoid the recursive nature of a tree diagram, although it appears that both algorithms possess the same level of computing complexity. This completes the proof of Theorem 1.1.

Let us now apply (1.4) to the calculation of c_5, which is given as 511/ $66 \cdot 9!$ or $73/3\,421\,440$ in Table 1.1. There are seven partitions summing to 5: {5}, {4, 1}, {3, 2}, {3, 1, 1}, {2, 2, 1}, {2, 1, 1, 1} and {1, 1, 1, 1, 1}. The multiplicities for each partition are displayed in Table 1.2. We also require the specific contribution made by each partition in the table. From Theorem 1.1 each part i is assigned a value of $(-1)^{i+1}/(2i+1)!$. Since each part occurs λ_i times, we need to take their assigned value to the power of λ_i or $(-1)^{(i+1)\lambda_i}/((2i+1)!)^{\lambda_i}$ in each partition. Thus, (1.13) yields

$$c_5 = \frac{1}{11!} - \frac{2!}{1!\cdot 1!}\frac{1}{3!\cdot 9!} - \frac{2!}{1!\cdot 1!}\frac{1}{5!\cdot 7!} + \frac{3!}{1!\cdot 2!}\frac{1}{3!^2\cdot 7!} + \frac{3!}{1!\cdot 2!}\frac{1}{3!\cdot 5!^2}$$
$$- \frac{4!}{3!\cdot 1!}\frac{1}{3!^3\cdot 5!} + \frac{5!}{5!}\frac{1}{3!^5} = \frac{511}{66\cdot 9!}. \tag{1.14}$$

By comparing (1.1) with (1.3), we immediately observe that

$$c_k = \frac{(-1)^{k+1}}{(2k)!}(2^{2k}-2)B_{2k}. \tag{1.15}$$

Moreover, according to No. 9.616 of Ref. [14], $B_{2k}/(2k)!$ is related to $\zeta(2k)$, where $\zeta(s)$ represents the Riemann zeta function. Hence the cosecant numbers can be written as

$$c_k = 2(1 - 2^{1-2k}) \frac{\zeta(2k)}{\pi^{2k}}. \tag{1.16}$$

Alternatively, we can express the Bernoulli numbers as a sum over partitions, viz.

$$B_{2k} = \frac{(2k)!}{2 - 2^{2k}} \sum_{\substack{\lambda_1, \lambda_2, \lambda_3, \ldots, \lambda_k = 0 \\ \sum_{i=1}^{k} i\lambda_i = k}}^{k, \lfloor k/2 \rfloor, \lfloor k/3 \rfloor, \ldots, 1} (-1)^{N_k} N_k! \prod_{i=1}^{k} \left(\frac{1}{(2i+1)!} \right)^{\lambda_i} \frac{1}{\lambda_i!}. \tag{1.17}$$

Hence we have a discrete method of evaluating the Bernoulli numbers or positive even integer values of the Riemann zeta function.

There is, however, one anomaly when (1.1) is compared with (1.3). We see that (1.1) is valid for $|x| < \pi$, while Theorem 1.1 states that the disk of absolute convergence is given at least by $|x| < \sqrt{6}$. This difference is not due to the fact that the partition method for a power series expansion yields different power series expansions to those derived by other methods such as the Taylor/Maclaurin series method, but is due to how the series in (1.5) has been handled, particularly the use of the inequality given as No. 3.7.29 in Ref. [23]. Had we identified the series instead as $(1 - \sin(x)/x)$, and expanded this form using the geometric series, we would have found that (1.3) was valid for $|1 - \sin(x)/x| < 1$ or $0 < \sin(x)/x < 2$. Since $\sin(x)/x$ is always less than or equal to unity, we only need to be concerned with the lower bound, which dips below zero when $|x| > \pi$. In other words, we obtain the same range of validity as (1.1). Therefore, we have seen that it may not be possible to obtain the correct range of absolute convergence directly from the partition method for a power series expansion.

We can, however, use the material in Appendix A to determine the regions in the complex plane where the power series expansion given by (1.3) is convergent and divergent as exemplified by the following corollary.

Corollary to Theorem 1.1. The power series given by (1.3) is absolutely convergent for $|x| < \pi$, conditionally convergent for $|x| > \pi$ and $|\Re x| < \pi$ and divergent everywhere else.

Proof. By introducing (1.16) into (1.3) we find that

$$\csc x \equiv \frac{2}{x} \sum_{k=0}^{\infty} (1 - 2^{1-2k}) \zeta(2k) \left(\frac{x}{2\pi} \right)^{2k}. \tag{1.18}$$

Next we introduce the Dirichlet series form for the Riemann zeta function and separate the two series over x. This yields

$$\csc x \equiv \frac{2}{x} \sum_{j=1}^{\infty} \sum_{k=0}^{\infty} \left(\left(\frac{x}{j\pi} \right)^{2k} - 2 \left(\frac{x}{2j\pi} \right)^{2k} \right). \tag{1.19}$$

Both the sums over k represent the geometric series, the first involving the variable $x/j\pi$ and the second, $x/2j\pi$. Consequently, we can use the material in Appendix A to determine where (1.3) is convergent and divergent. First, we note that the convergence of the entire rhs is determined by the behaviour of the $j=1$ value because the radius of absolute convergence is smallest for this value of j. That is, even though the series for other values of j may be convergent for values of x outside the region of convergence for the $j=1$ series, the divergence of the latter for these values of x means that the rhs is still divergent. Therefore, the convergence behaviour of the series $\sum_{k=0}^{\infty}(x/\pi)^{2k}$ determines the convergence behaviour of (1.3). From Appendix A we know that the radius of absolute convergence of the geometric series is $|x/\pi| < 1$. By writing the series as

$$\sum_{k=0}^{\infty} \left(\frac{x}{\pi} \right)^{2k} = \sum_{k=0}^{\infty} \frac{(1+(-1)^k)}{2} \left(\frac{x}{\pi} \right)^{k}, \tag{1.20}$$

we see that the series is composed of two separate geometric series. Again from Appendix A, we know that the first of these series is conditionally convergent for $|x| > \pi$ and $\Re x < \pi$, while the second is conditionally convergent for $|x| > \pi$ and $\Re x > -\pi$. For the remaining values of x both series are divergent. This means that both series possess a common strip where they will be conditionally convergent, which is given by $|x| > \pi$ and $|\Re x| < \pi$. For all other values of x lying outside the strip, the rhs of (1.18) is divergent. This completes the proof of the corollary.

It should also be mentioned that the cosecant numbers are shown in Corollary 2 to Theorem 1 of Ref. [5] to obey various recurrence relations. Two of these results are

$$c_k = \sum_{j=0}^{k-1} \frac{(-1)^{k-j-1}}{(2k-2j+1)!} c_j, \tag{1.21}$$

and

$$\frac{c_k}{1-2^{1-2k}} = \sum_{j=0}^{k} \frac{(-1)^{k-j-1}}{(2k-2j)!} c_j. \tag{1.22}$$

When developing the general theory of the partition method for a power series expansion in Chapter 4, we shall refer to the series given by $x^2/3! - x^4/5! + \cdots$ in the denominator of (1.5) as the inner series, while the general series resulting in (1.7) will be referred to as the outer series. In the above example we have seen that the coefficients of the inner series were instrumental for assigning values to the parts in the partitions, while the outer series was simply the geometric series.

As a consequence of Theorem 1.1, we are now in a position to derive an analogous power series expansion for $\sec x$ or $1/\cos x$. With the aid of the power series expansion for cosine, secant can be expressed as

$$\sec x = \frac{1}{1 - x^2/2! + x^4/4! - x^6/6! + x^8/8! - \cdots} . \tag{1.23}$$

By comparing (1.23) with (1.5), we observe immediately that the coefficients of x^{2i} are now $(-1)^{i+1}/(2i)!$, whereas previously they were given by $(-1)^{i+1}/(2i+1)!$. If we denote the coefficients of the power series expansion for secant as d_k, then all we need to do is replace $(2i+1)!$ in (1.4) by $(2i)!$ to obtain the new coefficients. Hence we arrive at

$$d_k = (-1)^k \sum_{\substack{\lambda_1,\lambda_2,\lambda_3,\ldots,\lambda_k=0 \\ \sum_{i=1}^{k} i\lambda_i=k}}^{k,\lfloor k/2 \rfloor, \lfloor k/3 \rfloor,\ldots,1} (-1)^{N_k} N_k! \prod_{i=1}^{k} \left(\frac{1}{(2i)!} \right)^{\lambda_i} \frac{1}{\lambda_i!}, \tag{1.24}$$

where

$$\sec x \equiv \sum_{k=0}^{\infty} d_k x^{2k} . \tag{1.25}$$

We shall refer to this special set of rational numbers, d_k, as the secant numbers. The first few of these numbers are: $d_0 = 1$, $d_1 = 1/2$, $d_2 = 5/24$, and $d_3 = 61/720$. The third column in Table 1.1 displays them up to $k = 12$. Although these numbers converge to zero as $k \to \infty$, they are not as rapidly converging as their cosecant counterparts.

Since (1.23) can be written as $1/(1 - (\cos x - 1))$ and the geometric series is again the outer series, we observe that (1.25) is absolutely convergent only when $|1 - \cos x| < 1$ or $0 < \cos x < 2$. In this case $\cos x$ dips below zero at $|x| = \pi/2$. Therefore, the disk of absolute convergence is given by $|x| < \pi/2$. For these values of x, we can replace the equivalence symbol in (1.25) by an equals sign. If we had applied the same inequality approach as in Theorem 1.1 to (1.23), then we would have found that the disk of absolute convergence would have been given by at least $|x| < \sqrt{2}$ as in Theorem 4 of Ref. [5].

According to No. 1.411 from Ref. [14], secant can be expressed in terms of the Euler numbers as

$$\sec x = \sum_{k=0}^{\infty} \frac{|E_k|}{(2k)!} x^{2k}, \tag{1.26}$$

provided $|x| < \pi/2$. For these values of x, we can replace the equivalence symbol in (1.25) by an equals sign. Moreover, by comparing the above result with (1.25), we see that the secant numbers are related to the Euler numbers, where $d_k = (-1)^k E_{2k}/(2k)!$. Consequently, the Euler numbers can be expressed as a sum over partitions, viz.

$$E_{2k} = (2k)! \sum_{\substack{\lambda_1,\lambda_2,\lambda_3,\dots,\lambda_k=0 \\ \sum_{i=1}^{k} i\lambda_i=k}}^{k,\lfloor k/2 \rfloor, \lfloor k/3 \rfloor,\dots,1} (-1)^{N_k} N_k! \prod_{i=1}^{k} \left(\frac{1}{(2i)!}\right)^{\lambda_i} \frac{1}{\lambda_i!}. \tag{1.27}$$

The secant numbers can also be related to the Hurwitz zeta function. According to No. 1.422 in Ref. [14], we have

$$\sec\left(\frac{\pi x}{2}\right) = \frac{4}{\pi} \sum_{k=1}^{\infty} (-1)^{k+1} \frac{2k-1}{(2k-1)^2 - x^2}, \tag{1.28}$$

provided $|x| < 1$. Since $|x| < 1$, we can substitute the lhs of the above result by (1.25) after the equivalence symbol is replaced by an equals sign. Then decomposing the resulting equation into partial fractions yields

$$\sum_{k=0}^{\infty} d_k \left(\frac{\pi x}{2}\right)^{2k} = \frac{2}{\pi} \sum_{k=1}^{\infty} \frac{(-1)^{k+1}}{2k-1}\left(\frac{1}{1 - x/(2k-1)} + \frac{1}{1 + x/(2k-1)}\right). \tag{1.29}$$

Because $|x/(2k-1)| < 1$, we can expand the denominators using the geometric series. In addition, x is arbitrary, which means that we can equate like powers of x on both sides of the resulting equation. Thus, we obtain

$$d_j = \frac{2^{2j+2}}{\pi^{2j+1}} \sum_{k=1}^{\infty} \frac{(-1)^{k+1}}{(2k-1)^{2j+1}}. \tag{1.30}$$

By separating the sum in (1.30) into even and odd values of k, taking out a factor of 4^{2j+1} and carrying out a little algebra, one finds that the secant numbers are given by

$$d_k = \frac{2}{(2\pi)^{2k+1}}\left(\zeta(2k+1, 1/4) - \zeta(2k+1, 3/4)\right). \tag{1.31}$$

As shown in Sec. 9 of Ref. [5] the secant numbers can also be determined from the following recurrence relation:

$$d_k = \sum_{j=0}^{k-1} \frac{(-1)^{k-j-1}}{(2k-2j)!} d_j . \tag{1.32}$$

This result is the analogue of (1.21), but seems to resemble (1.22) more closely.

Now that we have derived a power series expansion for $\csc x$ in terms of the cosecant numbers c_k, let us investigate what happens when we invert it and apply the method for a power series expansion again. Then we find that

$$\sin x \equiv \frac{x}{1 + c_1 x^2 + c_2 x^4 + c_6 x^6 + \cdots} . \tag{1.33}$$

Aside from the factor of x in the numerator, (1.33) is the same form as (1.5) except that the cosecant numbers have become the coefficients of the powers of x^2. In addition, we know that if $|x| < \pi$, then we can replace the equivalence symbol by an equals sign.

To obtain the power series expansion for $\sin x$ in terms of the cosecant numbers, we treat (1.33) as the geometric series as we did in Theorem 1.1, which yields

$$\sin x \equiv x \sum_{k=0}^{\infty} (-1)^k \left(c_1 x^2 + c_2 x^4 + c_6 x^6 + \cdots \right)^k . \tag{1.34}$$

In order to replace the equivalence symbol by an equals sign, we require that the series in brackets be less than unity. This means that $|x/\sin x - 1| < 1$ or $0 < x/\sin x < 2$. Hence $|x| < 2$, which is less than π, but still indicates that there is a region in the complex plane where the equivalence symbol can be replaced by an equals sign. To apply the partition method for a power series expansion to (1.33), all that is required is to assign a value of $-c_i$ to each part i instead of $(-1)^{i+1}/(2i+1)!$ as in the case of cosecant. Consequently, we obtain the power series expansion for $\sin x$ with the coefficients expressed as a sum over partitions involving $c_i^{\lambda_i}$. Furthermore, since we know that the coefficients of the power series for $\sin x$ are equal to $(-1)^k/(2k+1)!$ for x^{2k+1}, we arrive at

$$\frac{(-1)^k}{(2k+1)!} = \sum_{\substack{\lambda_1,\lambda_2,\lambda_3,\ldots,\lambda_k=0 \\ \lambda_1+2\lambda_2+3\lambda_3+\cdots+k\lambda_k=k}}^{k,\lfloor k/2 \rfloor,\lfloor k/3 \rfloor,\ldots,1} (-1)^{N_k} N_k! \prod_{i=1}^{k} c_i^{\lambda_i} \frac{1}{\lambda_i!} . \tag{1.35}$$

This is valid for all values of k, but it has been obtained by equating the power series expansion for $\sin x$ with the resulting power series obtained via the partition method for a power series expansion, which was found to be only valid for $|x| < 2$. Therefore, we see that alternative methods are required for determining the exact range of validity for results derived by the partition method for a power series expansion. Moreover, by inverting the power series for secant given by (1.25), one obtains

$$\frac{(-1)^k}{(2k)!} = \sum_{\substack{\lambda_1,\lambda_2,\lambda_3,\ldots,\lambda_k=0 \\ \sum_{i=1}^k i\lambda_i=k}}^{k,\lfloor k/2\rfloor,\lfloor k/3\rfloor,\ldots,1} (-1)^{N_k} N_k! \prod_{i=1}^{k} d_i^{\lambda_i} \frac{1}{\lambda_i!}, \qquad (1.36)$$

where we have used the fact that the coefficients of x^{2k} in the power series expansion for cosine are equal to $(-1)^k/(2k)!$. That is, $\cos x = \sum_{k=0}^{\infty}(-1)^k x^{2k}/(2k)!$. These last two results are examples where the partition method for a power series expansion has produced the standard Taylor/Maclaurin series, but from a different perspective. As a consequence, we have combined the two approaches to derive new identities for $1/(2k+1)!$ and $1/(2k)!$.

1.2 RECIPROCAL LOGARITHM NUMBERS

After the partition method for a power expansion had been applied to cosecant and secant, it was decided to apply it to an elementary function, where a power series had not been derived or at least, was not well-known. Consequently, the function $f(z) = 1/\ln(1+z)$ was selected. The coefficients A_k of the resulting power series expansion were referred to as the reciprocal logarithm numbers, but since the publication of Ref. [3], it has emerged that the moduli or signless form of the first few of these numbers had been calculated by Gregory and Cauchy. In fact, Apelblat states on p. 138 of Ref. [25] that they are known as the Gregory or Cauchy numbers, while more recently, Blagouchine [26] refers to them as the Gregory coefficients, but also mentions that they have been called various names in the past including (normalized) Cauchy numbers (of the first kind) and normalized (generalized) Bernoulli numbers. In addition, Weisstein [27] refers to them as logarithmic numbers, while their numerators and denominators appear respectively as sequences A002206/M5066 and A002207/M2017 in Ref. [28], where they are known as the Gregory numbers. When they are multiplied by $k!(k+1)!$, they also appear the integer sequence A009763 in the same reference. Notwithstanding, we shall refer to them as the reciprocal logarithm numbers in keeping with Ref. [3]. Moreover, they will be generalized in the next chapter, where they will be referred to as the generalized reciprocal numbers and denoted by $A_k(s)$.

Theorem 1.2. There exists a power series expansion for $1/\ln(1+z)$, which can be written as

$$\frac{1}{\ln(1+z)} \equiv \sum_{k=0}^{\infty} A_k z^{k-1} , \tag{1.37}$$

where the reciprocal logarithm numbers, A_k, represent an infinite set of special rational numbers and are displayed in Table 1.1 up to $k = 12$. Via the partition method for a power series expansion they are given by

$$A_k = (-1)^k \sum_{\substack{\lambda_1,\lambda_2,\lambda_3,\dots,\lambda_k=0 \\ \sum_{i=1}^{k} i\lambda_i=k}}^{k,\lfloor k/2 \rfloor,\lfloor k/3 \rfloor,\dots,1} (-1)^{N_k} N_k! \prod_{i=1}^{k} \frac{1}{\lambda_i!} \frac{1}{(i+1)^{\lambda_i}} . \tag{1.38}$$

Furthermore, the expansion represents a small $|z|$-expansion with a unit disk of absolute convergence, i.e., $|z| < 1$. For these values of z the equivalence symbol can be replaced by an equals sign.

Proof. With the aid of the Taylor/Maclaurin series for logarithm, which is only absolutely convergent for $|z| < 1$, we write the reciprocal of $\ln(1+z)$ as

$$\frac{1}{\ln(1+z)} = \frac{1}{z} \frac{1}{(1 - z/2 + z^2/3 - z^3/4 + z^4/5 - \cdots)}$$
$$\equiv \frac{1}{z} \sum_{k=0}^{\infty} \left(\frac{z}{2} - \frac{z^2}{3} + \frac{z^3}{4} - \frac{z^4}{5} + \cdots \right)^k . \tag{1.39}$$

In arriving at the final form we have treated the denominator in the intermediate member as the regularized value for the geometric series, which yet again represents the outer series. As in the proof of Theorem 1.1, we expand the first few powers of z in (1.39), thereby obtaining

$$\frac{1}{\ln(1+z)} \equiv \frac{1}{z} \left(1 + \frac{z}{2} + z^2 \left(\frac{1}{4} - \frac{1}{3} \right) + z^3 \left(\frac{1}{8} - \frac{1}{3} + \frac{1}{4} \right) + \right.$$
$$\left. + z^4 \left(\frac{1}{16} + \frac{1}{9} - \frac{1}{5} \right) + \cdots \right) = \sum_{k=0}^{\infty} A_k z^{k-1} . \tag{1.40}$$

From the above result we see that the coefficients, A_0, A_1, A_2, A_3 and A_4 are equal to 1, 1/2, −1/12, 1/24 and −19/720, respectively and thus,

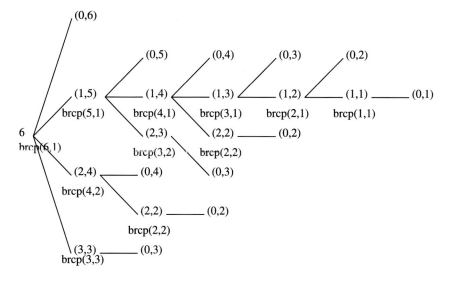

Figure 1.2 Partition tree of partitions summing to 6.

appear in Table 1.1. If we compare these values with the corresponding values in the other columns, then we observe that they do not converge as rapidly to zero as the cosecant and secant numbers. In fact, they are weakly-converging numbers.

We now turn our attention to deriving a general formula for the higher order coefficients of the above power series via the partition method for a power series expansion. As was done in Theorem 1.1, we shall create a tree diagram, but in this case we shall consider all partitions summing to the order k, not only those with even parts. From here on we shall refer to such diagrams as the partition trees since all partitions are involved. To create a partition tree, one draws branch lines to all pairs of numbers that can be summed to the seed number k, where the first member in the tuple is an integer less than or equal to $\lfloor k/2 \rfloor$. Again, $\lfloor x \rfloor$ denotes the floor function or the greatest integer less than or equal to x. For example, in Fig. 1.2, which displays the partition tree for partitions summing to 6, we draw branch lines to $(0, 6)$, $(1, 5)$, $(2, 4)$ and $(3, 3)$. As in the case of the tree diagram for the cosecant numbers, whenever a zero appears as the first member of a tuple, the path terminates, as can be seen by $(0, 6)$ or $(0, 3)$ in Fig. 1.2. For the remaining pairs, one draws another set of branch lines to all pairs with integers that sum to the second member of the preceding pair, but once again under the prescription that the first member of each new pair is less than or equal to half of the second member of the previous tuple. Hence for $(1, 5)$, we draw paths branching out to $(0, 5)$, $(1, 4)$ and $(2, 3)$. This recursive approach continues until all paths in the partition tree are terminated by a tuple possessing a zero.

This, however, is not all that is required to produce a partition tree for evaluating the reciprocal logarithm numbers. We still have the problem of dealing with duplicated paths or permutations of partitions. Previously, it was stated that a search algorithm was required to remove these partitions, but such an algorithm would not only result in much time being expended, it would also add significantly to the complexity of the partition method for a power series. We can avoid having to introduce a search algorithm by viewing partition trees from a different perspective. First we note that the entire tree emanating from $(1, 5)$ in Fig. 1.2 represents the same partition tree as if the seed number were 5 instead of 6. Similarly, the partition tree emanating from $(1, 4)$ is the same partition tree as if the seed number were equal to 4 instead of 6. This recursive behaviour continues all the way down to the last partition whose parts are only composed of ones. Second, only partitions with unity in them emanate from $(1, 5)$ in the partition tree, whereas the partitions emanating from $(2, 4)$ only possess parts greater than or equal to 2. Similarly, the partitions emanating from $(3, 3)$ only possess parts that are greater than or equal to 3. In fact, in the last instance since 3 is half of 6, there will only be threes involved along the path from $(3, 3)$. Thus, we see that the last path represents the central partition $\{3, 3\}$, which would have been $\{3, 4\}$ had we constructed a partition with the seed number equal to 7.

As a result of viewing partition trees in this manner, we note that broadly speaking the second member in a tuple decrements or decreases by unity with each rightward or horizontal movement, while the first member of each tuple increments with each downward or vertical movement. In other words, partition trees can be regarded as two–dimensional objects, which means in turn that two variables are required to construct them. As we shall see in Chapter 3, this is particularly interesting because it means that we are describing the rather rare instance of bivariate recursion. More importantly, we have eliminated the need to incorporate a search algorithm to remove duplicated partitions. Since it has been stated that the partition trees terminate at the central partition, we shall refer to the algorithm that generates the partition trees as the bivariate recursive central partition or BRCP, for short, algorithm.

Table 1.3 Multiplicities and lengths of partitions summing to 6

Partition	λ_1	λ_2	λ_3	λ_4	λ_5	λ_6	N_k
$\{6\}$						1	1
$\{1,5\}$	1				1		2
$\{1,1,4\}$	2			1			3
$\{1,1,1,3\}$	3		1				4
$\{1,1,1,1,2\}$	4	1					5
$\{1,1,1,1,1,1\}$	6						6
$\{1,2,3\}$	1	1	1				3
$\{2,4\}$		1		1			2
$\{2,2,2\}$		3					1
$\{1,1,2,2\}$	2	2					4
$\{3,3\}$			2				2

It should be emphasized that it is not necessary to create partition trees in order to apply the partition method for a power series expansion, but they do ensure that all partitions are captured, whereas there is a risk that partitions can be missed or omitted when one attempts to write them down directly. Another advantage of partition trees is that it becomes a simple matter to exclude specific parts from them. If one or more of the coefficients in the power series in the large parentheses in (1.40) were zero, then all branches where these powers appeared in a tuple would be excluded. For example, if the coefficient of z^2 had been zero instead of $-1/3$, then the partition tree could still be constructed in the same manner except that there would be no branches containing a two in any tuple whatsoever.

From the partition tree in Fig. 1.2 we see that there are eleven partitions summing to 6, which occur in the following order: $\{6\}$, $\{1,5\}$, $\{1,1,4\}$, $\{1,1,1,3\}$, $\{1,1,1,1,2\}$, $\{1,1,1,1,1,1\}$, $\{1,2,3\}$, $\{2,4\}$, $\{2,2,2\}$, $\{1,1,2,2\}$ and $\{3,3\}$. As for Theorem 1.1, we require the multiplicities of each partition, which are displayed in Table 1.3. Next each part must be assigned a value. This value is the negative value of the coefficient when the part is equal to the power of the corresponding term in the inner series. In this case the inner series is the series appearing in the denominator of the intermediate member of (1.39). Thus, whenever a two appears in a partition it is assigned a value of $-1/3$, while if a four appears, then it is assigned a value of $-1/5$. In general, a part k is assigned the value of $(-1)^{k+1}/(k+1)$. To obtain the contribution from a partition, we multiply all the assigned values by the multinomial factor, which is obtained from the factorial of the length of the partition or the total number of parts, $N_k!$ divided by the factorials of all the multiplicities for the partition, $\prod_{i=1}^{k} 1/\lambda_i!$. E.g., for the partition $\{1, 1, 2, 2\}$, the product of the assigned values is $(-1/2)^2(1/3)^2$, while the multinomial factor becomes $4!/2!2!$ since $N_k = 4$, $\lambda_1 = 2$ and $\lambda_2 = 2$. Finally, summing the contributions from all eleven partitions yields A_6, which is given by

$$
\begin{aligned}
A_6 = {} & -\frac{1!}{1!}\frac{1}{7} + \frac{2!}{1! \cdot 1!} \cdot \frac{1}{2}\frac{1}{6} - \frac{3!}{2! \cdot 1!}\left(\frac{1}{2}\right)^2 \frac{1}{5} + \frac{4!}{3! \cdot 1!}\left(\frac{1}{2}\right)^3 \frac{1}{4} \\
& - \frac{5!}{4! \cdot 1!}\left(\frac{1}{2}\right)^4 \frac{1}{3} + \frac{6!}{6!}\left(\frac{1}{2}\right)^6 - \frac{3!}{1! \cdot 1! \cdot 1!}\frac{1}{2}\frac{1}{3}\frac{1}{4} + \frac{2!}{1! \cdot 1!}\frac{1}{3}\frac{1}{5} \\
& - \frac{3!}{3!}\left(\frac{1}{3}\right)^3 + \frac{4!}{2! \cdot 2!}\left(\frac{1}{2}\right)^2\left(\frac{1}{3}\right)^2 + \frac{2!}{2!}\left(\frac{1}{4}\right)^2 = -\frac{863}{60480}.
\end{aligned}
\tag{1.41}
$$

This result also appears in the fourth column of Table 1.1.

We can generalize the above result to any order k by noting that we are basically summing over all partitions summing to k. As was done in Theorem 1.1, we sum over all values that the multiplicities, λ_1 to λ_k can take subject to the condition $\sum_{i=1}^{k} i\lambda_i = k$. Therefore, the generalization of (1.41) to A_k becomes

$$
A_k = \sum_{\substack{\lambda_1,\lambda_2,\lambda_3,\ldots,\lambda_k=0 \\ \sum_{i=1}^{k} i\lambda_i=k}}^{k,\lfloor k/2 \rfloor, \lfloor k/3 \rfloor, \ldots, 1} N_k! \prod_{i=1}^{k} \frac{1}{\lambda_i!}\left(\frac{(-1)^{i+1}}{(i+1)}\right)^{\lambda_i}.
\tag{1.42}
$$

Now the product $\prod_{i=1}^{k}(-1)^{i\lambda_i}$ yields $(-1)^k$, which can be taken outside all the sums, while $\prod_{i=1}^{k}(-1)^{\lambda_i}$ yields $(-1)^{N_k}$, which can be taken outside of the product. When this is done, we arrive at (1.38).

From (1.38) we see that the number of contributions for each A_k is finite and that the contributions themselves do not involve irrational numbers. Hence the reciprocal logarithm numbers are rational.

In the proof of Theorem 1.1 we found the radius of absolute convergence for the expansion of $\csc x$ was given at least by $|x^2| < 6$. Later this was extended to $|x| < \pi$ by replacing the inner series by $\sin(x)/x - 1$ and then applying the convergence properties of the outer or geometric series. This was possible because the inner series or the power series expansion for $\sin x$ is convergent for all values of x. Even though the power series for $\csc x$ in Theorem 1.1 has a disk of absolute convergence given by $|x| < \pi$, it was used to determine the standard power series for $\sin x$, the latter being valid for all values of x. Hence we observed that the partition method for a power series expansion may yield very limited or over-restricted results for the disk of absolute convergence and that alternative methods are often required to determine the exact values over which a power series expansion based on partitions is valid. We shall carry out a similar investigation for the power series expansion appearing in Theorem 1.2.

We begin by deriving an integral representation for the reciprocal logarithm numbers using the following simple integral:

$$(1+z) \int_0^1 dt\, e^{-t \ln(1+z)} = \frac{z}{\ln(1+z)}. \tag{1.43}$$

By noting that the integrand is equal to $(z+1)^{-t}$, we introduce the binomial series, which gives

$$\sum_{k=0}^{\infty} (-z)^k \int_0^1 dt\, \frac{\Gamma(k+t-1)}{k!\,\Gamma(t-1)} \equiv \frac{z}{\ln(1+z)}. \tag{1.44}$$

Note that the equivalence symbol has replaced the equals sign because the binomial series is only absolutely convergent for $|z| < 1$. For $\Re(-z) < 1$ and $|z| > 1$, the binomial series is conditionally convergent, while for the other values of z, it is divergent just like the geometric series, in which case the series needs to be regularized. In addition, like the geometric series its regularized value is the same limit value as when it is absolutely convergent. See Lemma A.1 in Appendix A for the proof of its regularized value.

Since the lhs's of (1.37) and (1.44) share the same regularized value and by definition the regularized value is unique as discussed in Appendix A, they are equal to another. Moreover, because z is arbitrary, we can equate like powers of z on both sides of the resulting equation. After making the change of variable, $y = 1 - t$, we find that

$$A_k = \frac{(-1)^k}{k!} \int_0^1 \frac{\Gamma(k+t-1)}{\Gamma(t-1)}\, dt = \frac{(-1)^k}{k!} \int_{-1}^0 (y)_k\, dy. \tag{1.45}$$

The above equation has been simplified with the introduction of the Pochhammer notation, $(y)_k$, for $\Gamma(k+y)/\Gamma(y)$. The integral in (1.45) is always negative except when $k = 0$. In addition, since $(1)_k = k!$, it is readily seen that the reciprocal logarithm numbers are normalized integrals of the Pochhammer polynomials [29].

The modulus or magnitude of the reciprocal logarithm numbers can also be expressed as

$$|A_k| = \frac{1}{k!} \left| \int_0^1 (k+t-2)(k+t-3)\cdots t(t-1)\, dt \right|. \tag{1.46}$$

Consequently, upper and lower bounds to this result are obtained by putting $t = 1$ and $t = 0$ respectively in all but the last three factors in the integrand.

Hence for $k > 1$, the signless or unsigned forms of the reciprocal logarithm numbers are bounded as follows:

$$\frac{1}{4k(k-1)} \leq |A_k| \leq \frac{1}{8k} . \tag{1.47}$$

Although sufficient for our purposes, these bounds can be narrowed even further by considering more than just the last three factors in the integrand of (1.46). From (1.47), however, we observe that as $k \to \infty$, $|A_k| \to 0$. Thus, the reciprocal logarithm numbers converge at a rate that is between linear and quadratic in $1/k$. Because this rate of the convergence is not exponentially small, they can be regarded as slowly or weakly converging rational numbers as mentioned previously.

Now we complete the proof of Theorem 1.2 by determining the radius of absolute convergence of (1.37) via d'Alembert's ratio test [31]. Putting u_k equal to $A_k z^{k-1}$ in this test yields

$$\left| \frac{u_{k+1}}{u_k} \right| = \left| \frac{A_{k+1} z}{A_k} \right| = \frac{|z|}{k+1} \left| \frac{\int_0^1 \Gamma(k+t)/\Gamma(t-1)\, dt}{\int_0^1 \Gamma(k+t-1)/\Gamma(t-1)\, dt} \right| , \tag{1.48}$$

where the integral representation for the reciprocal logarithm numbers given by (1.46) has been introduced. The integral in the numerator of (1.48) can be bounded by noting that

$$\left| \int_0^1 \Gamma(k+t)/\Gamma(t-1)\, dt \right| \leq k \left| \int_0^1 \Gamma(k+t-1)/\Gamma(t-1)\, dt \right| . \tag{1.49}$$

Therefore, the ratio of successive terms is bounded by

$$\left| \frac{u_{k+1}}{u_k} \right| \leq \frac{k\,|z|}{k+1} \frac{|A_k|}{|A_k|} \leq |z| , \tag{1.50}$$

which means in turn that the power series expansion given by (1.37) is absolutely convergent for $|z| < 1$. This completes the proof of Theorem 1.2.

As was done in the previous section, we can invert (1.37), thereby obtaining

$$\ln(1+z) \equiv \frac{z}{1 + \sum_{k=1}^{\infty} A_k z^k} . \tag{1.51}$$

Once again, we treat the denominator as the regularized value of the geometric series. Hence the above result can be expressed as

$$\ln(1+z) \equiv z \sum_{k=0}^{\infty} (-1)^k \left(A_1 z + A_2 z^2 + A_3 z^3 + \cdots \right)^k , \tag{1.52}$$

which is basically the same as (1.39) except that the coefficients have been replaced by the reciprocal logarithm numbers. Consequently, we can apply the partition method for a power series expansion to (1.52) with each part i assigned a value of $-A_i$ instead of $(-1)^{i+1}/(i+1)$. Then (1.52) becomes

$$\ln(1+z) \equiv z\left(1 + zA_1 + z^2\left(A_1^2 - A_2\right) + z^3\left(-A_1^3 + \frac{2!}{1! \cdot 1!}\right.\right.$$
$$\left.\left. \times (-A_1)(-A_2) - A_3\right) + \cdots\right). \tag{1.53}$$

This is another representation of the Taylor/Maclaurin series for $\ln(1+z)$, which is given by $\sum_{k=1}^{\infty}(-1)^{k+1}z^k/(k+1)$. Therefore, by equating them to each another, we find that

$$\frac{(-1)^k}{k+1} = \sum_{\substack{\lambda_1,\lambda_2,\lambda_3,\ldots,\lambda_k=0 \\ \sum_{i=1}^{k} i\lambda_i=k}}^{k,\lfloor k/2\rfloor,\lfloor k/3\rfloor,\ldots,1} (-1)^{N_k}N_k! \prod_{i=1}^{k}\frac{A_i^{\lambda_i}}{\lambda_i!}. \tag{1.54}$$

To complete this section, it should be mentioned that a recurrence or recursion relation for the reciprocal logarithm numbers can be derived simply by analysing the quotient of $\ln(1+z)$ over itself. By introducing the Taylor/Maclaurin series for $\ln(1+z)$ in the numerator and substituting the denominator by (1.37), we obtain a product of two infinite series. Because both series are absolutely convergent for $|z| < 1$, we can replace the equivalence symbol by an equals sign for these values of z. Moreover, since z is arbitrary, like powers of z on both sides can be equated to one another, despite the fact they vanish on the rhs of the resulting equation. Thus we arrive at

$$A_k = \sum_{j=0}^{k-1}\frac{(-1)^{k-j+1}}{k-j+1}A_j. \tag{1.55}$$

In Chapter 2 a master equation will be derived, which yields (1.55) when its variables are set to particular values, while a similar recurrence relation for the related Gregory numbers is derived essentially by the same approach in Ref. [30].

The reciprocal logarithm numbers can also be expressed in terms of Stirling numbers of the first kind. According to Chapters 24 and 18 of Refs. [23] and [29] respectively, the Pochhammer polynomials can be written as

$$(\gamma)_k = \frac{\Gamma(\gamma+k)}{\Gamma(\gamma)} = (-1)^k\sum_{j=0}^{k}(-1)^j s_k^{(j)}\gamma^j, \tag{1.56}$$

where the integers $s_k^{(j)}$ are known as the (signed) Stirling numbers of the first kind with $s_k^{(0)} = s_0^{(k)} = 0$ for $k \geq 1$ and $s_0^0 = 1$. It should be noted here that the Stirling numbers of the first kind are often represented as $s(k, j)$. They also satisfy the following recurrence relation:

$$s_{k+1}^{(j)} = s_k^{(j-1)} - k\, s_k^{(j)} . \tag{1.57}$$

In the appendix of Ref. [4], a general formula for these numbers is derived, which is given as

$$s_k^{(k-j)} = (-1)^j \sum_{i_j=j}^{k-1} i_j \sum_{i_{j-1}=j-1}^{i_j-1} \sum_{i_{j-2}=j-2}^{i_{j-1}-1} i_{j-2} \cdots \sum_{i_1=1}^{i_2-1} i_1 . \tag{1.58}$$

With this result we can calculate general formulas for these numbers for j ranging from 0 to 4. These are

$$s_k^{(k)} = 1, \quad s_k^{(k-1)} = -\binom{k}{2}, \quad s_k^{(k-2)} = \frac{(3k-1)}{4}\binom{k}{3},$$

$$s_k^{(k-3)} = -\binom{k}{2}\binom{k}{4}, \quad s_k^{(k-4)} = \frac{1}{48}(15k^3 - 30k^2 + 5k + 2)\binom{k}{5}. \tag{1.59}$$

Note that the denominator of the last result has been corrected here compared with Ref. [4], where 336 appears instead of 48. These results have been obtained by introducing the nested sum formula given by (1.58) as a module in Mathematica [32]. In general, one obtains a polynomial in k of degree $2j$ with one of the terms in the polynomials being equal to $(-1)^j\binom{k}{j+1}$. In Chapter 6 we shall derive an alternative formula for the Stirling numbers of the first kind, which will enable us to determine a general formula for $s_k^{(k-5)}$ more easily than using (1.58). Moreover, $s_k^{(k-\ell)}$ will be related to special polynomials $r_\ell(k)$, which will be tabulated up to $\ell = 10$.

If we introduce (1.56) into (1.45), then we find that the reciprocal logarithm numbers can be expressed in terms of the Stirling numbers of the first kind as

$$A_k = \frac{(-1)^k}{k!} \sum_{j=1}^{k} \frac{s_k^{(j)}}{j+1} . \tag{1.60}$$

When (1.60) is programmed in Mathematica [32], it becomes a very quick method of obtaining the values of the reciprocal logarithm numbers for large values of k. The above result should also be regarded as being complementary to other known relations involving the Stirling numbers of the first kind such as

$$\sum_{j=1}^{k} s_k^{(j)} = 0 \quad \text{for } k \geq 2, \quad \text{and} \quad \sum_{j=1}^{k} \left| s_k^{(j)} \right| = k! \quad \text{for } k \geq 1. \tag{1.61}$$

To conclude this section, it should be mentioned that a brief history of the Stirling numbers of the first kind is presented in Ref. [26] together with an extensive list of references for the interested reader to pursue. Also appearing in this reference are many infinite series involving the Stirling numbers of the first kind that are related to elementary functions such as cosine or hyperbolic tangent, but where the argument is equal to $\ln(1+z)$.

1.3 BERNOULLI AND RELATED POLYNOMIALS

In both the previous sections of this chapter we considered applications of the partition method for a power series expansion where the coefficients of the resulting power series expansion turned out to be numbers. It should, however, be emphasized that the method can be applied to situations where the coefficients need not necessarily be numerical, but may result in polynomials or involve special functions. To demonstrate this, we shall describe here how the method can be used to determine the well-known Bernoulli polynomials, $B_k(x)$.

When $x = 0$ in the Bernoulli polynomials, they reduce to the Bernoulli numbers, which we have already discussed in Chapter 1. There we derived a sum over partitions for them that is given by (1.17). This result was based on relating the Bernoulli numbers to the cosecant numbers, but was not obtained directly from the generating function or (1.2). Consequently, (1.17) has an unwieldy factor of $2 - 2^{2k}$ in the denominator, which, as we shall soon see, can be avoided by applying the method directly to the generating function for the Bernoulli polynomials. By putting $x = 0$ in this result, we shall be left with a more direct form of the sum over partitions for the Bernoulli numbers than (1.17).

The Bernoulli polynomials are defined simply by multiplying the lhs of (1.2) by $\exp(xt)$ and replacing B_k on the rhs by $B_k(x)$. Therefore, we arrive at

$$\frac{t\, e^{xt}}{e^t - 1} \equiv \sum_{k=0}^{\infty} a_k(x)\, t^k, \tag{1.62}$$

where $a_k(x) = B_k(x)/k!$. At this stage, all we have done is provide a definition. That is, we have not even established whether the quantity on the rhs of (1.62) is convergent or divergent and if so, for which values of x. Since the quantity on the rhs is a power series in t, it must be equal to the power series in t when we multiply the power series expansion for $\exp(xt)$ by the series in (1.2). Hence we find that

$$\sum_{k=0}^{\infty} a_k(x)\, t^k = \sum_{k=0}^{\infty} \frac{(xt)^k}{k!} \sum_{k=0}^{\infty} a_k\, t^k = \sum_{k=0}^{\infty} t^k \sum_{j=0}^{k} a_{k-j}\, \frac{x^j}{j!}\,, \qquad (1.63)$$

where $a_k = B_k/k!$. Equating like powers of t yields

$$a_k(x) = \sum_{j=0}^{k} a_{k-j}\, \frac{x^j}{j!}\,. \qquad (1.64)$$

By expressing a_k and $a_k(x)$ respectively in terms of the Bernoulli numbers and polynomials, we obtain the standard definition of the polynomials, namely

$$B_k(x) = \sum_{j=0}^{k} \binom{k}{j} B_{k-j}\, x^j\,. \qquad (1.65)$$

On the other hand, if we multiply both sides of (1.62) by $\exp(-xt)$ and introduce its power series representation on the rhs, then we obtain

$$\frac{t}{e^t - 1} \equiv \sum_{k=0}^{\infty} t^k \sum_{j=0}^{k} \frac{(-1)^{k-j}}{(k-j)!}\, a_j(x)\, x^{k-j}\,. \qquad (1.66)$$

Now we replace the lhs of the above equivalence by the rhs of (1.2). Since the lhs is a divergent power series in t, it is, therefore, equal to the rhs of the above equivalence. Moreover, t is arbitrary again. So, we can equate like powers of both sides. Then we find that

$$\sum_{j=0}^{k} \frac{(-1)^{k-j}}{(k-j)!}\, a_j(x)\, x^{k-j} = a_k\,. \qquad (1.67)$$

If we express the $a_j(x)$ and a_k in terms of the Bernoulli polynomials and numbers respectively, then (1.67) becomes

$$\sum_{j=0}^{k} (-1)^j \binom{k}{j} B_j(x)\, x^{-j} = (-1)^k B_k\, x^{-k}\,. \qquad (1.68)$$

From the preceding analysis we can determine the conditions under which (1.62) is convergent and divergent. In the derivation of (1.64) we have used the power series expansion for $\exp(xt)$, which is convergent for all values of x and t. On the other hand, we have used (1.17), which with the aid of No. 9.616 from Ref. [14] can be expressed in terms of the Riemann zeta function as

$$\frac{t}{e^t - 1} \equiv 1 - \frac{t}{2} + 2 \sum_{k=1}^{\infty} \frac{(-1)^{k+1}}{(2\pi)^{2k}} \zeta(2k) \, t^k . \tag{1.69}$$

By introducing the Dirichlet series for the Riemann zeta function, viz. $\zeta(k) = \sum_{l=1}^{\infty} 1/l^k$, one obtains

$$\frac{t}{e^t - 1} \equiv 1 - \frac{t}{2} - 2 \sum_{l=1}^{\infty} \sum_{k=1}^{\infty} \left(-\frac{t^2}{4\pi^2 l^2} \right)^k . \tag{1.70}$$

The inner series in the above result is merely the geometric series with the variable z in (1.6) equal to $-t^2/4\pi^2 l^2$. From Appendix A we know that the radius of absolute convergence each geometric series is $|t| < 2\pi l$. Therefore, all the geometric series and the lhs of (1.70) are absolutely convergent only when $|t| < 2\pi$. In fact, from the discussion in Appendix A, (1.70) can be written as

$$\sum_{k=0}^{\infty} a_k(x) \, t^k \begin{cases} = \dfrac{t\, e^{xt}}{e^t - 1} , & \Re\, t^2 > -4\pi^2 \text{ and } \forall\, x , \\[2mm] \equiv \dfrac{t\, e^{xt}}{e^t - 1} , & \Re\, t^2 \leq -4\pi^2 \text{ and } \forall\, x . \end{cases} \tag{1.71}$$

Now we are in a position to apply the partition method for a power series expansion to (1.62). By inverting the exponential factor on the lhs into the denominator and substituting the power series for the resulting exponential, we find that

$$\frac{t\, e^{xt}}{e^t - 1} = \frac{1}{1 - \displaystyle\sum_{k=1}^{\infty} \frac{(-t)^k}{(k+1)!} \left((x-1)^{k+1} - x^{k+1} \right)} . \tag{1.72}$$

The rhs of the above equation can be viewed again as the regularized value of the geometric series. Hence we arrive at

$$\frac{t\, e^{xt}}{e^t - 1} \equiv \sum_{k=0}^{\infty} \left(\sum_{j=1}^{\infty} \frac{(-1)^j}{(j+1)!} \left((x-1)^{j+1} - x^{j+1} \right) t^j \right)^k . \tag{1.73}$$

This result is identical to (87) in Ref. [4] except for the phase factor of $(-1)^j$, which is given incorrectly as $(-1)^{j+1}$. It can also be expressed as

$$\frac{t\, e^{xt}}{e^t - 1} \equiv 1 - \left(\frac{t}{2!} \left((x-1)^2 - x^2 \right) - \frac{t^2}{3!} \left((x-1)^3 - x^3 \right) + \cdots \right)$$
$$+ \left(\frac{t}{2!} \left((x-1)^2 - x^2 \right) - \frac{t^2}{3!} \left((x-1)^3 - x^3 \right) + \cdots \right)^2 - \cdots . \tag{1.74}$$

From this result we can easily determine the lowest order terms by expanding in powers of t. These are

$$\frac{t e^{xt}}{e^t - 1} \equiv 1 + \left(x - \frac{1}{2}\right) t + \left(\frac{x^2}{2} - \frac{x}{2} + \frac{1}{12}\right) t^2 + \left(\frac{x^3}{6} - \frac{x^2}{4} + \frac{x}{12}\right) t^3 + O(t^4).$$

(1.75)

The above statement, which represents the corrected version of (89) in Ref. [4], is now of the same form as (1.62). Again, since the regularized value is unique, the rhs's of (1.75) and (1.62) are equal to one another. Consequently, we observe that $a_1(x) = x - 1/2$, $a_2(x) = x^2/2 - x/2 + 1/12$ and $a_3(x) = x^3/6 - x^2/4 + x/12$, which are, of course, $B_1(x)$, $B_2(x)/2!$ and $B_3(x)/3!$, respectively.

To determine the higher order coefficients, we require the partition method for a power series expansion. In applying the method we note that the outer series is again the geometric series. Hence the multinomial factor multiplying the assigned values to the parts remains the same as in the previous examples presented in this chapter. In addition, we assigned a value to each part i by taking the negative value of the coefficient of the i-th power in the inner series, as for example in (1.40) and (1.42). In (1.72) the coefficient of i-th power of t is $(-1)^i ((x-1)^{i+1} - x^{i+1})/(i+1)!$, which means that the value we now assign to each part i in a partition is minus or the negative value of this coefficient. Therefore, to obtain $a_k(x)/k!$, all we have to do is replace $(-1)^{i+1}/(i+1)$ in the product of (1.42) by $(-1)^i((x-1)^{i+1} - x^{i+1})/(i+1)!$. Then we obtain

$$a_k(x) = (-1)^k \sum_{\substack{\lambda_1, \lambda_2, \lambda_3, \dots, \lambda_k = 0 \\ \sum_{i=1}^{k} i\lambda_i = k}}^{k, \lfloor k/2 \rfloor, \lfloor k/3 \rfloor, \dots, 1} N_k! \prod_{i=1}^{k} \frac{1}{\lambda_i!} \left(\frac{(x-1)^{i+1} - x^{i+1}}{(i+1)!}\right)^{\lambda_i}.$$

(1.76)

In particular, for the partition $\{1, 1, 3\}$ the multinomial factor is simply $3!/(2! \cdot 1!)$, whilst the values assigned to the ones and three are $-((x-1)^2 - x^2)/2!$ and $-((x-1)^4 - x^4)/4!$, respectively. Then the contribution to $a_5(x)$ from this partition is

$$C(\{1, 1, 3\}) = -\frac{3!}{2! \cdot 1!} \left(\frac{((x-1)^2 - x^2)}{2!}\right)^2 \left(\frac{((x-1)^4 - x^4)}{4!}\right)$$

$$= \frac{x^5}{2} - \frac{5}{4} x^4 + \frac{11}{8} x^3 - \frac{13}{16} x^2 + \frac{x}{4} - \frac{1}{32}.$$

(1.77)

We calculate $a_5(x)$ by evaluating the sum of the contributions from all seven partitions summing to 5, which are $\{5\}$, $\{1, 4\}$, $\{1, 1, 3\}$, $\{2, 3\}$,

Table 1.4 Coefficients of the products of the polynomials $h_i(x)$ in the calculation of $a_5(x)$

	x^5	x^4	x^3	x^2	x	x^0
$h_5(x)$	$-1/120$	$1/48$	$-1/36$	$1/48$	$-1/120$	$1/720$
$2h_1(x)h_4(x)$	$1/12$	$-5/24$	$1/4$	$-1/6$	$7/120$	$-1/120$
$3h_1(x)^2h_3(x)$	$-1/2$	$5/4$	$-11/8$	$13/16$	$-1/4$	$1/32$
$2h_2(x)h_3(x)$	$1/6$	$-5/12$	$17/36$	$-7/24$	$7/72$	$1/72$
$4h_1(x)^3h_2(x)$	2	-3	$31/6$	$-11/4$	$3/4$	$-1/12$
$3h_1(x)h_2(x)^2$	$-3/4$	$15/8$	-2	$9/8$	$-1/3$	$1/24$
$h_1(x)^5$	-1	$5/2$	$-5/2$	$5/4$	$-5/16$	$1/32$
Total	$-1/120$	$1/48$	$-1/72$	0	$1/720$	0

$\{1, 1, 1, 2\}$, $\{1, 2, 2\}$ and $\{1, 1, 1, 1, 1\}$. If we put $h_i(x) = ((x - 1)^{i+1} - x^{i+1})/(i + 1)!$, then (1.76) yields

$$a_5(x) = -\left(h_5(x) + 2h_1(x)h_4(x) + 3h_1(x)^2h_3(x) + 2h_2(x)h_3(x) \right.$$
$$\left. + 4h_1(x)^3h_2(x) + 3h_1(x)h_2(x)^2 + h_1(x)^5\right). \tag{1.78}$$

Each term on the rhs of (1.78) yields a fifth order polynomial in x whose coefficients are displayed in Table 1.4. Aside from the minus sign arising from the different definition of $h_i(x)$ here, Table 1.4 represents the corrected version of Table 2 in Ref. [4], which possesses a few typographical errors. The sums of the coefficients in each column appear in the bottom row of the table. From (1.76) these sums represent the negative values of the coefficients of $a_5(x)$. Hence $a_5(x)$ is given by

$$a_5(x) = \frac{x^5}{120} - \frac{x^4}{48} + \frac{x^3}{72} - \frac{x}{720} = \frac{1}{5!}\left(x^5 - \frac{5}{2}x^4 + \frac{5}{3}x^3 - \frac{x}{6}\right). \tag{1.79}$$

In other words, we find that $a_5(x) = B_5(x)/5!$. Furthermore, we can determine the Bernoulli numbers by setting $x = 0$ in (1.76), thereby obtaining

$$B_k = k! \sum_{\substack{\lambda_1,\lambda_2,\lambda_3,\dots,\lambda_k=0 \\ \sum_{i=1}^{k} i\lambda_i=k}}^{k,\lfloor k/2\rfloor,\lfloor k/3\rfloor,\dots,1} (-1)^{N_k} N_k! \prod_{i=1}^{k} \frac{1}{\lambda_i!}\left(\frac{1}{(i+1)!}\right)^{\lambda_i}. \tag{1.80}$$

The Bernoulli numbers are zero for odd values of k greater than unity. Therefore, the above result reduces to

$$\sum_{\substack{\lambda_1,\lambda_2,\lambda_3,\dots,\lambda_{2k+1}=0 \\ \sum_{i=1}^{2k+1} i\lambda_i=2k+1}}^{2k+1,k,\lfloor(2k+1)/3\rfloor,\dots,1} (-1)^{N_{2k+1}} N_{2k+1}! \prod_{i=1}^{2k+1} \frac{1}{\lambda_i!}\left(\frac{1}{(i+1)!}\right)^{\lambda_i} = 0, \tag{1.81}$$

and

$$B_{2k} = (2k)! \sum_{\substack{\lambda_1,\lambda_2,\lambda_3,\dots,\lambda_{2k}=0 \\ \sum_{i=1}^{2k} i\lambda_i = 2k}}^{2k,k,\lfloor 2k/3 \rfloor,\dots,1} (-1)^{N_{2k}} N_{2k}! \prod_{i=1}^{2k} \frac{1}{\lambda_i!} \left(\frac{1}{(i+1)!} \right)^{\lambda_i}. \tag{1.82}$$

Comparing the above result with (1.17), we see that there is no factor of $2 - 2^{2k}$ in the denominator, but it is considerably more complex because it involves all the parts up to $2k$, whereas (1.17) only involves parts up to k. In fact, equating both results yields

$$\sum_{\substack{\lambda_1,\lambda_2,\lambda_3,\dots,\lambda_k=0 \\ \sum_{i=1}^{k} i\lambda_i = k}}^{k,\lfloor k/2 \rfloor,\lfloor k/3 \rfloor,\dots,1} (-1)^{N_k} N_k! \prod_{i=1}^{k} \left(\frac{1}{(2i+1)!} \right)^{\lambda_i} \frac{1}{\lambda_i!} = (2 - 2^{2k})$$

$$\times \sum_{\substack{\lambda_1,\lambda_2,\lambda_3,\dots,\lambda_{2k}=0 \\ \sum_{i=1}^{2k} i\lambda_i = 2k}}^{2k,k,\lfloor 2k/3 \rfloor,\dots,1} (-1)^{N_{2k}} N_{2k}! \prod_{i=1}^{2k} \frac{1}{\lambda_i!} \left(\frac{1}{(i+1)!} \right)^{\lambda_i}. \tag{1.83}$$

This is an unusual result relating the sum over partitions summing to k with the sum over partitions summing to $2k$.

Just as the Bernoulli numbers give rise to Bernoulli polynomials, the cosecant numbers also give rise to their own polynomials except that there is now a trigonometric function in the generating function. These polynomials denoted by $c_k(x)$ are defined in Ref. [5] as

$$c_k(t) := \sum_{j=0}^{k} \frac{(-1)^j}{(2j)!} c_{k-j} t^{2j}, \tag{1.84}$$

and satisfy the following generating function:

$$\sum_{k=0}^{\infty} c_k(t) x^{2k} \equiv \frac{x \cos(xt)}{\sin x}. \tag{1.85}$$

Thus, we see that the polynomials are of degree $2k$, while the coefficients oscillate in sign with the constant term equal to c_k and the highest order coefficient equal to $1/(2k)!$. In addition to being even, they possess the following properties:

$$\frac{d^2}{dt^2} c_{k+1}(t) = -c_k(t),$$

$$c_k(1/2) = 2^{-2k} c_k,$$

$$c_k(1) = \frac{c_k}{2^{1-2k} - 1} \, ,$$

$$c_k(2) = c_k + \frac{2(-1)^k}{(2k-1)!} \, ,$$

$$c_k(3) = \frac{c_k}{2^{1-2k} - 1} + \frac{(-1)^k 2^{2k}}{(2k-1)!} \, . \tag{1.86}$$

The last two results can be obtained from the $t=1$ cases of the following identities:

$$c_k(t+1) - c_k(t-1) = \frac{2(-1)^k t^{2k-1}}{(2k-1)!} \, , \tag{1.87}$$

and

$$c_k(t+2) - c_k(t-2) = \frac{2(-1)^k}{(2k-1)!} \left((t+1)^{2k-1} + (t-1)^{2k-1} \right) . \tag{1.88}$$

Furthermore, from Theorem 11 in Ref. [5], they are related to the Bernoulli polynomials via

$$c_k(2t-1) = (-1)^k 2^{2k} \frac{B_{2k}(t)}{(2k)!} \, , \tag{1.89}$$

and

$$c_k'(2t-1) = (-1)^{k+1} 2^{2k+1} \frac{B_{2k+1}(t)}{(2k+1)!} \, . \tag{1.90}$$

In (1.90) the prime superscript denotes the derivative. The cosecant polynomials are also superior to the Bernoulli polynomials since their coefficients do not diverge at any stage. These interesting polynomials are displayed in Table 1.5 up to $k=15$.

It should also be mentioned that (1.85) has been derived by multiplying the power series for $\cos(zt)$ by (1.3). Since the power series for cosine is convergent for all values of its argument, it is the convergence behaviour of (1.3) that determines the convergence behaviour of (1.85). From the corollary to Theorem 1.1, (1.3) is convergent for $|\Re x| < \pi$ and divergent for all other values of x. Hence (1.85) is convergent for $|\Re t| < \pi$ and for all values of x. For t lying in this strip of the complex plane, the equivalence symbol can be replaced by an equals sign. For t situated outside the strip, it is divergent irrespective of the value of x and thus, the rhs represents the regularized value of the lhs. Furthermore, within the strip (1.85) is absolutely convergent for $|t| < \pi$ and conditionally convergent elsewhere.

Table 1.5 The cosecant polynomials $c_k(t)$

k	$c_k(t)$
0	1
1	$\frac{1}{6} - \frac{t^2}{2}$
2	$\frac{7}{360} - \frac{t^2}{12} + \frac{t^4}{24}$
3	$\frac{31}{15120} - \frac{7t^2}{720} + \frac{t^4}{144} - \frac{t^6}{720}$
4	$\frac{127}{604800} - \frac{31\,t^2}{30240} + \frac{7\,t^4}{8640} - \frac{t^6}{4320} + + \frac{t^6}{40320}$
5	$\frac{73}{3421440} - \frac{127\,t^2}{1209600} + \frac{31\,t^4}{362880} - \frac{7\,t^6}{259200} + \frac{t^8}{241920} - \frac{t^{10}}{3628800}$
6	$\frac{1414477}{653837184000} - \frac{73\,t^2}{6842880} + \frac{127\,t^4}{14515200} - \frac{31\,t^6}{10886400} + \frac{t^8}{2073600} - \frac{t^{10}}{21772800} + \frac{t^{12}}{479001600}$
7	$\frac{8191}{37362124800} - \frac{1414477\,t^2}{1307674368000} + \frac{73\,t^4}{82114560} - \frac{127\,t^6}{435456000} + \frac{31\,t^8}{609638400} - \frac{t^{10}}{186624000}$ $+ \frac{t^{12}}{2874009600} - \frac{t^{14}}{87178291200}$
8	$\frac{16931177}{762187345920000} - \frac{8191t^2}{74724249600} + \frac{1414477t^4}{15692092416000} - \frac{73t^6}{2463436800} + \frac{127t^8}{24385536000}$ $- \frac{31t^{10}}{54867456000} + \frac{t^{12}}{24634368000} - \frac{t^{14}}{523069747200} + \frac{t^{16}}{20922789888000}$
9	$\frac{5749691557}{2554547108585472000} - \frac{16931177t^2}{1524374691840000} + \frac{8191t^4}{896690995200} - \frac{1414477t^6}{470762772480000}$ $+ \frac{73t^8}{137952460800} - \frac{127t^{10}}{2194698240000} + \frac{31t^{12}}{7242504192000} - \frac{t^{14}}{4483454976000}$ $+ \frac{t^{16}}{125536739328000} - \frac{t^{18}}{6402373705728000}$
10	$\frac{91546277357}{401428831349145600000} - \frac{5749691557t^2}{5109094217170944000} + \frac{16931177t^4}{18292496302080000} - \frac{8191t^6}{26900729856000}$ $+ \frac{1414477t^8}{26362715258880000} - \frac{73t^{10}}{12415721472000} + \frac{127t^{12}}{289700167680000} - \frac{31t^{14}}{1318135762944000}$ $+ \frac{t^{16}}{1076029194240000} - \frac{t^{18}}{38414242234368000} + \frac{t^{20}}{2432902008176640000}$
11	$\frac{3324754717}{143888775912161280000} - \frac{91546277357t^2}{80285766269829120000} + \frac{5749691557t^4}{61309130606051328000}$ $- \frac{16931177t^6}{5487748890062400000} + \frac{8191t^8}{1506440871936000} - \frac{1414477t^{10}}{2372644373299200000} + \frac{73t^{12}}{1638875234304000}$ $- \frac{127t^{14}}{52725430517760000} + \frac{31t^{16}}{316352583106560000} - \frac{t^{18}}{329264933437440000}$ $+ \frac{t^{20}}{145974120490598400000} - \frac{t^{22}}{1124000727777607680000}$
12	$\frac{1982765468311237}{846912068365871834726400000} - \frac{3324754717t^2}{287777551824322560000} + \frac{91546277357t^4}{96342919523794944000000}$ $- \frac{5749691557t^6}{18392739181815398400000} + \frac{16931177t^8}{30731393787494400000} - \frac{8191t^{10}}{1355579678474240000}$ $+ \frac{1414477t^{12}}{3131890572754944000000} - \frac{73t^{14}}{298275292643328000} + \frac{127t^{16}}{12654103324262400000}$ $- \frac{31t^{18}}{9680389043060736000} + \frac{t^{20}}{125120674706227200000} - \frac{t^{22}}{674400436665646080000}$ $+ \frac{t^{24}}{620448401733239439360000}$
13	$\frac{22076500342261}{9306726025998591590400000000} - \frac{1982765468311237t^2}{169382413673174366945280000} + \frac{3324754717t^4}{345333062189187072000}$ $- \frac{91546277357t^6}{289028758571384832000000} + \frac{5749691557t^8}{10299933941816623104000} - \frac{16931177t^{10}}{27658254408744960000}$ $+ \frac{8191t^{12}}{17896517558599680000} - \frac{1414477t^{14}}{570004084241399808000000} + \frac{73t^{16}}{71586070234398720000}$ $- \frac{127t^{18}}{3872155617224294400000} + \frac{31t^{20}}{36785478363630796800000} - \frac{t^{22}}{578057517142769664000000}$ $+ \frac{t^{24}}{372269041039436636160000} - \frac{t^{26}}{403291461126605635584000000}$

(continued on next page)

Table 1.5 (continued)

k	$c_k(t)$

| 14 | |

$$\frac{65053034220152267}{27066618348182761085337600000000} - \frac{22076500342261t^2}{18613452051997183180800000}$$

$$+ \frac{1982765468311237t^4}{2032588964078092403343600000} - \frac{3324754717t^6}{103599918656756121600000}$$

$$+ \frac{91546277357t^8}{161856104799975505920000000} - \frac{5749691557t^{10}}{9269940547634960793600000}$$

$$+ \frac{16931177t^{12}}{365088958195433472000000} - \frac{8191t^{14}}{325716619565141760000} + \frac{1414477t^{16}}{136800980217935953920000000}$$

$$- \frac{73t^{18}}{2190533749172600832000} + \frac{127t^{20}}{14714191345452318720000000} - \frac{31t^{22}}{169948910039974281216000000}$$

$$+ \frac{t^{24}}{31908774946280885452800000} - \frac{t^{26}}{2419748766759633813504000000}$$

$$+ \frac{t^{28}}{304888344611713860501504000000}$$

| 15 | |

$$\frac{9251189109760413581111}{3798951458230200341789210050560000000} - \frac{65053034220152267t^2}{5413323669636552217067520000000}$$

$$+ \frac{22076500342261t^4}{223361424623966198169600000} - \frac{1982765468311237t^6}{60977668922342772100300800000}$$

$$+ \frac{3324754717t^8}{580159544477834280960000} - \frac{91546277357t^{10}}{14567049431997795532800000000}$$

$$+ \frac{5749691557t^{12}}{1223632152228781482475520000} - \frac{16931177t^{14}}{66446190391568891904000000}$$

$$+ \frac{8191t^{16}}{78171988695963402240000} - \frac{1414477t^{18}}{418610999466884018995200000}$$

$$+ \frac{73t^{20}}{832402824685588316160000} - \frac{127t^{22}}{67979564015989712486400000}$$

$$+ \frac{31t^{24}}{9381179834206580323123200000} - \frac{t^{26}}{2074070371508257554432000000}$$

$$+ \frac{t^{28}}{18293300676702831630090240000000} - \frac{t^{30}}{265252859812191058636308480000000}$$

If we put $t = (x+1)/2$ and replace the rhs of (1.89) by (1.76), then we find that

$$c_k(x) = (-1)^k \sum_{\substack{\lambda_1,\lambda_2,\lambda_3,\ldots,\lambda_{2k}=0 \\ \sum_{i=1}^{2k} i\lambda_i=2k}}^{2k,k,\lfloor 2k/3\rfloor,\ldots,1} 2^{-N_{2k}} N_{2k}!$$

$$\times \prod_{i=1}^{2k} \frac{1}{\lambda_i!} \left(\frac{(x-1)^{i+1} - (x+1)^{i+1}}{(i+1)!} \right)^{\lambda_i}. \tag{1.91}$$

Similarly for (1.90), we find that

$$c_k'(x) = (-1)^k \sum_{\substack{\lambda_1,\lambda_2,\lambda_3,\ldots,\lambda_{2k+1}=0 \\ \sum_{i=1}^{2k+1} i\lambda_i=2k+1}}^{2k+1,k,\lfloor (2k+1)/3\rfloor,\ldots,1} 2^{-N_{2k+1}} N_{2k+1}!$$

$$\times \prod_{i=1}^{2k+1} \frac{1}{\lambda_i!} \left(\frac{(x-1)^{i+1} - (x+1)^{i+1}}{(i+1)!} \right)^{\lambda_i}. \tag{1.92}$$

By putting $x = 1$ in the above results and using (1.89) and (1.90), we re-derive (1.81) and (1.82). Moreover, according to (27) of Ref. [33], we have

$$B_{2k}(1/3) = -\frac{1}{2}(1 - 3^{-2k})B_{2k}.$$
(1.93)

If we put $t = 1/3$ in (1.89) and (1.91), then we obtain

$$\sum_{\substack{\lambda_1,\lambda_2,\lambda_3,\dots,\lambda_{2k}=0 \\ \sum_{i=1}^{2k} i\lambda_i=2k}}^{2k,k,\lfloor 2k/3 \rfloor,\dots,1} \left(\frac{1}{3}\right)^{N_{2k}} N_{2k}! \prod_{i=1}^{2k} \frac{1}{\lambda_i!}\left(\frac{(-2)^{i+1}-1}{(i+1)!}\right)^{\lambda_i} = \left(\frac{3-3^{2k}}{2(2k)!}\right) B_{2k}.$$

(1.94)

Since we have shown how the Bernoulli polynomials can be derived via the partition method for a power series expansion, we shall conclude this chapter by applying the method to the generating function whose coefficients are the Euler polynomials. This appears as (2) in Ref. [34] and is given by

$$\sum_{k=0}^{\infty} E_k(x) \frac{t^k}{k!} \equiv \frac{2e^{xt}}{e^t + 1}.$$
(1.95)

As in the case of the generating function with Bernoulli polynomial coefficients, it has been written as an equivalence because the series on the lhs can become divergent, while the rhs is finite. This can be observed by taking out a factor of $\exp(t/2)$ in the denominator on the rhs and writing it with the aid of (1.25) as

$$\frac{e^{(x-1/2)t}}{\cos(it/2)} \equiv \sum_{k=0}^{\infty}(x-1/2)^k \frac{t^k}{k!} \sum_{k=0}^{\infty}(-1)^k d_k(t/2)^{2k}.$$
(1.96)

The first series on the rhs of (1.96) is not divergent. Therefore, we concentrate on the second series, which is equivalent to $\mathrm{sech}(t/2)$ according to (1.25). By introducing (1.30) into the second series and interchanging the summations, we obtain

$$\mathrm{sech}(t/2) \equiv \frac{4}{\pi} \sum_{j=1}^{\infty} \frac{(-1)^{j+1}}{(2j-1)} \sum_{k=0}^{\infty}(-1)^k \left(\frac{t}{(2j-1)\pi}\right)^{2k}.$$
(1.97)

The second series in (1.97) is simply the geometric series. From Appendix A we know that this series is only convergent for $\Re(-t^2) < (2j-1)^2\pi^2$. Since the $j = 1$ case incorporates all other values of j, the condition for convergence becomes $\Re(it)^2 < \pi^2$ or $-\pi < \Im t < \pi$. Outside this strip, the series is divergent.

If we divide the exponential factor in the numerator on the rhs of (1.95) into denominator and expand both exponentials into their power series form, then we find that

$$\frac{2e^{xt}}{e^t + 1} = \frac{1}{1 + \sum_{k=1}^{\infty} \frac{(-t)^k}{2\,k!}\left((x-1)^k + x^k\right)}.$$ (1.98)

As with the previous examples in this chapter, the outer series is again the geometric series. Comparing (1.98) with (1.73), we see that: (1) there is now a minus sign outside the summation over k, (2) the power of t^k has a factor of $k!$ in its denominator rather than $(k+1)!$, and $(x-1)^{k+1} - x^{k+1}$ has been replaced by $(x-1)^k + x^k$. The first of these differences introduces a factor of $(-1)^{N_k}$ into the sum over partitions, while the second and third differences mean that instead of using $h_i(x)$ they become modified to $g_i(x) = ((x-1)^i + x^i)/i!$ inside the product. Introducing these modifications into (1.76) yields the following result for the Euler polynomials:

$$E_k(x) = (-1)^k\,k! \sum_{\substack{\lambda_1,\lambda_2,\lambda_3,\dots,\lambda_k=0 \\ \sum_{i=1}^{k} i\lambda_i = k}}^{k,\lfloor k/2\rfloor,\lfloor k/3\rfloor,\dots,1} \left(-\frac{1}{2}\right)^{N_k} N_k! \prod_{i=1}^{k} \frac{1}{\lambda_i!}\left(\frac{(x-1)^i + x^i}{i!}\right)^{\lambda_i}.$$ (1.99)

The analogue of the cosecant polynomials are the secant polynomials, which are defined in Ref. [5] as

$$d_k(t) := \sum_{j=0}^{k} \frac{(-1)^j}{(2j+1)!}\, d_{k-j}\, t^{2j+1}.$$ (1.100)

The trigonometric function that produces the generating function with these odd polynomials as their coefficients is given by

$$\sum_{k=0}^{\infty} d_k(t)x^{2k} \equiv \frac{\sin(xt)}{x\cos x}.$$ (1.101)

In addition, they possess the following properties:

$$\frac{d^2}{dt^2}\, d_{k+1}(t) = -d_k(t),$$
$$d_k(1) = -(2^{2k+2} - 1)\, c_{k+1}(1),$$
$$d_k(2) = \frac{2(-1)^k}{(2k+1)!}.$$ (1.102)

They are displayed up to $k = 15$ in Table 1.6.

Table 1.6 The secant polynomials $d_k(t)$

k	$d_k(t)$
0	t
1	$\frac{t}{2} - \frac{t^3}{6}$
2	$\frac{5t}{24} - \frac{t^3}{12} + \frac{t^5}{120}$
3	$\frac{61t}{720} - \frac{5t^3}{144} + \frac{t^5}{240} - \frac{t^7}{5040}$
4	$\frac{277t}{8064} - \frac{61t^3}{4320} + \frac{t^5}{576} - \frac{t^7}{10080} + +\frac{t^9}{362880}$
5	$\frac{50521t}{3628800} - \frac{277t^3}{48384} + \frac{61t^5}{86400} - \frac{t^7}{24192} + \frac{t^9}{725760} - \frac{t^{11}}{39916800}$
6	$\frac{540553t}{95800320} - \frac{50521t^3}{21772800} + \frac{277t^5}{967680} - \frac{61t^7}{3628800} + \frac{t^9}{1741824} - \frac{t^{11}}{79833600} + \frac{t^{13}}{6227020800}$
7	$\frac{199360981t}{87178291200} - \frac{540553t^3}{574801920} + \frac{50521t^5}{435456000} - \frac{277t^7}{40642560} + \frac{61t^9}{261273600} - \frac{t^{11}}{191600640}$ $- \frac{t^{13}}{12454041600} - \frac{t^{15}}{1307674368000}$
8	$\frac{3878302429t}{4184557977600} - \frac{199360981t^3}{523069747200} + \frac{540553t^5}{11496038400} - \frac{50521t^7}{18289152000} + \frac{277t^9}{2926264320}$ $- \frac{61t^{11}}{28740096000} + \frac{t^{13}}{29889699840} - \frac{t^{15}}{2615348736000} + \frac{t^{17}}{355687428096000}$
9	$\frac{2404879675441t}{6402373705728000} - \frac{3878302429t^3}{25107347865600} + \frac{199360981t^5}{10461394944000} - \frac{540553t^7}{482833612800} + \frac{50521t^9}{1316818944000}$ $- \frac{277t^{11}}{321889075200} + \frac{61t^{13}}{4483454976000} - \frac{t^{15}}{6276836966400} + \frac{t^{17}}{711374856192000}$ $- \frac{t^{19}}{121645100408832000}$
10	$\frac{14814847529501t}{973160803270656000} - \frac{2404879675441t^3}{38414242234368000} + \frac{3878302429t^5}{502146957312000} - \frac{199360981t^7}{439378587648000}$ $+ \frac{540553t^9}{34764020121600} - \frac{50521t^{11}}{144850083840000} + \frac{277t^{13}}{50214695731200} - \frac{61t^{15}}{941525544960000}$ $+ \frac{t^{17}}{1707299654860800} - \frac{t^{19}}{243290200817664000} + \frac{t^{21}}{51090942171709440000}$
11	$\frac{69348874393137901t}{1124000727777607680000} - \frac{14814847529501t^3}{583896481962393600} + \frac{2404879675441t^5}{768284844687360000}$ $- \frac{3878302429t^7}{21090172207104000} + \frac{199360981t^9}{31635258310656000} - \frac{540553t^{11}}{3824042213376000} + \frac{50521t^{13}}{22596613079040000}$ $- \frac{277t^{15}}{10545086103552000} + \frac{61t^{17}}{256094948229120000} - \frac{t^{19}}{583896481962393600}$ $+ \frac{t^{21}}{102181884343418880000} - \frac{t^{23}}{25852016738884976640000}$
12	$\frac{238685140977801337t}{9545360026665221144000} - \frac{69348874393137901t^3}{6744004366665646080000} + \frac{14814847529501t^5}{11677929639247872000}$ $- \frac{2404879675441t^7}{32267963476869120000} + \frac{3878302429t^9}{1518492398911488000} - \frac{199360981t^{11}}{3479878414172160000}$ $+ \frac{41581t^{13}}{45888506560512000} - \frac{50521t^{15}}{4745288746598400000} + \frac{277t^{17}}{2868263420166144000}$ $- \frac{61t^{19}}{87584472294359040000} + \frac{t^{21}}{24523652224205312000} - \frac{t^{23}}{51704033477769953280000}$ $+ \frac{t^{25}}{15511210043330985984000000}$
13	$\frac{4087072509293123892361t}{403291461126605635584000000} - \frac{238685140977801337t^3}{57272160159991332864000} + \frac{69348874393137901t^5}{13488008733331292921600000}$ $- \frac{14814847529501t^7}{490473044848410624000} + \frac{2404879675441t^9}{23232933703345766640000} - \frac{3878302429t^{11}}{16703416388026368000}$ $+ \frac{199360981t^{13}}{542861032610856960000} - \frac{41581t^{15}}{9636586377707520000} + \frac{50521t^{17}}{1290718539074764800000}$ $- \frac{277t^{19}}{980946089696821248000} + \frac{61t^{21}}{3678547836363079680000} - \frac{t^{23}}{1240896803466478878720000}$ $+ \frac{t^{25}}{310224200866619719680000000} - \frac{t^{27}}{10888869450418352160768000000}$

(continued on next page)

Table 1.6 (*continued*)

k	$d_k(t)$

14
$$\frac{1318168043582768279403\,t}{3209350995912777478963200000} - \frac{4087072509293123892361\,t^3}{24197487667596338135040000000}$$
$$+ \frac{238685140977801337\,t^5}{1145443203199826657280000} - \frac{69348874393137901\,t^7}{566496366799914270720000} + \frac{14814847529501\,t^9}{3531405922908556492800}$$
$$- \frac{2404879675441\,t^{11}}{2555622707368034304000000} + \frac{3878302429t^{13}}{2605732956532113408000000} - \frac{199360981\,t^{15}}{114000816848279961600000}$$
$$+ \frac{41581\,t^{17}}{262115149473644544000000} - \frac{2659t^{19}}{23232933703345766400000} + \frac{277t^{21}}{4119973576726649241600000}$$
$$- \frac{61\,t^{23}}{18613452051997183180800000} + \frac{t^{25}}{7445380820798873272320000000}$$
$$- \frac{t^{27}}{217777389008367043215360000000} + \frac{t^{29}}{8841761993739701954543616000000}$$

15
$$\frac{44154389324902310455368282\,t}{265252859812191058636308480000000} - \frac{1318168043582768279403\,t^3}{192561059754766648737792000000}$$
$$+ \frac{4087072509293123892361\,t^5}{483949753351926762700800000000} - \frac{238685140977801337\,t^7}{48108614534392719605760000}$$
$$+ \frac{69348874393137901\,t^9}{40787738409593827491840000} - \frac{14814847529501\,t^{11}}{3884546515199412142080000}$$
$$+ \frac{2404879675441\,t^{13}}{398677142349413351424000} - \frac{3878302429\,t^{15}}{547203920871743815680000}$$
$$+ \frac{199360981\,t^{17}}{3100822218273214955520000} - \frac{41581\,t^{19}}{896433811199864340480000}$$
$$+ \frac{2659\,t^{21}}{9757832155405221888000000} - \frac{277\,t^{23}}{208470662982368451624960000}$$
$$+ \frac{61\,t^{25}}{11168071231198309908480000000} - \frac{t^{27}}{5226657336200809037168640000}$$
$$+ \frac{t^{29}}{176835239874794039090872320000000} - \frac{t^{31}}{822283865417792281772556288000000}$$

In Theorem 14 of Ref. [5], the secant polynomials are related to the Euler polynomials by

$$d_k(2t-1) = \frac{(-1)^k 2^{2k+1}}{(2k+1)!}\, E_{2k+1}(t)\,, \tag{1.103}$$

and

$$d_k'(2t-1) = \frac{(-1)^k 2^{2k}}{(2k)!}\, E_{2k}(t)\,. \tag{1.104}$$

If we substitute t by $(x+1)/2$ after introducing (1.99) in these results, then we arrive at

$$d_k(x) = (-1)^{k+1} \sum_{\substack{\lambda_1,\lambda_2,\lambda_3,\ldots,\lambda_{2k+1}=0 \\ \sum_{i=1}^{2k+1} i\lambda_i = 2k+1}}^{2k+1,k,\lfloor (2k+1)/3\rfloor,\ldots,1} \left(-\frac{1}{2}\right)^{N_{2k+1}} N_{2k+1}!$$
$$\times \prod_{i=1}^{2k+1} \frac{1}{\lambda_i!}\left(\frac{(x-1)^i + (x+1)^i}{i!}\right)^{\lambda_i}\,, \tag{1.105}$$

and

$$d'_k(x) = (-1)^k \sum_{\substack{\lambda_1,\lambda_2,\lambda_3,\ldots,\lambda_{2k}=0 \\ \sum_{i=1}^{2k} i\lambda_i=2k}}^{2k,k,\lfloor 2k/3\rfloor,\ldots,1} \left(-\frac{1}{2}\right)^{N_{2k}} N_{2k}!$$

$$\times \prod_{i=1}^{2k} \frac{1}{\lambda_i!} \left(\frac{(x-1)^i + (x+1)^i}{i!}\right)^{\lambda_i}. \tag{1.106}$$

By putting $x = 1$ in (1.105) and using the third and second results in (1.86) and (1.102) respectively, we find after a little algebra that

$$c_k = (-1)^k \left(\frac{2^{2k-1}-1}{2^{2k}-1}\right) \sum_{\substack{\lambda_1,\lambda_2,\lambda_3,\ldots,\lambda_{2k-1}=0 \\ \sum_{i=1}^{2k-1} i\lambda_i=2k-1}}^{2k-1,k-1,\lfloor(2k-1)/3\rfloor,\ldots,1} \left(-\frac{1}{2}\right)^{N_{2k-1}}$$

$$\times N_{2k-1}! \prod_{i=1}^{2k-1} \frac{1}{\lambda_i!(i!)^{\lambda_i}}. \tag{1.107}$$

The above result is a different sum over partitions for the cosecant numbers compared with (1.13). In this case the sum is over all partitions that sum to $2k - 1$, which means that this form for the cosecant numbers is more complex. On the other hand, if we put $x = 2$ in (1.105) and use the third result in (1.102), then we obtain

$$\frac{(-1)^{k+1}}{(2k+1)!} = \sum_{\substack{\lambda_1,\lambda_2,\lambda_3,\ldots,\lambda_{2k+1}=0 \\ \sum_{i=1}^{2k+1} i\lambda_i=2k+1}}^{2k+1,k,\lfloor(2k+1)/3\rfloor,\ldots,1} \left(-\frac{1}{2}\right)^{N_{2k+1}} N_{2k+1}! \prod_{i=1}^{2k+1} \frac{1}{\lambda_i!} \left(\frac{1+3^i}{i!}\right)^{\lambda_i}. \tag{1.108}$$

This is the alternative form of (1.34) with the sum over partitions extended to $2k + 1$ instead of k.

To conclude this chapter, we write the rhs of (1.101) as

$$\frac{\sin(xt)}{x\cos x} = \frac{t}{xt\csc(xt)\cos x}. \tag{1.109}$$

Next we replace $\csc(xt)$ by (1.8), which we have seen is valid provided that $|xt| < \pi$. Introducing the power series for $\cos x$ results in

$$\frac{\sin(xt)}{x\cos x} = \frac{t}{1 + \sum_{k=1}^{\infty}(xt)^{2k} \sum_{j=0}^{k} \frac{(-1)^j}{(2j)!} c_{k-j} t^{-2j}}. \tag{1.110}$$

The summation over j is merely $c_k(1/t)$ according to (1.84). Hence the above equation becomes

$$\frac{\sin(xt)}{x\cos x} = \frac{t}{1 + \sum\limits_{k=1}^{\infty} c_k(1/t)\left(xt\right)^{2k}}. \tag{1.111}$$

Now we are in a position to apply the partition method for a power series expansion. In this case the inner series has coefficients $c_k(1/t)$, while the outer series is the geometric series. The resulting power series is in powers of $(xt)^2$. In fact, the situation is identical to (1.33) except that x and c_k are replaced by xt and $c_k(1/t)$, respectively. Therefore, we find that

$$d_k(t) = t^{2k+1} \sum_{\substack{\lambda_1,\lambda_2,\lambda_3,\dots,\lambda_k=0 \\ \sum_{i=0}^{k} i\lambda_i=k}}^{k,\lfloor k/2 \rfloor, \lfloor k/3 \rfloor,\dots,1} (-1)^{N_k} N_k! \prod_{i=1}^{k} c_i\left(1/t\right)^{\lambda_i} \frac{1}{\lambda_i!}. \tag{1.112}$$

A similar type of analysis can be carried out for the generating function of the cosecant polynomials. Eventually, one obtains

$$c_k(t) = t^{2k} \sum_{\substack{\lambda_1,\lambda_2,\lambda_3,\dots,\lambda_k=0 \\ \sum_{i=1}^{k} i\lambda_i=k}}^{k,\lfloor k/2 \rfloor, \lfloor k/3 \rfloor,\dots,1} (-1)^{N_k} N_k!\, t^{N_k} \prod_{i=1}^{k} d_i\left(1/t\right)^{\lambda_i} \frac{1}{\lambda_i!}. \tag{1.113}$$

The major difference between both the above results is the appearance of the factor of t^{N_k} in the sum over partitions for the cosecant polynomials.

The results presented in this section have shown that the partition method for a power series expansion can be applied to situations where the coefficients of the resulting power series are not necessarily numerical. In the next chapter we shall investigate more advanced applications by adapting the method to problems where the outer series is no longer limited to the geometric series as has been the case throughout this chapter.

CHAPTER 2

More Advanced Applications

In the introductory chapter the partition method for a power series expansion was applied to situations involving elementary transcendental functions. Various power series expansions including a not so well-known one for the reciprocal of the logarithm function, were derived by representing the coefficients at order k as a sum over all the partitions summing to that order. It was seen that the method involves identifying an inner series, whose coefficients are directly related to the values each part is assigned when calculating the contribution to a coefficient from a partition, and an outer series, which is responsible for the multinomial factor for each partition in these calculations. All the examples, however, involved the same outer series, the geometric series. In this chapter we aim to investigate more advanced examples where the outer series is no longer the geometric series. Consequently, the multinomial factor of Chapter 1, which was represented by the factorial of the total number of parts or the length of a partition, viz. $N_k!$, divided by the factorials of the multiplicities, $\lambda_i!$, will take on different forms here. This will enable a more general theory of the method to be developed in Chapter 4.

2.1 BELL POLYNOMIALS OF THE FIRST KIND

There are actually two types of Bell polynomials. The first type, which are investigated in this section, are the resulting polynomials derived from the generating function of the exponential, $\exp(x(e^t - 1)x)$. They are normally denoted by $B_k(x)$ in the literature, but to avoid confusion with the more famous Bernoulli polynomials, they will be denoted here by $\mathcal{B}_k(x)$.

The second type of Bell polynomials, which are sometimes referred to as Bell polynomials of the second kind [35] or partition polynomials, are more general quantities that are derived by calculating sums over partitions in a similar manner to the partition method for a power series expansion [36]. They also involve the same multinomial factor appearing in the various examples presented in Chapter 1. Because of this, one can be easily misled into believing that both methods are identical to each other, an issue which will be discussed in greater detail in Chapter 4. In fact, by considering examples with a different multinomial factor as in this chapter it will become

The Partition Method for a Power Series Expansion.
DOI: http://dx.doi.org/10.1016/B978-0-12-804466-7.00002-4

evident that the partition method for a power series expansion is far more versatile.

As mentioned above, Bell polynomials of the first kind or $\mathcal{B}_k(x)$ appear as the following generating function:

$$\sum_{k=0}^{\infty} \mathcal{B}_k(x) \frac{t^k}{k!} = \exp\left(x(e^t - 1)\right) . \tag{2.1}$$

If we expand the exponential term $\exp(t)$ into its standard power series and do likewise to the resulting exponential, then we obtain

$$\sum_{k=0}^{\infty} \mathcal{B}_k(x) \frac{t^k}{k!} = \sum_{k=0}^{\infty} \left(t + t^2/2! + t^3/3! + t^4/4! + \cdots\right)^k \frac{x^k}{k!} . \tag{2.2}$$

Note that there is no equivalence symbol in the above result, because the exponentials have been expanded into their power series form, which is convergent for all values in the complex plane. Expanding (2.2) in powers of t yields

$$\sum_{k=0}^{\infty} \mathcal{B}_k(x) \frac{t^k}{k!} = 1 + xt + \left(x + x^2\right) \frac{t^2}{2!} + \left(x + 3x^3 + x^3\right) \frac{t^3}{3!}$$

$$+ \left(x + 7x^2 + 6x^3 + x^4\right) \frac{t^4}{4!} + \cdots . \tag{2.3}$$

From the above we see that higher order polynomials become unwieldy. Therefore, let us apply the method for a power series expansion as in Chapter 1. If we compare (2.2) with the corresponding form for the reciprocal logarithm numbers, viz. (1.39), then we see that the coefficients of the inner series are $1/k!$, whereas for the reciprocal logarithm numbers they are given by $(-1)^{k+1}/(k+1)$. Hence each part i in a partition needs to be assigned a value of $1/i!$ in the case of the $\mathcal{B}_k(x)$. In addition, there is an extra factor of $x^k/k!$ appearing in the summand of (2.2). This factor affects the multinomial factor of $N_k! \prod_{i=1}^{k} 1/\lambda_i!$, which appeared so prominently in the previous chapter. In the partition method for a power series expansion the contribution made by a partition with N_k parts is determined from the $k = N_k$ term in the summation over k on the rhs of (2.2). Since there is an extra factor of $x^k/k!$ due to the fact that we have expanded an exponential, $x^{N_k}/N_k!$ must now be carried through. This factor cancels $N_k!$ in the original multinomial factor leaving only $x^{N_k} \prod_{i=1}^{k} 1/\lambda_i!$. As a consequence, the

Bell polynomials of the first kind are given by

$$\mathcal{B}_k(x) = k! \sum_{\substack{\lambda_1,\lambda_2,\lambda_3,\dots,\lambda_k=0 \\ \sum_{i=1}^{k} i\lambda_i = k}}^{k,\lfloor k/2 \rfloor,\lfloor k/3 \rfloor,\dots,1} x^{N_k} \prod_{i=1}^{k} \frac{1}{\lambda_i!} \left(\frac{1}{i!}\right)^{\lambda_i}. \qquad (2.4)$$

Just as the Bernoulli polynomials yield the Bernoulli numbers for $x = 1$, a similar situation exists with the Bell polynomials yielding Bell numbers $\mathcal{B}_k$ when $x = 1$. From (2.4) we obtain

$$\mathcal{B}_k = k! \sum_{\substack{\lambda_1,\lambda_2,\lambda_3,\dots,\lambda_k=0 \\ \sum_{i=1}^{k} i\lambda_i = k}}^{k,\lfloor k/2 \rfloor,\lfloor k/3 \rfloor,\dots,1} \prod_{i=1}^{k} \frac{1}{\lambda_i!} \left(\frac{1}{i!}\right)^{\lambda_i}. \qquad (2.5)$$

According to Ref. [35], Bell numbers represent the number of ways that n elements can be partitioned into non-empty subsets. The first few Bell numbers are $1, 2, 5, 15, 52, \dots$. Via Dobinski's formula they can be expressed in series form as

$$\mathcal{B}_k = \frac{1}{e} \sum_{j=0}^{\infty} \frac{j^k}{j!}. \qquad (2.6)$$

In addition, they satisfy the recurrence relation of

$$\mathcal{B}_k = \sum_{j=0}^{k-1} \binom{k-1}{j} \mathcal{B}_j, \qquad (2.7)$$

and can be expressed in terms of the Stirling numbers of the second kind, $S(k,j)$, as

$$\mathcal{B}_k = \sum_{j=1}^{k} S(k,j). \qquad (2.8)$$

In fact, the Stirling numbers of the second kind, which represent the number of ways of partitioning a group of k objects into j subgroups, are the coefficients of Bell polynomials. E.g., $S(k, 3)$, which is the number of ways of dividing k elements into three groups, is the coefficient of the third order term or x^3 in the Bell polynomial of the first kind, $\mathcal{B}_k(x)$. With the aid of (2.4), however, we can also define them by

$$S(k,j) = k! \sum_{\substack{\lambda_1,\lambda_2,\lambda_3,\dots,\lambda_k=0 \\ \sum_{i=1}^{k} i\lambda_i = k, \ \sum_{i=1}^{k} \lambda_i = j}}^{k,\lfloor k/2 \rfloor,\lfloor k/3 \rfloor,\dots,1} \prod_{i=1}^{k} \frac{1}{\lambda_i!} \left(\frac{1}{i!}\right)^{\lambda_i}. \qquad (2.9)$$

This fascinating result, which demonstrates that the number of subgroups j is dependent upon there being only j parts in the partitions summing to k, appears in a different form as (8.25) in Ref. [37]. Moreover, earlier in the same chapter, namely in Theorem 8.4, Charalambides presents another fascinating expression for the Stirling numbers of the second kind, which is given as

$$S(k,j) = \frac{1}{j!} \sum_{i=0}^{j} (-1)^{j-i} \binom{j}{i} i^k . \tag{2.10}$$

This result is referred to as Euler's formula for Stirling numbers of the second kind in Ref. [38]. In actual fact, (2.10) is missing an extra term since for $k=j=0$, it yields zero, whereas $S(0,0)=1$. Hence (2.10) should be written as

$$S(k,j) = \frac{1}{j!} \sum_{i=0}^{j} (-1)^{j-i} \binom{j}{i} i^k + \delta_k , \tag{2.11}$$

where δ_k represents the Kronecker delta. The above result is, however, easily verified by introducing it into the standard recurrence relation for the Stirling numbers of the second kind, viz.

$$S(k,j) = S(k-1,j-1) + jS(k-1,j) . \tag{2.12}$$

If we introduce (2.11) into (2.8), then we obtain a general formula for the Bell numbers in terms of a finite double sum, which is given by

$$\mathcal{B}_k = 1 + \sum_{j=0}^{k-1} \frac{(-1)^j}{j!} \sum_{i=1}^{j} (-1)^i \binom{j}{i} i^k . \tag{2.13}$$

Let us examine how (2.9) can be implemented to calculate the Stirling numbers of the second kind. In this equation j corresponds to a fixed number of branches in the paths emanating from the seed number k in a partition tree. For example, if we wish to determine the number of ways a group of six elements can be arranged into two subgroups, then we have to consider all the paths that terminate with two branches in Fig. 1.2 or possess a zero as the first member of their tuple on the second branch. That means the tuples of $(0,5)$, $(0,4)$ and $(0,3)$, which correspond to the partitions, $\{1,5\}$, $\{2,4\}$ and $\{3,3\}$. For the first of these partitions $\lambda_1 = 1$ and $\lambda_5 = 1$, while all the remaining multiplicities are equal to zero. In the second partition the only non-zero multiplicities are λ_2 and λ_4, both of which are equal to unity. For the partition $\{3,3\}$, the only non-zero multiplicity is

λ_3, which equals 2. Therefore, introducing these results in (2.9) gives

$$S(6,2) = 6!\left(\frac{1}{1!}\frac{1}{1!}\frac{1}{1!}\frac{1}{5!} + \frac{1}{1!}\frac{1}{1!}\frac{1}{2!}\frac{1}{4!} + \frac{1}{2!}\left(\frac{1}{3!}\right)^2\right) = 31. \tag{2.14}$$

Similarly, the number of ways of arranging of a group of six elements into three subgroups can be evaluated from the paths terminating after three branches in Fig. 1.2 or the partitions, $\{1,1,4\}$, $\{1,2,3\}$ and $\{2,2,2\}$. By introducing the multiplicities for these partitions into (2.9), one finds that $S(6,3) = 90$.

We can also use (2.9) to derive general formulas for the Stirling numbers of the second kind. The highest order term or $S(k,k)$ represents the case where the most number of parts occur in a partition, i.e., k ones. In this case $\lambda_1 = k$, while the remaining λ_i equal zero. Hence the sum over partitions reduces to a sum over one partition, where $1/(\lambda_i! i!) = 1/k!$. Multiplying by $k!$ means that $S(k,k) = 1$ for all values of k. The next highest order term $S(k,k-1)$ is determined by the partitions with $(k-1)$ parts. There is only one such partition, viz. the partition with $k-2$ ones and one two. Since the sum over partitions in (2.9) reduces to a sum over one partition again, we find that

$$S(k,k-1) = k!\left(\frac{1}{(k-2)!}\frac{1}{2!}\right) = \frac{k(k-1)}{2}. \tag{2.15}$$

The above result agrees with (8) in Ref. [39], but we can proceed still further.

The third highest order term, $S(k,k-2)$, is determined by considering all the partitions with only $(k-2)$ parts. These are the partitions with $(k-3)$ ones and a three and $(k-4)$ ones and two twos. We shall use the shorthand notation of $\{1_{k-3},3\}$ and $\{1_{k-3},2_2\}$ to denote these partitions. Note that other references, e.g., Chapter 1 of Ref. [40] or Chapter 1 of Ref. [41], represent these partitions as $(1^{k-3},3)$ and $(1^{k-4},2^2)$, which we have avoided since confusion arises if the part needs to be represented as an actual power as will occur in Chapter 6 when we discuss perfect partitions. For both of these partitions (2.9) yields

$$S(k,k-2) = k!\left(\frac{1}{(k-3)!}\frac{1}{3!} + \frac{1}{(k-4)!}\frac{1}{2!}\left(\frac{1}{2!}\right)^2\right)$$
$$= \frac{(3k-5)}{24}k(k-1)(k-2). \tag{2.16}$$

Similarly, the fourth highest order term, $S(k,k-3)$, is represented by the sum over those partitions with $k-3$ parts. In this case there are three partitions: $\{1_{k-4}, 4\}$, $\{1_{k-5}, 2, 3\}$, and $\{1_{k-6}, 3_2\}$. Then one finds from (2.9) that

$$S(k, k-3) = k! \left(\frac{1}{(k-4)!} \frac{1}{4!} + \frac{1}{(k-5)!} \frac{1}{2!} \frac{1}{3!} + \frac{1}{(k-6)!} \frac{1}{3!} \left(\frac{1}{2!} \right)^3 \right)$$

$$= \frac{k}{48} (k-1)(k-2)^2 (k-3)^2 . \tag{2.17}$$

The next highest order term or $S(k, k-4)$ is evaluated by summing over the partitions with $k-4$ parts. In this instance there are five such partitions: $\{1_{k-5}, 5\}$, $\{1_{k-6}, 3_2\}$, $\{1_{k-6}, 2, 4\}$, $\{1_{k-7}, 2_2, 3\}$ and $\{1_{k-8}, 2_4\}$. Consequently, introducing the parts and their multiplicities into (2.9) gives

$$S(k, k-4) = \frac{k!}{(k-5)!} \frac{1}{5!} + \frac{k!}{(k-6)!} \frac{1}{2!} \left(\frac{1}{4!} + \frac{1}{3!^2} \right) + \frac{k!}{(k-7)!} \frac{1}{3! \cdot 2!^3}$$

$$+ \frac{k!}{(k-8)!} \frac{1}{4! \cdot 2!^4} = \frac{1}{48} \binom{k}{5} (15k^3 - 150k^2 + 485k - 502) . \tag{2.18}$$

For $S(k, k-5)$ there are 7 seven partitions with $k-5$ parts, which are: $\{1_{k-6}, 6\}$, $\{1_{k-7}, 2, 5\}$, $\{1_{k-7}, 3, 4\}$, $\{1_{k-8}, 2_2, 4\}$, $\{1_{k-8}, 2, 3_2\}$, $\{1_{k-9}, 2_3, 3\}$ and $\{1_{k-10}, 2_5\}$. Hence according to (2.9), we obtain

$$S(k, k-5) = \frac{k!}{(k-6)!} \frac{1}{6!} + \frac{k!}{(k-7)!} \left(\frac{1}{2! \cdot 5!} + \frac{1}{3! \cdot 4!} \right) + \frac{k!}{(k-8)!} \left(\frac{1}{4! \cdot 2!^3} \right.$$

$$\left. + \frac{1}{2!^2 \cdot 3!^2} \right) + \frac{k!}{(k-9)!} \frac{1}{3!^2 \cdot 2!^3} + \frac{k!}{(k-10)!} \frac{1}{2!^5 \cdot 5!} = \binom{k}{6} \frac{(k-4)}{16}$$

$$\times (k-5)(3k^2 - 23k + 38) . \tag{2.19}$$

In each calculation of $S(k, k-\ell)$, there are $p(\ell)$ distinct contributions, where $p(k)$ is the partition–number function mentioned at the beginning of Chapter 1 or the number of partitions summing to k. As we shall see, this follows from the discussion around (3.1) in the next chapter. This function, which is implemented as the PartitionsP[n] routine in Mathematica [32], will become very important in the later chapters of the book.

It should also be mentioned that all the results for $S(k, k-\ell)$ given above agree with the results on pp. 115–116 of Ref. [38], which in addition to requiring closed forms for the sum $\sum_{k=0}^{n-r} k^p$, have been obtained by putting $\alpha = n - \ell$ in the following recurrence relation:

$$S(n, \alpha) = \sum_{k=1}^{\alpha} (\alpha - k + 1) S(n - k, \alpha - k + 1) , \tag{2.20}$$

where $\alpha \geq 1$ and $n \neq \alpha$. Consequently, (2.9) is a more direct approach even though as ℓ increases, an ever increasing number of partitions needs to be considered. More results yielding the Stirling numbers of the second kind, where the second argument is equal to $n - \ell$ will be determined by the above method shortly.

From the above analysis, we see that the leading order term in k for $S(k, k - \ell)$ comes from the term with the least number of ones, viz. $\{1_{k-2l}, 2_l\}$. This partition yields a contribution of $1/2^l l!$, which in turn represents the coefficient of the leading term. The next leading order term is the sum of two contributions. First, we must determine the coefficient of the $k^{2\ell - 1}$ term in the contribution from the partition with the least number of ones. If we denote the contribution from this partition as $C_{\{1_{k-2\ell},2_\ell\}}$, then it is given by

$$C_{\{1_{k-2\ell},2_\ell\}} = \frac{k(k-1)\cdots(k-2\ell+1)}{2^\ell \ell!} = \frac{1}{2^\ell \ell!}\left(k^{2\ell} - (2\ell-1)\ell\, k^{2\ell-1}\right.$$
$$\left. + (\ell - 1/6)(2\ell - 1)\ell(\ell - 1)\, k^{2\ell-2} + \cdots\right). \tag{2.21}$$

The second contribution is the leading order term in the contribution from the partition with the second least number of ones, namely $\{1_{k-2\ell+1}, 2_{\ell-2}, 3\}$. The contribution from this partition is given by

$$C_{\{1_{k-2\ell+1},2_{\ell-2},3\}} = \frac{1}{3! \cdot 2^{\ell-2}} \frac{k!}{(k-2\ell+1)!(\ell-2)!} = \frac{1}{3! \cdot 2^{\ell-2}(\ell-2)!}\left(k^{2\ell-1}\right.$$
$$\left. - (2\ell-1)(\ell-1)\, k^{2\ell-2} + \cdots\right). \tag{2.22}$$

The coefficient of the leading order term in (2.22) is $1/(3! \cdot 2^{2\ell-8}(l-2)!)$. Therefore, to obtain the coefficient of the second leading order term in $S(k, k-\ell)$, we sum this term with the coefficient of the $k^{2\ell-1}$ term in (2.21). The third leading order term is obtained from the coefficients of the $k^{2\ell-2}$ terms in (2.21) and (2.22) plus the leading order terms of the partitions with $k - 2\ell - 2$ ones in them. There are two such partitions that possess a total of $k - \ell$ parts. These are $\{1_{k-2l+2}, 2_{\ell-3}, 4\}$ and $\{1_{k-2l+2}, 2_{\ell-4}, 3_2\}$. According to (2.9) the contribution from the first partition is given by

$$C_{\{1_{k-2\ell+2},2_{\ell-3},4\}} = \frac{1}{4! \cdot 2^{\ell-3}} \frac{k!}{(k-2\ell+2)!(\ell-3)!}, \tag{2.23}$$

while the contribution from the second partition is

$$C_{\{1_{k-2\ell+2},2_{\ell-4},3_2\}} = \frac{1}{2! \cdot 3^2 \cdot 2^{\ell-4}} \frac{k!}{(k-2\ell+2)!(\ell-4)!}. \tag{2.24}$$

In the above results the leading order term is $k^{2\ell-2}$, which arises from $k!/(k-2\ell+2)!$. Then the other factors contribute to the coefficient of the leading order term. Thus, to obtain the coefficient of the $k^{2\ell-2}$ term

Table 2.1 The polynomials $R_\ell(k)$

ℓ	$R_\ell(k)$
1	1
2	$\frac{1}{4}(3k-5)$
3	$\frac{1}{2}(k-2)(k-3)$
4	$\frac{1}{48}(15k^3 - 150k^2 + 485k - 502)$
5	$\frac{1}{96}(k-4)(k-5)(3k^2 - 23k + 38)$
6	$\frac{1}{576}(63k^5 - 1575k^4 + 15435k^3 - 73801k^2 + 171150k - 152696)$
7	$\frac{1}{144}(k-6)(k-7)(9k^4 - 198k^3 + 1563k^2 - 5182k + 6008)$
8	$\frac{1}{3840}(135k^7 - 6300k^6 + 124110k^5 - 1334760k^4 + 8437975k^3 - 31231500k^2$ $+ 62333204k - 51360816)$
9	$\frac{1}{768}(k-8)(k-9)(15k^6 - 645k^5 + 11265k^4 - 101807k^3 + 499176k^2$ $- 1249444k + 1234224)$
10	$\frac{1}{9216}(99k^9 - 7425k^8 + 244530k^7 - 4634322k^6 + 55598235k^5 - 436886945k^4$ $+ 2242194592k^3 - 7220722828k^2 + 13175306672k - 10307425152)$

in $S(k, k-\ell)$, we sum these factors together with the coefficient of the $k^{2\ell-2}$ terms in (2.21) and (2.22). After some algebra, we find that the three highest order terms in $S(k, k-\ell)$ in terms of k become

$$S(k, k-\ell) = \frac{1}{2^\ell \ell!} k^{2\ell} - \frac{(4l-1)}{3 \cdot 2^\ell (\ell-1)!} k^{2\ell-1}$$
$$+ \frac{(16\ell^2 - 2\ell + 3)}{9 \cdot 2^{\ell+1}(\ell-2)!} k^{2\ell-2} + O\left(k^{2\ell-3}\right). \qquad (2.25)$$

From these results we see that the Stirling numbers of the second kind can be expressed as $S(k, k-\ell) = f(k)\binom{k}{\ell+1}$, where $f(k)$ is a polynomial of degree $\ell - 1$ in k with each coefficient dependent only upon ℓ. In fact, they have the same form as the Stirling numbers of the first kind or $s_k^{(k-\ell)}$ as given by (1.59) except the polynomial $f(k)$ has different coefficients. There is also a phase factor of $(-1)^\ell$ that appears in the latter, whereas the Stirling numbers of the second kind are always positive. Furthermore, we can combine the preceding results with (2.11) to arrive at the following result:

$$\sum_{j=0}^{k-\ell}(-1)^{k-\ell-j}\binom{k-\ell}{j}j^k = R_\ell(k)\frac{(k-\ell)\,k!}{(l+1)!}. \qquad (2.26)$$

In other words, $S(k, k-\ell) = \binom{k}{\ell+1}R_\ell(k)$, where the $R_\ell(k)$ are polynomials of degree $\ell - 1$ and are displayed in Table 2.1 up to $\ell = 10$. An interesting property of these polynomials is that the sign of the coefficients toggles between positive and negative as the order of k is decremented with the highest order term always being positive. In addition, when ℓ is an odd integer, the polynomials can be simplified further by taking the factor of $(k - \ell)(k - \ell + 1)$ outside.

The lowest value of j in $S(k, j)$ given by (2.9) is represented by the partition with a single part, namely $\{k\}$. Since there is only one partition in this instance, (2.9) yields $S(k, 1) = k!/k! = 1$. The next lowest value of j is determined by all the partitions with two parts in them. The sum over partitions in (2.9) now depends upon whether k is even or not. If it is odd, i.e. $k = 2N + 1$, then we have N partitions with two parts in them, $\{1, 2N\}$, $\{2, 2N - 1\}, \ldots, \{N, N + 1\}$. Thus, (2.9) reduces to

$$S(2N + 1, 2) = \sum_{j=1}^{N} \frac{(2N + 1)!}{j!\,(2N + 1 - j)!} = \sum_{j=1}^{N} \binom{2N + 1}{j} = 2^{2N} - 1 \,, \quad (2.27)$$

where the last result has been obtained using No. 4.2.1.6. in Prudnikov et al. [42]. When $k = 2N$, we have the partitions $\{1, 2N - 1\}, \{2, 2N - 2\}$, $\ldots, \{N, N\}$, but now the multiplicity for the last partition, λ_N, is equal to 2. Consequently, (2.9) becomes

$$S(2N, 2) = (2N)! \sum_{j=1}^{N-1} \frac{1}{j!} \frac{1}{(2N - j)!} + (2N)! \frac{1}{2!} \left(\frac{1}{N!} \right)^2$$

$$= \sum_{j=1}^{N} \binom{2N}{j} - \frac{1}{2} \binom{2N}{N} = 2^{2N-1} - 1 \,. \quad (2.28)$$

In other words, we obtain the same result as when k is an odd number. Therefore, we have

$$S(k, 2) = 2^{k-1} - 1 \,, \quad (2.29)$$

for all values of k, which agrees with (6) in Ref. [39].

The third lowest value of j or $S(k, 3)$ is more difficult to evaluate, but is still manageable. In this case, we fix one of the partitions and apply the knowledge gained from considering the sum over two part partitions given above. E.g., consider first all the parts with at least one part equal to unity. This means that we apply (2.9) to the two-part partitions summing to $k - 1$ with the other part equal to unity. That is, we sum over the partitions over $\{i, k - 1 - i\}$, where i ranges from 1 to $\lfloor (k - 1)/2 \rfloor$. Two problems, however, arise. When i equals unity, we have the partition $\{1_2, k - 2\}$. According to (2.9), we need to take into account that there are two parts equal to unity, which in turn means that we need to ensure that $\lambda_1 = 2$ for this partition, whereas for all other two-part partitions $\lambda_1 = 1$. The second problem arises whenever $k - 1$ is even. Then we need to take into account that there will be two parts equal to $\lfloor (k - 1)/2 \rfloor$. Again, there will be a factor of $2!$ appearing in the denominator of the product in (2.9), while the sum over the other two-part partitions does not require this factor. Therefore, we shall carry out the sum over all two-part partitions as if the pairs were distinct except when dealing with $\{1_2, k - 2\}$ and when $k - 1$ is even, where we need to subtract half value of the contribution due to $\{1, ((k - 1)/2)_2\}$. Hence the contribution to $S(k, 3)$ from the partitions where at least one part is unity, which we denote as $S_1(k, 3)$, becomes

$$S_1(k, 3) = k! \left(\sum_{i=1}^{\lfloor (k-1)/2 \rfloor} \frac{1}{i!(k - 1 - i)!} - \frac{1}{2(k - 2)!} - \frac{(1 + (-1)^{k-1})}{2 \cdot 2! 1!(((k - 1)/2)!)^2} \right). \tag{2.30}$$

The above result needs to be amended for $k = 3$, which will be discussed shortly. Furthermore, there are 3-part partitions that do not possess unity once $k > 5$. These will begin with the partitions where all the parts are equal to 2, viz. $k = 6$. So now the summation is over the 2-part partitions summing to $k - 2$, i.e. the partitions $\{i, (k - 2 - i)\}$, where i ranges from 2 to $\lfloor (k - 2)/2 \rfloor$. In this case we need to subtract half the contribution due to the partition $\{2_2, k - 4\}$ and half the contribution due to $\{2, (k/2 - 1)_2\}$, when k is even. We continue by applying the same process to those partitions with at least a value of three and so on until we finally reach the partitions where the parts are at least equal to $\lfloor k/3 \rfloor$. This, however, is not the end of the calculation. When $k \equiv 0 \pmod 3$, we need to consider the partition $\{(k/3)_3\}$ separately because its contribution is cancelled in the preceding sums. To ensure that there is only a contribution at multiples of three we need to put $\ell = 3$ in the following identity:

$$\sum_{j=1}^{\ell} e^{2\pi i j k/\ell} = \begin{cases} \ell, & k \equiv 0 \pmod \ell, \\ 0, & k, \text{ otherwise.} \end{cases} \tag{2.31}$$

Hence the contribution to $S(k, 3)$ where all three parts are equal to one another can be expressed as

$$S_{\{(k/3)_3\}}(k, 3) = \frac{(1 + 2(-1)^k \cos(k\pi/3))}{3 \cdot 3!(k/3)!^3}. \tag{2.32}$$

By combining all the results from (2.30) onwards, we obtain $S(k, 3)$, which becomes

$$S(k, 3) = k! \left[\sum_{j=1}^{\lfloor k/3 \rfloor} \left(\sum_{i=j}^{\lfloor (k-j)/2 \rfloor} \frac{1}{i!(k-j-i)!j!} - \frac{1}{2(j!)^2} \frac{1}{(k-2j)!} \right. \right.$$
$$\left. \left. - \frac{(1+(-1)^{k-j})}{2 \cdot 2!j!(\lfloor (k-j)/2 \rfloor!)^2} \right) + \frac{(1+2(-1)^k \cos(k\pi/3))}{18((k/3)!)^3} \right]. \qquad (2.33)$$

In more compact notation (2.33) can be written as

$$S(k, 3) = \sum_{j=1}^{\lfloor k/3 \rfloor} \binom{k}{j} \left(\sum_{i=j}^{\lfloor (k-j)/2 \rfloor} \binom{k-j}{i} - \frac{1}{2} \binom{k-j}{j} - \frac{(1+(-1)^{k-j})}{4} \right.$$
$$\left. \times \binom{k-j}{\lfloor (k-j)/2 \rfloor} \right) + \frac{(1+2(-1)^k \cos(k\pi/3))k!}{18((k/3)!)^3}. \qquad (2.34)$$

On the other hand, Weisstein presents an elegant formula for the above result, which appears as (7) in Ref. [39]. This result, which can be obtained from Euler's formula for the Stirling numbers of the second kind, viz. (2.11), is

$$S(k, 3) = \frac{1}{6} \left(3^k - 3 \cdot 2^k + 3 \right). \qquad (2.35)$$

Both results for $S(k, 3)$ are easily implemented in Mathematica [32] where it is found that they not only give identical values, but they also agree with the values obtained via the intrinsic StirlingS2 routine.

We now turn our attention to inverting (2.1), which gives

$$\exp\left(-x(e^t - 1) \right) = \frac{1}{1 + B_1(x)\,t + B_2(x)\,t^2/2! + B_3(x)\,t^3/3! + \cdots}. \qquad (2.36)$$

By treating the rhs as the regularized value of the geometric series, we can express the above result as

$$\exp\left(-x(e^t - 1) \right) \equiv \sum_{k=0}^{\infty} (-1)^k \left(B_1(x)\,t + B_2(x)\,\frac{t^2}{2!} + B_3(x)\,\frac{t^3}{3!} + \cdots \right)^k. \qquad (2.37)$$

This result is in a similar form to the examples presented in Chapter 1. In fact, the series is virtually identical to that in (1.52) except the reciprocal logarithm numbers A_i have been replaced by the Bell polynomials divided by a factorial, i.e. $B_i(x)/i!$. This means that all we need to do is make the appropriate changes to (1.54) in order to apply the partition method for a power series expansion to the above result. Furthermore, we may replace the equivalence symbol by an equals sign because the lhs is merely (2.1) with $\mathcal{B}_k(x)$ replaced by $\mathcal{B}_k(-x)$. Therefore, we arrive at

$$\mathcal{B}_k(-x) = k! \sum_{\substack{\lambda_1,\lambda_2,\lambda_3,\ldots,\lambda_k=0 \\ \sum_{i=1}^{k} i\lambda_i=k}}^{k,\lfloor k/2 \rfloor,\lfloor k/3 \rfloor,\ldots,1} (-1)^{N_k} N_k! \prod_{i=1}^{k} \left(\frac{\mathcal{B}_i(x)}{i!}\right)^{\lambda_i} \frac{1}{\lambda_i!}. \qquad (2.38)$$

The rhs of (2.38) possesses a term with $\mathcal{B}_k(x)$, which needs to be separated from the sum over partitions and placed on the lhs. This term corresponds to the case where $\lambda_k = 1$ and all the other λ_i are zero. Hence we find that

$$\mathcal{B}_k(-x) + \mathcal{B}_k(x) = k! \sum_{\substack{\lambda_1,\lambda_2,\lambda_3,\ldots,\lambda_{k-1}=0 \\ \sum_{i=1}^{k-1} i\lambda_i=k}}^{k,\lfloor k/2 \rfloor,\lfloor k/3 \rfloor,\ldots,1} (-1)^{N_k} N_k! \prod_{i=1}^{k-1} \left(\frac{\mathcal{B}_i(\pm x)}{i!}\right)^{\lambda_i} \frac{1}{\lambda_i!}. \qquad (2.39)$$

Note that the constraint remains the same except that i ranges from 1 to $k - 1$, while the lhs remains the same irrespective of whether the upper or lower sign of x is used in the Bell polynomials on the rhs provided the same sign is used for all of them. Moreover, the lhs represents double all the even-powered terms of the Bell polynomials. For example, if $k = 5$, then the lhs reduces to

$$\mathcal{B}_5(-x) + \mathcal{B}_5(x) = 2\Big(S(5,2)x^2 + S(5,4)x^4\Big) = 30x^2 + 20x^4, \qquad (2.40)$$

while the rhs becomes

$$5!\left(-\mathcal{B}_1(-x)^5 + 4\mathcal{B}_1(-x)^3\frac{\mathcal{B}_2(-x)}{2!} - 3\mathcal{B}_1(-x)\left(\frac{\mathcal{B}_2(-x)}{2!}\right)^2\right.$$
$$\left. - 3\mathcal{B}_1(-x)^2\frac{\mathcal{B}_3(-x)}{3!} + 2\frac{\mathcal{B}_2(-x)}{2!}\frac{\mathcal{B}_3(-x)}{3!} + 2\mathcal{B}_1(-x)\frac{\mathcal{B}_4(-x)}{4!}\right). \qquad (2.41)$$

Introducing the first four Bell polynomials or the coefficients of $t^k/k!$ in (2.3) into the above result gives (2.40).

Suppose our aim had been to derive a power series expansion for $\exp(xt/(e^t - 1))$ instead of the exponential on the rhs of (2.1). If we introduce (1.2) into this exponential, then we obtain

$$\exp\left(\frac{xt}{e^t - 1}\right) \equiv e^x \sum_{k=0}^{\infty} \frac{x^k}{k!}\left(B_1 t + B_2 \frac{t^2}{2!} + B_3 \frac{t^3}{3!} + B_4 \frac{t^4}{4!} + \cdots\right)^k. \qquad (2.42)$$

Note the appearance of the equivalence symbol since we have seen that (1.2) is divergent for $|t| > 2\pi$. Alternatively, for $|t| < 2\pi$, we can replace the equivalence symbol by an equals sign. The rhs of (2.37) is of the same form as the rhs of (2.2). As a consequence, we can apply the partition method

Table 2.2 The polynomials $\mathcal{C}_k(x)$ as defined by (2.43)

k	$\mathcal{C}_k(x)$
0	1
1	$-\frac{x}{2}$
2	$\frac{x}{6} + \frac{x^2}{4}$
3	$-\frac{x^2}{4} - \frac{x^3}{8}$
4	$-\frac{x}{30} + \frac{x^2}{12} + \frac{x^3}{4} + \frac{x^4}{16}$
5	$\frac{x^2}{12} - \frac{5x^3}{24} - \frac{5x^4}{24} - \frac{x^5}{32}$
6	$\frac{x}{42} - \frac{x^2}{12} - \frac{x^3}{18} + \frac{5x^4}{16} + \frac{5x^5}{32} + \frac{x^6}{64}$
7	$-\frac{x^2}{12} + \frac{7x^3}{24} - \frac{7x^4}{72} - \frac{35x^5}{96} - \frac{7x^6}{64} - \frac{x^7}{128}$
8	$-\frac{x}{30} + \frac{3x^2}{20} - \frac{x^3}{36} - \frac{217x^4}{432} + \frac{49x^5}{144} + \frac{35x^6}{96} + \frac{7x^7}{96} + \frac{x^8}{256}$
9	$\frac{3x^2}{20} - \frac{27x^3}{40} + \frac{5x^4}{8} + \frac{49x^5}{96} - \frac{287x^6}{480} - \frac{21x^7}{64} - \frac{3x^8}{64} - \frac{x^9}{512}$
10	$\frac{5x}{66} - \frac{5x^2}{12} + \frac{x^3}{3} + \frac{173x^4}{144} - \frac{505x^5}{288} - \frac{35x^6}{192} + \frac{77x^7}{96} + \frac{35x^8}{128} + \frac{15x^9}{512} + \frac{x^{10}}{1024}$

for a power series expansion, but in this instance the value assigned to each part i is $B_i/i!$ instead of $1/i!$.

Let us define the polynomials $\mathcal{C}_k(x)$ by

$$\sum_{k=0}^{\infty} \mathcal{C}_k(x) \frac{t^k}{k!} := \sum_{k=0}^{\infty} \frac{x^k}{k!} \left(B_1 t + B_2 \frac{t^2}{2!} + B_3 \frac{t^3}{3!} + B_4 \frac{t^4}{4!} + \cdots \right)^k. \qquad (2.43)$$

Applying the partition method for a power series expansion in the same manner as (2.2), we arrive at

$$\mathcal{C}_k(x) = k! \sum_{\substack{\lambda_1,\lambda_2,\lambda_3,\ldots,\lambda_k=0 \\ \sum_{i=1}^{k} i\lambda_i = k}}^{k,\lfloor k/2 \rfloor,\lfloor k/3 \rfloor,\ldots,1} x^{N_k} \prod_{i=1}^{k} \frac{1}{\lambda_i!} \left(\frac{B_i}{i!} \right)^{\lambda_i}. \qquad (2.44)$$

Hence we see that the Bernoulli numbers appear as the coded value for each part i in a partition. Note that many of the contributions will vanish because for odd indices greater than unity the Bernoulli numbers are equal to zero. The polynomials $\mathcal{C}_k(x)$ are displayed in Table 2.2 up to $k = 10$. There we see that the degree of these polynomials is k with the coefficient of this term equal to $(-1/2)^k$ since $B_1 = -1/2$. The second highest order term is the coding of the partition with $k-2$ ones and a two. If we denote this coefficient by $\mathcal{C}_{k,k-1}$, then it is found to be

$$\mathcal{C}_{k,k-1} = \frac{k!}{(k-2)!} \frac{B_1^{k-2} B_2}{2!} = \frac{k(k-1)}{12} \left(-\frac{1}{2} \right)^{k-2}. \qquad (2.45)$$

On the other hand, for even values of k the lowest order term is linear, while for odd values of k it is quadratic, again because the Bernoulli numbers vanish for odd indices greater than unity. If we denote the lowest order term by $\mathcal{C}_{k,1}$, then we have

$$\mathcal{C}_{2k,1} = B_{2k}, \tag{2.46}$$

while $\mathcal{C}_{2k+1,1} = 0$. The quadratic terms represent the sum of the contributions of the pairs of numbers summing to k. Hence we find that

$$\mathcal{C}_{k,2} = \sum_{j=1}^{\lfloor k/2 \rfloor} \binom{k}{j} B_j B_{k-j} - \frac{1}{4}\left(1 + (-1)^k\right)\binom{k}{k/2} B_{k/2}^2, \tag{2.47}$$

where $\lfloor x \rfloor$ again denotes the floor function or the greatest integer lower than or equal to x. Note that the second term after the sum on the rhs of (2.47) only contributes for even values of k. In terms of the partition method for a power series expansion, the coefficients $\mathcal{C}_{k,j}$ can be expressed as

$$\mathcal{C}_{k,j} = k! \sum_{\substack{\lambda_1,\lambda_2,\lambda_3,\dots,\lambda_k=0 \\ \sum_{i=1}^{k} i\lambda_i = k, \ \sum_{i=1}^{k} \lambda_i = j}}^{k,\lfloor k/2 \rfloor,\lfloor k/3 \rfloor,\dots,1} \prod_{i=1}^{k} \frac{1}{\lambda_i!}\left(\frac{B_i}{i!}\right)^{\lambda_i}. \tag{2.48}$$

We can also derive an interesting result involving the coefficients of the $\mathcal{C}_k(x)$ polynomials. For $|t| < 2\pi$ and x replaced by $-x$, the generating function becomes

$$\exp\left(-\frac{xt}{e^t - 1}\right) = e^{-x} \sum_{k=0}^{\infty} \frac{t^k}{k!} \sum_{j=1}^{k} (-1)^j \mathcal{C}_{k,j} \, x^j. \tag{2.49}$$

Integrating both sides of (2.49) gives

$$\frac{1}{t}\left(e^t - 1\right) = \sum_{k=0}^{\infty} \frac{t^k}{(k+1)!} = \sum_{k=0}^{\infty} \frac{t^k}{k!} \sum_{j=1}^{k} (-1)^j j! \, \mathcal{C}_{k,j}. \tag{2.50}$$

Since t is arbitrary, we can equate like powers of t on both sides of the above result, thereby obtaining

$$\sum_{j=1}^{k} (-1)^j j! \, \mathcal{C}_{k,j} = \frac{1}{k+1}. \tag{2.51}$$

The sum over j in (2.51) is effectively over all partitions summing to k when (2.48) is introduced and it is realized that j corresponds to N_k in the partition method for a power series expansion. This means that the above result can be represented in a similar manner as in previous applications of the method. Thus, (2.51) can be expressed alternatively as

$$\frac{1}{(k+1)!} = \sum_{\substack{\lambda_1,\lambda_2,\lambda_3,\ldots,\lambda_k=0 \\ \sum_{i=1}^{k} i\lambda_i=k}}^{k,\lfloor k/2\rfloor,\lfloor k/3\rfloor,\ldots,1} (-1)^{N_k} N_k! \prod_{i=1}^{k} \frac{1}{\lambda_i!} \left(\frac{B_i}{i!}\right)^{\lambda_i}. \tag{2.52}$$

The above result can be regarded as being complementary to (1.35) and (1.36).

2.2 GENERALIZED COSECANT AND SECANT NUMBERS

The cosecant and secant numbers have already been discussed in great detail in Chapter 1. In this section we aim to generalize both sets of numbers by introducing a power of ρ, which can be complex, to the trigonometric functions that spawn them. To derive the coefficients of the power series expansions in these cases, we shall adapt the partition method for a power series expansion as presented in Theorem 1.1. Consequently, the generalized coefficients in the new expansions will turn out to be polynomials of order k, which reduce to the cosecant and secant numbers when ρ is set equal to unity.

Theorem 2.1. There exists a power series expansion for cosecant raised to an arbitrary power of ρ, which is given by

$$\csc^\rho z = \frac{1}{\sin^\rho z} \equiv z^{-\rho} \sum_{k=0}^{\infty} c_{\rho,k}\, z^{2k}. \tag{2.53}$$

In (2.53) the coefficients $c_{\rho,k}$ are referred to as the generalized cosecant numbers, and are polynomials of order k in ρ. Via the partition method for a power series expansion they are expressed as

$$c_{\rho,k} = (-1)^k \sum_{\substack{\lambda_1,\lambda_2,\lambda_3,\ldots,\lambda_k=0 \\ \sum_{i=1}^{k} i\lambda_i=k}}^{k,\lfloor k/2\rfloor,\lfloor k/3\rfloor,\ldots,1} (-1)^{N_k} (\rho)_{N_k} \prod_{i=1}^{k} \left(\frac{1}{(2i+1)!}\right)^{\lambda_i} \frac{1}{\lambda_i!}, \tag{2.54}$$

where, as before, $(\rho)_{N_k}$ represents the Pochhammer notation for $\Gamma(\rho + N_k)/\Gamma(\rho)$. Moreover, the generalized cosecant numbers satisfy the following recurrence relations:

$$\rho\, c_{\rho+1,k} = \sum_{j=0}^{k} \left(\rho - 2k + 2j\right) d_j\, c_{\rho,k-j}, \tag{2.55}$$

and

$$c_{\mu+\nu,k} = \sum_{j=0}^{k} c_{\mu,j}\, c_{\nu,k-j}\,.\qquad(2.56)$$

For ρ not equal to a negative integer, the power series expansion is absolutely convergent when $|z| < \pi$, while for $|z| \geq \pi$ and $|\Re z| < \pi$, it is conditionally convergent. For both these cases the equivalence symbol can be replaced by an equals sign. For all other values of z the power series expansion is divergent and therefore, only the equivalence symbol applies. For ρ equal to a negative integer, the power series expansion is convergent for all values of z.

Proof. By introducing the power series for sine into $\csc^\rho z$, we obtain

$$z^\rho\, \csc^\rho z = \left(1 - \frac{z^2}{3!} + \frac{z^4}{5!} - \frac{z^6}{7!} + \cdots\right)^{-\rho}.\qquad(2.57)$$

Comparing the above result with (1.5), which was used in the proof of Theorem 1.1, we see that although the coefficients of the powers of z are still the same, there is now a power of ρ overall. This means that the denominator can no longer be regarded as the regularized value of the geometric series. Instead, the denominator can only be regarded as the regularized value of the binomial series, which is discussed extensively in Appendix A and Ref. [3]. From Lemma A.1, (2.57) can be written as

$$z^\rho\, \csc^\rho z \equiv \sum_{l=0}^{\infty} \frac{\Gamma(l+\rho)}{\Gamma(\rho)\, l!} \left(\frac{z^2}{3!} - \frac{z^4}{5!} + \frac{z^6}{7!} - \frac{z^8}{9!} + \cdots\right)^l$$

$$= \sum_{k=0}^{\infty} c_{\rho,k}\, z^{2k}\,.\qquad(2.58)$$

Like the geometric series the binomial series is absolutely convergent when the variable lies within the unit disk in the complex plane. Moreover, since the same series appears inside the large brackets in the same manner as in the proof of Theorem 1.1, (1.9) applies here too. Hence (2.58) is at least absolutely convergent for $|z^2| < 6$, in which case we can replace the equivalence symbol by an equals sign for these values of z. Unfortunately, we have seen previously that analysing the convergence of a power series via (1.9) is too restrictive, but we have also observed in Chapter 1 that the radius of absolute convergence is determined by the nearest singularity to the origin of the complex plane. From the infinite product for sine, viz.

$$\sin z = z \prod_{k=1}^{\infty}\left(1 - \frac{z^2}{k^2\pi^2}\right), \tag{2.59}$$

we see that the convergence behaviour of the power series expansion for $z^\rho \csc^\rho z$ will be determined by $(1 - z^2/\pi^2)^{-\rho}$, which in turn represents the regularized value for the binomial series, viz. $\sum_{k=0}^{\infty}(\rho)_k(z/\pi)^{2k}/k!$. From Appendix A and Ref. [3], this series is absolutely convergent for $|z| < \pi$ and conditionally convergent for $|z| > \pi$ and $|\Re z| < \pi$. For these values of z the equivalence symbol in (2.58) can be replaced by an equals sign. In addition, the series in (2.59) is divergent when $|\Re z| \geq \pi$, which means that for these values of z, (2.58) is divergent. Then only the equivalence symbol applies. For the special case of ρ equal to a negative integer, the binomial series is finite. Therefore, the equivalence symbol can be replaced by an equals sign for all values of z in this case.

As seen in Chapter 1, the partition method for a power series expansion depends upon two steps: (1) determining all the partitions summing to each order k and (2) coding the partitions, which includes the evaluation of a multinomial factor based on the multiplicities of the parts in each partition. To determine the generalized cosecant numbers from (2.58), we need another or third step or else the situation would not be different from Theorem 1.1. Whilst each part i in a partition is assigned a value of $(-1)^{i+1}/(2i+1)!$ and the multinomial factor equals $N_k!/\lambda_1!\lambda_2!\cdots\lambda_k!$ as in the proof of Theorem 1.1, the extra step that is now required is to introduce the dependence upon ρ into the proof. This will, in turn, result in a generalization of the cosecant numbers.

Basically, the extra step/modification is the introduction of a factor that accounts for the coefficients $\Gamma(\rho+l)/\Gamma(\rho)\,l!$ appearing in the binomial theorem. Since the index l in the intermediate member of (2.58) corresponds to the number of parts N_k in the partition method for a power series expansion, an extra factor of $\Gamma(\rho+N_k)/\Gamma(\rho)\,N_k!$ is required to evaluate the contribution from each partition. For example, consider the partitions summing to 5, which are $\{1_5\}$, $\{1_3,2\}$, $\{1,2_2\}$, $\{1_2,3\}$, $\{2,3\}$, $\{1,4\}$ and $\{5\}$. The partition with only ones emanates from the $l=5$ term in the intermediate member of (2.58). Therefore, the contribution to $c_{\rho,5}$ from this partition possesses an extra factor of $\Gamma(\rho+5)/\Gamma(\rho)\cdot5!$ or $(\rho)_5/5!$ using Pochhammer notation. For the partition $\{1_3,2\}$ the extra factor is $\Gamma(\rho+4)/\Gamma(\rho)\cdot4!$ or $(\rho)_4/4!$, while for the partitions $\{1_2,3\}$ and $\{1,2_2\}$ it is $\Gamma(\rho+3)/\Gamma(\rho)\cdot3!$

or $(\rho)_3/3!$. These factors are, of course, the Pochhammer polynomials [29] that have been discussed in Chapter 1 with an extra factor of $l!$ in the denominator. As a consequence, the contribution due to each partition becomes

$$C\left(\lambda_1, \lambda_2, \ldots, \lambda_k\right) = \frac{(\rho)_{N_k}}{N_k!} \prod_{i=1}^{k} \frac{(-1)^{\lambda_i(i+1)}}{\lambda_i!\,((2i+1)!)^{\lambda_i}}. \tag{2.60}$$

For example, according to (2.60), the contribution to $c_{\rho,5}$ from the partition $\{1_2, 3\}$ is $(\rho)_3/(3!)^3 \cdot 7!$. Finally, by summing over all the partitions summing to k, we arrive at (2.54).

From (2.54), $c_{\rho,5}$ is given by

$$\begin{aligned}
c_{\rho,5} = {} & \frac{(\rho)_5}{5!}\frac{1}{3!^5} - \frac{(\rho)_4}{4!}\frac{4!}{3!^4\cdot 5!} + \frac{(\rho)_3}{3!}\frac{3!}{2!\cdot 3!^2 \cdot 7!} + \frac{(\rho)_3}{2!}\frac{1}{5!^2\cdot 3!} \\
& - \frac{(\rho)_2}{2!}\frac{2!}{3!\cdot 9!} - \frac{(\rho)_2}{2!}\frac{2!}{5!\cdot 7!} + (\rho)_1\frac{1}{11!} = \frac{\rho}{359251200}\big(768 \\
& + 2288\rho + 2684\rho^2 + 1540\rho^3 + 385\rho^4\big).
\end{aligned} \tag{2.61}$$

Thus, we observe that $c_{\rho,5}$ is a polynomial in the exponent ρ of degree 5 where the highest order term is determined by the partition with k ones resulting in a factor of $(\rho)_k$. This means that the generalized cosecant numbers $c_{\rho,k}$ are polynomials in ρ of degree k. Those up to $k=15$ are presented in Table 2.3. As expected, when ρ is set equal to unity, they reduce to the cosecant numbers displayed in Table 1.1, while for $\rho=-1$, we find that $c_{-1,k} = (-1)^k/(2k+1)!$, i.e., the coefficients of the convergent power series expansion for $\sin z$. The interesting property of the results in Table 2.3 is that the coefficients remain invariant regardless of the value of the exponent ρ. Hence the generalized cosecant numbers can be expressed as $c_{\rho,k} = \sum_{i=1}^{k} C_{k,i}\rho^i$, where the $C_{k,i}$ represent rational numbers that have no dependence upon ρ.

The appearance of the power or exponent, ρ, in (2.53) makes the problem of deriving recurrence relations for the generalized cosecant numbers coefficients more formidable. Consequently, many of the methods used to derive recurrence relations for the cosecant numbers in Ref. [5] cannot be applied to the generalized case because ρ is no longer an integer. Nevertheless, by differentiating $z^\rho \csc^\rho z$, we find that

$$\frac{1}{\cos z}\frac{d}{dz}\left(\frac{z^\rho}{\sin^\rho z}\right) = \frac{1}{\cos z}\frac{\rho}{z}\frac{z^\rho}{\sin^\rho z} - \frac{\rho}{z}\frac{z^{\rho+1}}{\sin^{\rho+1} z}. \tag{2.62}$$

Table 2.3 Generalized cosecant numbers $c_{\rho,k}$

k	$c_{\rho,k}$
0	1
1	$\frac{1}{3!}\rho$
2	$\frac{2}{6!}\left(2\rho + 5\rho^2\right)$
3	$\frac{8}{9!}\left(16\rho + 42\rho^2 + 35\rho^3\right)$
4	$\frac{2}{3\cdot10!}\left(144\rho + 404\rho^2 + 420\rho^3 + 175\rho^4\right)$
5	$\frac{4}{3\cdot12!}\left(768\rho + 2288\rho^2 + 2684\rho^3 + 1540\rho^4 + 385\rho^5\right)$
6	$\frac{2}{9\cdot15!}\left(1061376\rho + 3327584\rho^2 + 4252248\rho^3 + 2862860\rho^4 + 1051050\rho^5 + 175175\rho^6\right)$
7	$\frac{1}{27\cdot15!}\left(552960\rho + 1810176\rho^2 + 2471456\rho^3 + 1849848\rho^4 + 820820\rho^5 + 210210\rho^6 \right.$ $\left. + 25025\rho^7\right)$
8	$\frac{2}{45\cdot18!}\left(200005632\rho + 679395072\rho^2 + 978649472\rho^3 + 792548432\rho^4 + 397517120\rho^5 \right.$ $\left. + 125925800\rho^6 + 23823800\rho^7 + 2127125\rho^8\right)$
9	$\frac{4}{81\cdot21!}\left(129369047040\rho + 453757851648\rho^2 + 683526873856\rho^3 + 589153364352\rho^4 \right.$ $\left. + 323159810064\rho^5 + 117327450240\rho^6 + 27973905960\rho^7 + 4073869800\rho^8 \right.$ $\left. + 282907625\rho^9\right)$
10	$\frac{2}{6075\cdot22!}\left(38930128699392\rho + 140441050828800\rho^2 + 219792161825280\rho^3 \right.$ $\left. + 199416835425280\rho^4 + 117302530691808\rho^5 + 47005085727600\rho^6 \right.$ $\left. + 12995644662000\rho^7 + 2422012593000\rho^8 + 280078548750\rho^9 + 15559919375\rho^{10}\right)$
11	$\frac{8}{243\cdot25!}\left(494848416153600\rho + 1830317979303936\rho^2 + 2961137042841600\rho^3 \right.$ $\left. + 2805729689044480\rho^4 + 1747214980192000\rho^5 + 755817391389984\rho^6 \right.$ $\left. + 232489541684400\rho^7 + 50749166067600\rho^8 + 7607466867000\rho^9 + 715756291250\rho^{10} \right.$ $\left. + 32534376875\rho^{11}\right)$
12	$\frac{2}{2835\cdot27!}\left(1505662706987827200\rho + 5695207005856038912\rho^2 + 9487372599204065280\rho^3 \right.$ $\left. + 9332354263294766080\rho^4 + 6096633539052376320\rho^5 + 2806128331871953088\rho^6 \right.$ $\left. + 937291839756592320\rho^7 + 229239926321406000\rho^8 + 40598842049766000\rho^9 \right.$ $\left. + 5005999501002500\rho^{10} + 390802935022500\rho^{11} + 14803141478125\rho^{12}\right)$
13	$\frac{232}{81\cdot30!}\left(844922884529848320\rho + 3261358271400247296\rho^2 + 5576528334428209152\rho^3 \right.$ $\left. + 5668465199488266240\rho^4 + 3858582205451484160\rho^5 + 1870620248833400064\rho^6 \right.$ $\left. + 667822651436228288\rho^7 + 178292330746770240\rho^8 + 35600276746834800\rho^9 \right.$ $\left. + 5225593531158000\rho^{10} + 539680243602500\rho^{11} + 35527539547500\rho^{12} \right.$ $\left. + 1138703190625\rho^{13}\right)$
14	$\frac{2}{1215\cdot30!}\left(138319015041155727360\rho + 543855095595477762048\rho^2 \right.$ $\left. + 952027796641042464768\rho^3 + 996352286992030556160\rho^4 + 703040965960031795200\rho^5 \right.$ $\left. + 356312537387839432192\rho^6 + 134466795172062184832\rho^7 + 38526945410311117760\rho^8 \right.$ $\left. + 8436987713444690400\rho^9 + 1404048942958662000\rho^{10} + 173777038440005000\rho^{11} \right.$ $\left. + 15258232341852500\rho^{12} + 858582205731250\rho^{13} + 23587423234375\rho^{14}\right)$
15	$\frac{1088}{729\cdot35!}\left(56200973964769840087040\rho + 224751194159631176407449\overline{6}\rho^2 \right.$ $\left. + 401910837930690543983001\overline{6}\rho^3 + 431774592520807259425996\overline{8}\rho^4 \right.$ $\left. + 314516377667793942941696\overline{0}\rho^5 + 165691720353903234153062\overline{4}\rho^6 \right.$ $\left. + 6556439193644205860234\overline{24}\rho^7 + 199227919419039256217472\rho^8 \right.$ $\left. + 46995751664475880185920\rho^9 + 8614026107092938211680\rho^{10} \right.$ $\left. + 1214778349162323946000\rho^{11} + 128587452922193265000\rho^{12} \right.$ $\left. + 9720180867524627500\rho^{13} + 472946705787806250\rho^{14} + 11260635852090625\rho^{15}\right)$

Introducing the power series expansions for $z^\rho \csc^\rho z$ as given above and for $\sec z$ in Chapter 1, viz. (1.25), one eventually arrives at

$$\sum_{k=0}^{\infty} d_k z^{2k} \sum_{k=0}^{\infty} 2k c_{\rho,k} z^{2k} \equiv \rho \sum_{k=0}^{\infty} d_k z^{2k} \sum_{k=0}^{\infty} c_{\rho,k} z^{2k} - \rho \sum_{k=0}^{\infty} c_{\rho+1,k} z^{2k} . \quad (2.63)$$

We have seen that the power series expansion for $\csc^\rho z$ is absolutely convergent for $|z| < \pi$, while from Chapter 1 the disk of absolute convergence for the power series expansion of $\sec z$ is $|z| < \pi/2$. Therefore, all the series in (2.63) are convergent for $|z| < \pi/2$, which means in turn that the equivalence symbol can be replaced by an equals sign for these values of z. Because z is still fairly arbitrary, we can equate like powers of z on both sides of (2.63), which results in (2.55).

The other recurrence relation in Theorem 2.1 can be derived by noting that

$$\frac{z^{\mu+\nu}}{\sin^{\mu+\nu} z} = \frac{z^\mu}{\sin^\mu z} \frac{z^\nu}{\sin^\nu z} . \quad (2.64)$$

Introducing (2.53) into the above identity gives

$$\sum_{k=0}^{\infty} c_{\mu+\nu,k} z^{2k} \equiv \sum_{k=0}^{\infty} c_{\mu,k} z^{2k} \sum_{k=0}^{\infty} c_{\nu,k} z^{2k} = \sum_{k=0}^{\infty} z^{2k} \sum_{j=0}^{k} c_{\nu,j} c_{\nu,k-j} . \quad (2.65)$$

For $|z| < \pi$, the equivalence symbol can be replaced by an equals sign. Because z is again fairly arbitrary, like powers of z can be equated on both sides of the resulting equation. Hence we obtain (2.56), thereby completing the proof of Theorem 2.1.

We can derive another interesting result by expressing the lhs of (2.53) as

$$\frac{z^\rho}{\sin^\rho z} = \frac{1}{(z^{-\rho}/\sin^{-\rho} z)} . \quad (2.66)$$

By replacing both sides of (2.66) with the corresponding forms given on the rhs of (2.53), we obtain

$$\sum_{k=0}^{\infty} c_{\rho,k} z^{2k} \equiv \frac{1}{1 + \sum_{k=1}^{\infty} c_{-\rho,k} z^{2k}} . \quad (2.67)$$

As in the examples of Chapter 1, the rhs of (2.67) can be regarded as the regularized value of the geometric series. Note also that the equivalence symbol can be replaced by an equals sign for $|\Re z| < \pi$. Thus, we can apply the partition method for a power series expansion again. Moreover, by applying the partition method for a power series expansion to the rhs, we are transforming the equivalence statement into an equation. In addition, we have seen that when applying the method, the negative values of the coefficients of the powers of z^2 in the denominator become the values that are assigned to the parts in the partitions. That is, each part is assigned a value of $-c_{-\rho,i}$. Equating like powers of z in the resulting equation yields

$$c_{\pm\rho,k} = \sum_{\substack{\lambda_1,\lambda_2,\lambda_3,\dots,\lambda_k=0 \\ \sum_{i=1}^{k} i\lambda_i=k}}^{k,\lfloor k/2\rfloor,\lfloor k/3\rfloor,\dots,1} (-1)^{N_k} N_k! \prod_{i=1}^{k} \frac{1}{\lambda_i!} c_{\mp\rho,i}^{\lambda_i}. \tag{2.68}$$

If we put ρ equal unity in this result, then it reduces to (1.17) when B_{2k} replaces c_k via (1.15), while if ρ is set equal to -1, then we obtain (1.35) again. Alternatively, (2.68) can be written as

$$c_{\pm\rho,k} + c_{\mp\rho,k} = \sum_{\substack{\lambda_1,\lambda_2,\lambda_3,\dots,\lambda_{k-1}=0 \\ \sum_{i=1}^{k-1} i\lambda_i=k}}^{k,\lfloor k/2\rfloor,\lfloor k/3\rfloor,\dots,1} (-1)^{N_k} N_k! \prod_{i=1}^{k-1} \frac{1}{\lambda_i!} c_{\mp\rho,i}^{\lambda_i}. \tag{2.69}$$

We can also use the partition method for a power series expansion to determine general formulas for the highest order values of the $C_{k,i}$. We have already noted that the highest order term in the generalized cosecant numbers is determined by the partition with the most number of ones. This partition yields a contribution of $(\rho)_k/(3!)^k k!$ to $c_{\rho,k}$. Therefore, the highest order term is the coefficient of ρ^k in the Pochhammer factor. Hence we find that

$$C_{k,k} = \frac{1}{(3!)^k k!}. \tag{2.70}$$

The next highest order term, namely $C_{k,k-1}$, is the sum of the contributions from two partitions. First, there is the ρ^{k-1} term from the partition with k ones and second, there is the highest order term from the partition with $k-2$ ones and one two. The first term represents the coefficient of ρ^{k-1} in the Pochhammer factor in the previous calculation, while the second term represents the ρ^{k-1} power in $-(\rho)_{k-1}/(5! \cdot (3!)^{k-1}(k-2)!)$. By combining the contributions, we obtain

$$C_{k,k-1} = \frac{1}{(3!)^k k!} \sum_{i=1}^{k-1} i - \frac{1}{5 \cdot (3!)^{k-2}(k-2)!} = \frac{1}{5 \cdot (3!)^k (k-2)!}. \tag{2.71}$$

The next highest order term or $C_{k,k-2}$ is the sum of the contributions from four partitions, viz. the ρ^{k-2} power from the partition with k ones, the second leading order term from the partition with $k-2$ ones and one two and the leading order terms in the partitions with $k-3$ ones and one three and $k-4$ ones and two twos. To evaluate each of these contributions in general form, we require the formula that gives the coefficient of ρ to an arbitrary power in each Pochhammer factor, i.e., (1.56). Hence $C_{k,k-2}$ becomes

$$C_{k,k-2} = \frac{s_k^{(k-2)}}{(3!)^k k!} + \frac{s_{k-1}^{(k-2)}}{5! \cdot (3!)^{k-2}(k-2)!} + \frac{s_{k-2}^{(k-2)}}{7! \cdot (3!)^{k-3}(k-3)!}$$
$$+ \frac{s_{k-2}^{(k-2)}}{2! \cdot (5!)^2 (3!)^{k-4}(k-4)!} . \tag{2.72}$$

Introducing the results from (1.59), we find that the coefficients reduce to

$$C_{k,k-2} = \frac{21k+17}{175\,(3!)^{k+1}\,(k-3)!} . \tag{2.73}$$

A pattern is developing here. The power of 3! seems to be increasing as the power of ρ decreases while the factorial is decrementing by unity. That is, $C_{k,k-1}$ goes as $1/(3!)^k(k-2)!$, while $C_{k,k-2}$ goes as $1/(3!)^{k+1}(k-3)!$. Moreover, the numerator for $C_{k,k-1}$ is constant, whereas it is linear in k for $C_{k,k-2}$. So let us conjecture that

$$C_{k,k-3} = \frac{ak^2 + bk + c}{(3!)^{k+2}(k-4)!} . \tag{2.74}$$

The logic behind choosing $1/(k-l-1)!$ in the denominator for $C_{k,k-l}$ is that the first value of these coefficients only appears when $k=l+1$. Therefore, we can put $k=4$ in the above result and equate it to $C_{4,1}$ in Table 2.3, which equals 144/5443200. Similarly, we put $k=5$ and $k=6$ in (2.74) and equate them respectively to $C_{5,2}$ and $C_{6,3}$ in the table. Then we obtain the following set of equations:

$$16a + 4b + c = \frac{216}{175} , \tag{2.75}$$

$$25a + 5b + c = \frac{312}{175} , \tag{2.76}$$

$$36a + 6b + c = \frac{2124}{875} . \tag{2.77}$$

The solution to the above set of equations is $a=6/125$, $b=102/875$ and $c=0$. Hence $C_{k,k-3}$ is given by

$$C_{k,k-3} = \frac{k^2 + 17k/7}{125(3!)^{k+1}(k-4)!} . \tag{2.78}$$

Putting $k = 8$ in this formula yields $73/26453952000$, which agrees with the coefficient of 397517120 in Table 2.3 for $k = 8$ when it is multiplied by the external factor of $2/45 \cdot 18!$. With this method it does not matter whether our conjecture has the exact power of 3! in the denominator provided the dependence upon k is correct. However, it is crucial that the correct factorial appears in the denominator. In the case of $C_{k,k-4}$ one would conjecture that

$$C_{k,k-4} = \frac{ak^3 + bk^2 + ck + d}{(3!)^{k+4}(k-5)!}.$$ (2.79)

In this case a set of four linear equations is required with k ranging from 5 to 8. Solving the equations using the LinearSolve routine in Mathematica [32] yields

$$C_{k,k-4} = \frac{3(3k^3 + 102k^2/7 + 289k/49 - 11170/539)}{625\,(3!)^{k+3}\,(k-5)!}.$$ (2.80)

Finally, putting $k = 9$ into the above result yields a value of $229051/733303549440000$, which agrees with the coefficient of ρ^5 for $k = 9$ in Table 2.3.

Determining general formulas for the coefficients of the lowest order terms in ρ in the generalized cosecant numbers is a far more formidable problem. If we consider the preceding methods, then to obtain the equivalent of (2.71) for the lowest order coefficient, viz. $C_{k,1}$, we need to evaluate the contributions from all partitions, whereas previously we only needed a fixed number, e.g., four for determining $C_{k,k-2}$. This is clearly not possible since the number of partitions or $p(k)$ is not fixed, but increases dramatically. On the other hand, the empirical approach discussed above is only effective when k appears in the second subscript of the coefficients, e.g. $C_{k,k-\ell}$ with ℓ set to fixed values. In the case of $C_{k,1}$, $\ell = k - 1$. Therefore, one would conjecture that

$$C_{k,1} = (3!)^{-k}\left(a_1 k^{k-2} + a_2 k^{k-3} + \cdots + a_{k-2}k + a_{k-1}\right).$$ (2.81)

Note that there is no factorial appearing in the denominator since it reduces to unity. However, the problem is that we cannot set the above polynomial to any of the values in Table 2.3 in order to determine the coefficients, a_i.

It should be noted that formulas for the generalized cosecant numbers can be derived when ρ is a negative integer. This is of particular interest because we have observed that the coefficients in the generalized cosecant numbers are invariant irrespective of the value of ρ. From Appendix I.1.9

of Ref. [42] we have

$$\sin^{2n}(x) = \frac{1}{2^{2n-1}} \sum_{k=0}^{n-1} (-1)^{n-k} \binom{2n}{k} \cos\big((2n-2k)x\big) + 2^{-2n} \binom{2n}{n}. \quad (2.82)$$

According to Theorem 2.1 we can express the lhs in terms of the generalized cosecant numbers, $c_{-2n,k}$, while the cosine on the rhs can be expanded as a power series. Note that since we are dealing with negative values of ρ we can replace the equivalence symbol by an equals sign. Thus we obtain

$$\sum_{l=0}^{\infty} c_{-2n,l} x^{2l+2n} = \frac{1}{2^{2n-1}} \sum_{k=0}^{n-1} (-1)^{n-k} \binom{2n}{k} \sum_{j=0}^{\infty} \frac{(-1)^j}{(2j)!} \big((2n-2k)x\big)^{2j}$$

$$+ 2^{-2n} \binom{2n}{n}. \quad (2.83)$$

Next we equate like powers of x on both sides of the above result. For $j < n$, the coefficients of x^{2j} on the rhs vanish. Consequently, we arrive at the following interesting result:

$$\frac{1}{(2j)!} \sum_{k=0}^{n-1} (-1)^{n-k} \binom{2n}{k} (2n-2k)^{2j} = \begin{cases} 0, & j < n, \\ (-1)^n 2^{2n-1}, & j = n. \end{cases} \quad (2.84)$$

For $j > n$, the generalized cosecant numbers appear on the lhs of (2.83). Equating like powers of x yields

$$c_{-2n,l} = \sum_{i=1}^{l} C_{l,i}(-2n)^i = \frac{1}{2^{2n-1}} \sum_{k=0}^{n-1} (-1)^{k+l} \binom{2n}{k} \frac{(2n-2k)^{2n+2l}}{(2n+2l)!}. \quad (2.85)$$

This result is fascinating because the rhs is highly combinatorial, while the lhs yields polynomials in n of degree l. It is also mysterious because the lhs indicates that we should obtain a polynomial of degree l in n, but the rhs includes terms such as $2n^{2n+2l}$ and $2^{2n+2l-1}$ when the power term is expanded in powers of n. In fact, the combination of the binomial term with the power can be written to lowest order in n as

$$\binom{2n}{k}(2n-2k)^{2n+2l} = \frac{1}{k!}\big(2n(-1)^{k+1}(k-1)! + (2n)^2(k-1)!s_k^{(2)} + O(n^2)\big)$$

$$\times \big((-2k)^{2n+2l} + (2n+2l)2n(-2k)^{2n+2l-1} + O(n^2)\big)$$

$$= 2n(-1)^{l+1} 2^{2n+2l} k^{2n+2l-1} + \cdots, \quad (2.86)$$

where we have used the fact from Ref. [3] that $\sum_{j=1}^{k-1} 1/j = (-1)^k s_k^{(2)}/$ $(k-1)!$. Therefore, it appears that the lowest order term is $O(n)$, but this is deceptive once we evaluate the sum over k in (2.85). Denoting the sum over k as G, we find that

$$G = \frac{(-4)^{l+1} n}{(2n+2l)!} \sum_{k=0}^{n-1} k^{2n+2l-1} = \frac{(-4)^{l+1} n}{(2n+2l)!(2n+2l)} \left(B_{2n+2l}(n) - B_{2n+2l} \right).$$

(2.87)

The above result has been obtained by using No. 4.1.1.3 in Ref. [42]. The lowest order term in the bracketed term on the rhs is $O(n^2)$. So we have gone from a first order term in n to a third order term in n as the lowest order term. Moreover, the coefficients still maintain a dependence upon n because of the $1/(2n+2l)(2n+2l)!$ factor. In actual fact, the $1/(2n+2l)!$ factor can be removed by expressing $B_{2n+2l}(n)$ and B_{2n+2l} in terms of the cosecant polynomial $c_{2n+2l}(2n-1)$ and cosecant number c_{2n+2l} via (1.89) and (1.15) respectively. However, we are left with the $2n+2l$ factor. By partial fraction decomposition, we can obtain a linear term in n. For example,

$$\frac{n^3}{n+l} = n^2 - nl + l^2 - \frac{l^3}{n+l}.$$

(2.88)

The problem is that the decomposition will have to be applied to all the powers in $B_{2n+2l}(n) - B_{2n+2l}$. Worse still, it may have to be applied to all the neglected terms in (2.86), which emphasizes the intractability of calculating the lowest order terms in the generalized cosecant numbers.

Alternatively, we can express (2.85) as

$$\sum_{j=0}^{n-1} (-1)^j \binom{2n}{j} (2n-2j)^{2n+2k} = (-1)^k 2^{2n-1} (2n+2k)! \, c_{-2n,k} ,$$

(2.89)

whereupon if we introduce the first three results listed in Table 2.3, then we obtain the following combinatorial identities:

$$\sum_{j=0}^{n-1} (-1)^j \binom{2n}{j} (2n-2j)^{2n+2} = \frac{2^{2n-1}}{3} n (2n+2)! ,$$

(2.90)

$$\sum_{j=0}^{n-1} (-1)^j \binom{2n}{j} (2n-2j)^{2n+4} = \frac{2^{2n-1}}{360} (20n^2 - 4n) (2n+4)! ,$$

(2.91)

and

$$\sum_{j=0}^{n-1} (-1)^j \binom{2n}{j} (2n-2j)^{2n+6} = \frac{2^{2n-1}}{9 \cdot 7!} (280n^3 - 168n^2 + 32n) (2n+6)! .$$

(2.92)

These results have been checked in Mathematica for various values of n. They also supplement the result given by No. 4.2.2.33 on p. 610 of Ref. [42], which is

$$\sum_{j=0}^{n}(-1)^j\binom{2n}{j}(2n-2j)^{2n} = 2^{2n-1}\,(2n)!\,, \qquad (2.93)$$

If the power of $2n$ is replaced by $2m$ in the sum, then (2.93) vanishes for all values of $m < n$.

We can also adopt a similar approach to the second result in Appendix I.1.9 of the same reference, which is

$$\sin^{2n+1}(x) = \frac{1}{2^{2n}}\sum_{k=0}^{n}(-1)^{n-k}\binom{2n+1}{k}\sin\big((2n-2k+1)x\big)\,. \qquad (2.94)$$

Then we find that

$$\frac{1}{(2j+1)!}\sum_{k=0}^{n}(-1)^{n-k-j}\binom{2n+1}{k}(2n-2k+1)^{2j+1} = \begin{cases} 0\,, & j<n\,, \\ 2^{2n}\,, & n\,, \end{cases} \qquad (2.95)$$

and

$$c_{-2n-1,l} = \sum_{i=1}^{l}C_{l,i}(l)(-2n-1)^i = \frac{1}{2^{2n}}\sum_{k=0}^{n}(-1)^{k+l}\binom{2n+1}{k}$$
$$\times \frac{(2n-2k+1)^{2n+2l+1}}{(2n+2l+1)!}\,. \qquad (2.96)$$

Again, we see that the combinatorial form of the rhs yields a polynomial of degree l in n. In addition, the above equation is valid for $n = 0$, where it gives the coefficients of the power series expansion for $\sin(z)/z$, viz. $(-1)^l/(2l+1)!$. It can also be expressed as

$$\sum_{j=0}^{n}(-1)^j\binom{2n+1}{j}(2n-2j+1)^{2n+2k+1} = (-1)^k2^{2n}(2n+2k+1)!\,c_{-2n-1,k}\,.$$

$$(2.97)$$

By using the first three results listed in Table 2.3 again, we arrive at

$$\sum_{j=0}^{n}(-1)^j\binom{2n+1}{j}(2n-2j+1)^{2n+3} = \frac{2^{2n-1}}{3}\,(2n+1)\,(2n+3)!\,, \qquad (2.98)$$

$$\sum_{j=0}^{n}(-1)^j\binom{2n+1}{j}(2n-2j+1)^{2n+5} = \frac{2^{2n}}{360}\,(2n+5)!(20n^2+16n+3)\,,$$

$$(2.99)$$

and

$$\sum_{j=0}^{n}(-1)^j\binom{2n+1}{j}(2n-2j+1)^{2n+7}=\frac{2^{2n}}{9\cdot 7!}(2n+7)!$$
$$\times(280n^3+252n^2+74n+9)\,.\tag{2.100}$$

These results have also been checked in Mathematica for numerous values of n.

In Section 10 of Ref. [5] it is shown that

$$\frac{1}{1-\cos z}\equiv 2\sum_{k=0}^{\infty}p_k\,z^{2k-2}\,,\tag{2.101}$$

where

$$p_k=\frac{(k-1/2)}{2^{2k-1}-1}\,c_k\,.\tag{2.102}$$

Since $1-\cos z=2\sin^2(z/2)$, we see immediately that $c_{2,k}=2^{2k}p_k$. Thus, with the aid of (2.54) we find that

$$c_k=(-1)^k\left(\frac{1-2^{1-2k}}{2k-1}\right)\sum_{\substack{\lambda_1,\lambda_2,\lambda_3,\dots,\lambda_k=0\\\sum_{i=1}^k i\lambda_i=k}}^{k,\lfloor k/2\rfloor,\lfloor k/3\rfloor,\dots,1}(-1)^{N_k}(N_k+1)!$$
$$\times\prod_{i=1}^{k}\left(\frac{1}{(2i+1)!}\right)^{\lambda_i}\frac{1}{\lambda_i!}\,.\tag{2.103}$$

The above result is an alternative to (1.4) in which the numerator of the multinomial factor has the total number of parts for each partition, viz. N_k, incremented by unity. Separating the factor of N_k+1 in the factorial, we find that (2.103) reduces to

$$\sum_{\substack{\lambda_1,\lambda_2,\lambda_3,\dots,\lambda_k=0\\\sum_{i=1}^k i\lambda_i=k}}^{k,\lfloor k/2\rfloor,\lfloor k/3\rfloor,\dots,1}(-1)^{N_k}N_k!\,N_k\prod_{i=1}^{k}\left(\frac{1}{(2i+1)!}\right)^{\lambda_i}\frac{1}{\lambda_i!}$$
$$=\left((-1)^k\frac{(2k-1)}{1-2^{1-2k}}-1\right)c_k\,.\tag{2.104}$$

Comparing the lhs of (2.104) with (1.4), we see that the introduction of N_k or the lengths of the partitions into the sum over the multiplicities has resulted in the factor preceding the cosecant numbers on the rhs.

For $\rho=2\nu$, where ν is a positive integer, it is shown in Ref. [7] that the generalized cosecant numbers can be expressed as

$$c_{2\nu,k}=2^{2k}\frac{\Gamma(2\nu-2k)}{\Gamma(2\nu)}s(\nu,k)\,,\tag{2.105}$$

where $k < v$ and $s(v, n)$ represents the n-th elementary symmetric polynomial derived by summing quadratic powers, viz. $1^2, 2^2, \ldots, (v-1)^2$. That is, they are defined as

$$s(v, n) = \sum_{1 \le i_1 < i_2 < \cdots < i_n < v-1} x_{i_1} x_{i_2} \cdots x_{i_n} , \qquad (2.106)$$

where $x_{i_1} < x_{i_2} < \cdots < x_{i_n}$ and each x_{i_j} is equal to at least one value in the set $\{1, 2^2, 3^2, \ldots, (v-1)^2\}$. For the three lowest values of n, the symmetric polynomials are given by

$$s(v, 0) = 1 , \quad s(v, 1) = (v-1)v(2v-1)/6 ,$$

and

$$s(v, 2) = \frac{(5v+1)}{4 \cdot 6!} (2v-4)_5 . \qquad (2.107)$$

On the other hand, for the three highest values of n, they are given by

$$s(v, v-1) = (v-1)!^2 , \quad s(v, v-2) = (v-1)!^2 \left(\zeta(2) - \zeta(2, v) \right) ,$$

and

$$s(v, v-3) = \frac{(v-1)!^2}{2} \left((\zeta(2) - \zeta(2, v))^2 + \zeta(4, v) - \zeta(4) \right) . \qquad (2.108)$$

If we replace ρ by $2v$ in (2.54) and let $k = v-1$, then we find that

$$\sum_{\substack{\lambda_1, \lambda_2, \lambda_3, \ldots, \lambda_{v-1}=0 \\ \sum_{i=1}^{v-1} i\lambda_i = v-1}}^{v-1, \lfloor (v-1)/2 \rfloor, \lfloor (v-1)/3 \rfloor, \ldots, 1} (-1)^{N_{v-1}} (2v)_{N_{v-1}} \prod_{i=1}^{v-1} \left(\frac{1}{(2i+1)!} \right)^{\lambda_i} \frac{1}{\lambda_i!}$$

$$= \frac{2^{2v-2}}{\Gamma(2v)} s(v, v-1) = \frac{\Gamma(3/2)\Gamma(v)}{\Gamma(v+1/2)} , \qquad (2.109)$$

where the first result in (2.108) and the duplication formula for the gamma function, No. 8.335(1) in Ref. [14], have been introduced. Replacing $v-1$ by k, we arrive at

$$\sum_{\substack{\lambda_1, \lambda_2, \lambda_3, \ldots, \lambda_k=0 \\ \sum_{i=1}^{k} i\lambda_i = k}}^{k, \lfloor k/2 \rfloor, \lfloor k/3 \rfloor, \ldots, 1} (-1)^{N_k} (2k+2)_{N_k} \prod_{i=1}^{k} \left(\frac{1}{(2i+1)!} \right)^{\lambda_i} \frac{1}{\lambda_i!}$$

$$= (-1)^k \frac{k!}{(3/2)_k} . \qquad (2.110)$$

Similarly, we can repeat the derivation with k replaced by $\nu - 2$ and $\nu - 3$ and use the other results in (2.108). Then we obtain

$$\sum_{\substack{\lambda_1,\lambda_2,\lambda_3,\dots,\lambda_k=0 \\ \sum_{i=1}^{k} i\lambda_i=k}}^{k,\lfloor k/2\rfloor,\lfloor k/3\rfloor,\dots,1} (-1)^{N_k} (2k+4)_{N_k} \prod_{i=1}^{k} \left(\frac{1}{(2i+1)!}\right)^{\lambda_i} \frac{1}{\lambda_i!}$$

$$= (-1)^k \frac{\Gamma(k+2)}{(5/2)_k} \left(\zeta(2) - \zeta(2, k+2)\right), \tag{2.111}$$

and

$$\sum_{\substack{\lambda_1,\lambda_2,\lambda_3,\dots,\lambda_k=0 \\ \sum_{i=1}^{k} i\lambda_i=k}}^{k,\lfloor k/2\rfloor,\lfloor k/3\rfloor,\dots,1} (-1)^{N_k} (2k+6)_{N_k} \prod_{i=1}^{k} \left(\frac{1}{(2i+1)!}\right)^{\lambda_i} \frac{1}{\lambda_i!}$$

$$= (-1)^k \frac{\Gamma(k+3)}{(7/2)_k} \left(\left(\zeta(2) - \zeta(2, k+3)\right)^2 + \zeta(4, k+3) - \zeta(4)\right). \tag{2.112}$$

We now turn our attention to the derivation of a power series expansion for secant raised to an arbitrary power with the following theorem.

Theorem 2.2. There exists a power series expansion for cosecant raised to an arbitrary power ρ, which can be expressed as

$$\sec^\rho z = \cos^{-\rho} z \equiv \sum_{k=0}^{\infty} d_{\rho,k}\, z^{2k}. \tag{2.113}$$

For ρ not equal to a negative integer, the power series expansion on the rhs is convergent when $|\Re z| < \pi/2$, in which case the equivalence symbol can be replaced by an equals sign. On the other hand, the power series is divergent when $|\Re z| \geq \pi/2$. Then only the equivalence symbol applies. For ρ equal to a negative integer, the power series is convergent for all values of z. As in the case of the generalized cosecant numbers, the coefficients $d_{\rho,k}$ or generalized secant numbers are polynomials of degree k in ρ and are given by

$$d_{\rho,k} = (-1)^k \sum_{\substack{\lambda_1,\lambda_2,\lambda_3,\dots,\lambda_k=0 \\ \sum_{i=1}^{k} i\lambda_i=k}}^{k,\lfloor k/2\rfloor,\lfloor k/3\rfloor,\dots,1} (-1)^{N_k} (\rho)_{N_k} \prod_{i=1}^{k} \left(\frac{1}{(2i)!}\right)^{\lambda_i} \frac{1}{\lambda_i!}. \tag{2.114}$$

Furthermore, they satisfy the following recurrence relations:

$$\rho\, d_{\rho+1,k} = \sum_{j=0}^{k} (2k - 2j + 2)\, c_j\, d_{\rho,k-j+1}, \tag{2.115}$$

and

$$d_{\mu+v,k} = \sum_{j=0}^{k} d_{\mu,j}\, d_{v,k-j}\,. \tag{2.116}$$

Proof. Introducing the power series expansion for cosine into $\sec^\rho z$ gives

$$\sec^\rho z = \left(1 - \frac{z^2}{2!} + \frac{z^4}{4!} - \frac{z^6}{6!} + \frac{z^8}{8!} + \cdots\right)^{-\rho}. \tag{2.117}$$

Comparing this equation with (1.23), we see that although the coefficients for the powers of z are identical, a power of ρ now appears in the denominator. Therefore, we now regard the denominator as the regularized value of the binomial series instead of the geometric series. Consequently, (2.117) becomes

$$\sec^\rho z \equiv \sum_{l=0}^{\infty} \frac{\Gamma(l+\rho)}{\Gamma(\rho)\, l!} \left(\frac{z^2}{2!} - \frac{z^4}{4!} + \frac{z^6}{6!} - \frac{z^8}{8!} + \cdots\right)^{l}$$

$$= \sum_{k=0}^{\infty} d_{\rho,k}\, z^{2k}\,, \tag{2.118}$$

where $d_{\rho,k}$ represent the generalization of the secant numbers presented in Chapter 1. In particular, it is obvious that $d_{1,k} = d_k$.

As in the proof of Theorem 2.1 we consider the infinite product for cosine, which from No. 1.431(3) in Ref. [14] can be written as

$$\sec z = \prod_{k=0}^{\infty} \left(1 - \left(\frac{z}{(k+1/2)\pi}\right)^2\right)^{-1}. \tag{2.119}$$

This means that when we take the ρ-th power of secant, where ρ is not a non-positive integer, we have an infinite product of regularized values of the binomial series. If we introduce the series expansion for each term in the product, then the convergence of the resulting expression is dictated by the convergence of the nearest singularity to the origin or the convergence behaviour of the $k = 0$ term. Hence the power series expansion for $\sec^\rho z$ will be absolutely convergent for $|z| < \pi/2$, and conditionally convergent for $|z| \geq \pi/2$ and $|\Re z| < \pi/2$. In both cases the equivalence symbol can be replaced by an equals sign, while for all other values of z the power series expansion is divergent in which case the lhs represents the regularized value of the power series expansion. For ρ equal to a non-positive integer, all the terms in the product reduce to finite series and thus, there is no singularity resulting in divergence. Hence the power series expansion is convergent for all values of z in these cases.

As in the case of the generalized cosecant numbers, we introduce an extra factor to account for the coefficients of $\Gamma(\rho + l)/\Gamma(\rho)\, l!$ appearing in the binomial theorem. Since we have seen that l in the intermediate form of (2.118) corresponds to the number of parts N_k or $\sum_{i=1}^{k}\lambda_i$ in the partition method for a power series expansion, a factor of $\Gamma(\rho + N_k)/\Gamma(\rho)\, N_k!$ must appear when evaluating the contribution from each partition. As for the secant numbers in Chapter 1, each part i in a partition is assigned a value of $(-1)^{i+1}/(2i)!$, while the multinomial factor once again equals $N_k!/\lambda_1!\lambda_2!\cdots\lambda_k!$. Therefore, the contribution due to each partition summing to k is given by

$$C\big(\lambda_1, \lambda_2, \ldots, \lambda_k\big) = (\rho)_{N_k} \prod_{i=1}^{k} \frac{(-1)^{\lambda_i(i+1)}}{\lambda_i!\,((2i)!)^{\lambda_i}} . \tag{2.120}$$

If we sum over all the partitions summing to k, then we obtain (2.114). Compared with (2.60) the only difference is that $(2i)!$ appears in the denominator of the product instead of $(2i+1)!$.

To derive the first recurrence relation, we differentiate $\sec^\rho z$, thereby obtaining

$$\frac{1}{\sin z}\frac{d}{dz}\sec^\rho z = \rho\,\sec^{\rho+1} z . \tag{2.121}$$

Introducing (1.3) and (2.113) in the above result, we arrive at

$$\rho\sum_{k=0}^{\infty} d_{\rho+1,k}\, x^{2k} = \sum_{k=0}^{\infty} x^{2k} \sum_{j=0}^{k}(2k - 2j + 2)\, c_j\, d_{\rho,k-j+1} . \tag{2.122}$$

Note that (2.122) is an equation because both sides are convergent and divergent for the same values of z. Then equating like powers of z on both sides yields (2.114).

The second recurrence relation for the generalized secant numbers is derived simply by writing

$$\sec^{\mu+\nu} z = \sec^\mu z\,\sec^\nu z$$

and introducing (2.113) into the above result. This yields

$$\sum_{k=0}^{\infty} d_{\mu+\nu,k}\, z^{2k} = \sum_{k=0}^{\infty} d_{\mu,k}\, z^{2k} \sum_{k=0}^{\infty} d_{\nu,k}\, z^{2k} . \tag{2.123}$$

The above result is again an equation because it is convergent and divergent for the same values of z. By equating like powers of z on both sides we derive the second recurrence relation, thereby completing the proof of Theorem 2.2.

As was found for the generalized cosecant numbers, the generalized secant numbers $d_{\rho,k}$ are polynomials of degree k in ρ, because the highest order term arises from the Pochhammer polynomial $(\rho)_k$. Therefore, they can be expressed in a similar manner to the generalized cosecant numbers, viz. $d_{\rho,k} = \sum_{j=1}^{k} D_{k,j}(k)\,\rho^j$. Table 2.4 displays them up to $k = 15$. As expected, when ρ is set equal to unity, they reduce to the secant numbers or d_k given in Table 1.1. For $\rho = 2$, they are related to the secant-squared numbers q_k of Ref. [5] by

$$d_{2,k} = 2^{2k} q_k \, . \tag{2.124}$$

The secant-squared numbers can be determined from either

$$q_k = (-1)^k \sum_{\substack{\lambda_1,\lambda_2,\lambda_3,\dots,\lambda_k=0 \\ \sum_{i=1}^{k} i\lambda_i=k}}^{k,\lfloor k/2\rfloor,\lfloor k/3\rfloor,\dots,1} \left(-\frac{1}{2}\right)^{N_k} N_k! \prod_{i=1}^{k} \left(\frac{1}{(2i)!}\right)^{\lambda_i} \frac{1}{\lambda_i!} \, , \tag{2.125}$$

or

$$q_k = 2^{-2k} \sum_{j=0}^{k} d_j\, d_{k-j} = \left(\frac{2^{2k+2}-1}{2^{2k+1}-1}\right)(4k+2)\, c_{k+1} \, . \tag{2.126}$$

On the other hand, for $\rho = -1$, $d_{-1,k} = (-1)^k/(2k)!$, which represent the coefficients of the convergent power or Taylor series expansion for $\cos z$. Furthermore, by using Nos. 1.412(2) and 1.412(4) in Ref. [14], we note that $d_{-2,k} = (-1)^k 2^{2k-1}/(2k)!$ and $d_{-3,k} = (-1)^k (3^{2k}+3)/4 \cdot (2k)!$.

As in the case of the generalized cosecant numbers, we can use the partition method for a power series expansion to derive general formulas for the coefficients $D_{k,j}$ of the generalized secant numbers. We have already noted that the highest order term in the $d_{\rho,k}$ comes from the term with the Pochhammer polynomial $(\rho)_k$, which corresponds in turn to the partition with k ones. This is analogous to (2.70) with the difference occurring in the value we assign to the parts in the partitions. That is, instead of assigning a value of $1/3!$, we now assign a value of $1/2!$ to each part that is a one. Thus, the highest order coefficient for the generalized secant numbers becomes

$$D_{k,k}(k) = \frac{1}{(2!)^k k!} \, . \tag{2.127}$$

Similarly, the next highest order term, which has the coefficient $D_{k,k-1}$, is the sum of the two contributions: (1) the ρ^{k-1} term from the partition with k ones and (2) the highest order term from the partition with $k - 2$ ones and one two. Instead of assigning a value of $-1/5!$ for the two, we now

Table 2.4 Generalized secant numbers $d_{\rho,k}$

k	$d_{\rho,k}$
0	1
1	$\frac{1}{2!}\,\rho$
2	$\frac{1}{4!}\left(2\rho + 3\rho^2\right)$
3	$\frac{1}{6!}\left(16\rho + 30\rho^2 + 15\rho^3\right)$
4	$\frac{1}{8!}\left(272\rho + 588\rho^2 + 420\rho^3 + 105\rho^4\right)$
5	$\frac{1}{10!}\left(7936\rho + 18960\rho^2 + 16380\rho^3 + 6300\rho^4 + 945\rho^5\right)$
6	$\frac{1}{12!}\left(353792\rho + 911328\rho^2 + 893640\rho^3 + 429660\rho^4 + 103950\rho^5 + 10395\rho^6\right)$
7	$\frac{1}{14!}\big(22368256\rho + 61152000\rho^2 + 65825760\rho^3 + 36636600\rho^4 + 11351340\rho^5$ $+ 1891890\rho^6 + 135135\rho^7\big)$
8	$\frac{1}{16!}\big(1903757312\rho + 5464904448\rho^2 + 6327135360\rho^3 + 3918554640\rho^4 + 1427025600\rho^5$ $+ 310269960\rho^6 + 37837800\rho^7 + 2027025\rho^8\big)$
9	$\frac{1}{18!}\big(209865346036\rho + 627708979200\rho^2 + 771099169560\rho^3 + 518915776140\rho^4$ $+ 212564111760\rho^5 + 54988053120\rho^6 + 8876747880\rho^7 + 827026200\rho^8$ $+ 34459425\rho^9\big)$
10	$\frac{1}{20!}\big(29088885112832\rho + 90133968949248\rho^2 + 116351757473280\rho^3$ $+ 83750172011520\rho^4 + 37584782071200\rho^5 + 11033668966320\rho^6 + 2140527589200\rho^7$ $+ 267129462600\rho^8 + 19641872250\rho^9 + 654729075\rho^{10}\big)$
11	$\frac{1}{22!}\big(4947139780649991\rho + 15814534765170218\rho^2 + 21304894383760218\rho^3$ $+ 16224432502183290\rho^4 + 7839772870777155\rho^5 + 2538741145580640\rho^6$ $+ 563078876920560\rho^7 + 85156682410800\rho^8 + 8469575314200\rho^9$ $+ 504141387750\rho^{10} + 13749310575\rho^{11}\big)$
12	$\frac{1}{24!}\big(1015423886506852352\rho + 3334995367266582528\rho^2 + 4660452027922944000\rho^3$ $+ 3722709536929152000\rho^4 + 1912883605551072000\rho^5 + 670567768280625600\rho^6$ $+ 165008446061659200\rho^7 + 28711168921234800\rho^8 + 3485201433714000\rho^9$ $+ 282924146805300\rho^{10} + 13914302301900\rho^{11} + 316234143225\rho^{12}\big)$
13	$\frac{1}{26!}\big(246921480190207983616\rho + 831075714033875681280\rho^2 + 1199816883168896778240\rho^3$ $+ 999152056361851392000\rho^4 + 541075658301443865600\rho^5 + 202638575299038278400\rho^6$ $+ 54234696715000401600\rho^7 + 10522205018757432000\rho^8 + 1477579639013478000\rho^9$ $+ 147292828652970000\rho^{10} + 9948726145858500\rho^{11} + 411104386192500\rho^{12}$ $+ 7905853580625\rho^{13}\big)$
14	$\frac{1}{28!}\big(70251601603943959887872\rho + 241739105025518063321088\rho^2$ $+ 359285730111528382955520\rho^3 + 310374268041902641274880\rho^4$ $+ 175911626797846587648000\rho^5 + 69699511706049902476800\rho^6$ $+ 20007718263237296592000\rho^7 + 4239318795668748792000\rho^8$ $+ 666689627505922020000\rho^9 + 77249145233642382000\rho^{10} + 6441347964618567000\rho^{11}$ $+ 367773983887810500\rho^{12} + 12949788165063750\rho^{13} + 213458046676875\rho^{14}\big)$
15	$\frac{1}{30!}\big(23119184187809597841473536\rho + 81173430481947309385973760\rho^2$ $+ 123843697685956391723335680\rho^3 + 110537942778566327962828800\rho^4$ $+ 65212319968676584133713920\rho^5 + 27132631482870606005913600\rho^6$ $+ 8267637211077679417382400\rho^7 + 1885316838557296448304000\rho^8$ $+ 324973012921416916056000\rho^9 + 42332829556495137660000\rho^{10}$ $+ 4120281213040990998000\rho^{11} + 291779693682582105000\rho^{12}$ $+ 14270666557900252500\rho^{13} + 433319834754056250\rho^{14} + 6190283353629375\rho^{15}\big)$

assign a value of $-1/4!$. Therefore, combining the contributions yields

$$D_{k,k-1}(k) = \frac{1}{(2!)^k k!} \sum_{i=1}^{k-1} i - \frac{1}{4! \cdot (2!)^{k-2}(k-2)!} = \frac{1}{3(2!)^k(k-2)!} . \tag{2.128}$$

The next highest order term $D_{k,k-2}(k)$ is the sum of the contributions from four partitions, viz. the ρ^{k-2} term from the partition with k ones, the second leading order term from the partition with $k-2$ ones and one two and the leading order terms in the partitions with $k-3$ ones and one three and $k-4$ ones and two twos. This is identical to the situation for $C_{k,k-2}$. Therefore, all we need to do is modify (2.72) so that the assigned values to the each part i is $(-1)^{i+1}/(2i)!$ instead of $(-1)^{i+1}/(2i+1)!$. Then we find that $D_{k,k-2}(k)$ is given by

$$D_{k,k-2}(k) = \frac{s_k^{(k-2)}}{(2!)^k k!} + \frac{s_{k-1}^{(k-2)}}{4! \cdot (2!)^{k-2}(k-2)!} + \frac{s_{k-2}^{(k-2)}}{6! \cdot (2!)^{k-3}(k-3)!}$$
$$+ \frac{s_{k-2}^{(k-2)}}{2! \cdot (4!)^2 (2!)^{k-4}(k-4)!} . \tag{2.129}$$

Introducing the results for the Stirling numbers of the first kind listed in (1.59) simplifies (2.129) to

$$D_{k,k-2}(k) = \frac{k+1}{9 \cdot 2^{k+1}(k-3)!} . \tag{2.130}$$

Again, a pattern is emerging, particularly in regard to the factorial in the denominator. It seems the next highest order coefficient will be dependent upon $1/(k-4)!$. In addition, we expect a quadratic term in k to appear in the numerator. So let us conjecture that

$$D_{k,k-3}(k) = \frac{a_1 k^2 + b_1 k + c_1}{2^k(k-4)!} . \tag{2.131}$$

We now put $k=4$ in the above result and equate it to $D_{4,1}(k)$ in Table 2.4, which equals $272/8!$. Similarly, we put $k=5$ and $k=6$ in the equation and equate both results respectively to $D_{5,2}(k)$ and $D_{6,3}(k)$ in the table. Therefore, we arrive at the following set of equations:

$$16a_1 + 4b_1 + c_1 = \frac{34}{315} , \tag{2.132}$$

$$25a_1 + 5b_1 + c_1 = \frac{158}{945} , \tag{2.133}$$

$$36a_1 + 6b_1 + c_1 = \frac{677}{2835} . \tag{2.134}$$

The solution to these equations is $a_1 = 1/162$, $b_1 = 1/270$ and $c_1 = -16/2835$. Hence $D_{k,k-3}(k)$ can be expressed as

$$D_{k,k-3}(k) = \frac{k^2/162 + k/207 - 16/2835}{2^k(k-4)!}. \tag{2.135}$$

Putting $k = 9$ in this formula yields $187/21772800$, which agrees with the coefficient of $54988053120/18!$ in Table 2.4. In the case of $D_{k,k-4}(k)$ we conjecture that

$$D_{k,k-4}(k) = \frac{a_1 k^3 + b_1 k^3 + c_1 k^2 + d_1}{2^k(k-5)!}. \tag{2.136}$$

In this case a set of four linear equations is required with k ranging from 5 to 8. Once again, the set is easily solved via the LinearSolve routine in Mathematica [32]. This yields

$$D_{k,k-4}(k) = \frac{k^3 + 6k^2/5 - 619k/175 - 222/175}{1944 \cdot 2^k (k-5)!}. \tag{2.137}$$

Finally, putting $k = 11$ into the above result yields a value of $10471/20901888000$, which agrees with the coefficient of ρ^7 for $k = 11$ in Table 2.4.

If we write $\sec^\rho z$ as $1/\sec^{-\rho} z$ and introduce (2.113) into it, then we obtain an inverted result that is similar to (2.67). Hence we find that

$$\sum_{k=0}^{\infty} d_{\rho,k} z^{2k} \equiv \frac{1}{1 + \sum_{k=1}^{\infty} d_{-\rho,k} z^{2k}}. \tag{2.138}$$

Again, the rhs of the above result can be regarded as the regularized value of the geometric series. Moreover, for $|\Re z| < \pi/2$ the equivalence symbol can be replaced by an equals sign. That is, for these values of z, the equivalence statement becomes an equation. Consequently, the partition method for a power series expansion can be applied to rhs of (2.138). From the previous examples of the partition method for a power series expansion, we note that the negative values of the coefficients of the powers of z^2 in the denominator represent the values that are assigned to the parts in the partitions. Hence each part i is now assigned a value of $-d_{-\rho,i}$. Equating like powers of z gives

$$d_{\rho,k} = \sum_{\substack{\lambda_1,\lambda_2,\lambda_3,\dots,\lambda_k=0 \\ \sum_{i=1}^{k} i\lambda_i=k}}^{k,\lfloor k/2 \rfloor, \lfloor k/3 \rfloor, \dots, 1} (-1)^{N_k} N_k! \prod_{i=1}^{k} \frac{(d_{-\rho,i})^{\lambda_i}}{\lambda_i!}. \tag{2.139}$$

Putting ρ equal to unity in (2.139) yields (1.24), while for ρ equal to -1, one obtains (1.36).

To conclude this section, some remarks are required about evaluating the generalized cosecant and secant numbers from the recurrence relations presented in Theorems 2.1 and 2.2. In order to determine the $c_{\rho+1,k}$ from the first recurrence relation in Theorem 2.1, we need to know all the values of the secant numbers from $j = 1$ to k. This is not a major problem since there are many recurrence relations presented in Ref. [5] for calculating them. However, we also need all the values of the generalized cosecant numbers $c_{\rho,j}$ ranging from $j = 1$ to k. These cannot be determined when ρ is not equal to an integer unless (2.54) is used, but using the latter already gives the values of $c_{\rho+1,k}$. Thus, the recurrence relation in Theorem 2.1 appears to be redundant at this stage. A similar situation occurs for the $d_{\rho,k}$ and the recurrence relations in Theorem 2.2. In this instance we need to know all the c_j and $d_{\rho,j}$ for $j = 1$ to k. Whilst there are many recurrence relations in Ref. [5] for the cosecant numbers, there is only one for the $d_{\rho,j}$, viz. (2.115). However, we can use (2.114) directly by putting $\rho = \rho + 1$. Therefore, the recurrence relations for both the generalized cosecant and secant numbers in the theorems are not going to be of much value unless complementary methods can be developed. This is the type or kind of situation where the partition method for a power series expansion comes into its own over other methods for deriving power series expansions.

2.3 GENERALIZED RECIPROCAL LOGARITHM NUMBERS

In the previous section we were concerned with adapting the partition method for a power series expansion where the coefficients became polynomials of the arbitrary power on the trigonometric function. Here we consider the problem of deriving the power series expansion via the method for a non-trigonometric function, namely $\ln(1+z)^{-s}$, where s can be complex. Thus, we aim to generalize the reciprocal logarithm numbers, A_k, presented in Chapter 1. As expected, we shall obtain polynomial coefficients for the resulting power series, which will be denoted by $A_k(s)$ and referred to as the generalized reciprocal logarithm numbers.

Theorem 2.3. For s complex, the power series expansion for $z^s/\ln^s(1+z)$ is given by

$$z^s \ln^{-s}(1+z) \equiv \sum_{k=0}^{\infty} A_k(s)\, z^k . \tag{2.140}$$

In (2.140) the coefficients $A_k(s)$, which represent the generalization of the reciprocal logarithm numbers $A_k(1)$ or simply A_k of Chapter 1 and Ref. [3],

Table 2.5 Generalized reciprocal logarithm numbers $A_k(s)$

k	$A_k(s)$
0	1
1	$\frac{1}{2}s$
2	$\frac{1}{2^2 \cdot 2!}\left(s^2 - 5s/3\right)$
3	$\frac{1}{2^3 \cdot 3!}\left(s^3 - 5s^2 + 6s\right)$
4	$\frac{1}{2^4 \cdot 4!}\left(s^4 - 10s^3 + 97s^2/3 - 502s/15\right)$
5	$\frac{1}{2^5 \cdot 5!}\left(s^5 - 50s^4/3 + 305s^3/3 - 153984s^2/576 + 760s/3\right)$
6	$\frac{1}{2^6 \cdot 6!}\left(s^6 - 25s^5 + 245s^4 - 10543s^3/9 + 8150s^2/3 - 152696s/63\right)$
7	$\frac{1}{2^7 \cdot 7!}\left(s^7 - 35s^6 + 1505s^5/3 - 33817s^4/9 + 46340s^3/3 - 295748s^2/9 + 84112s/3\right)$
8	$\frac{1}{2^8 \cdot 8!}\left(s^8 - 140s^7/3 + 2758s^6/3 - 88984s^5/9 + 1687595s^4/27 - 2082100s^3/9 \right.$ $\left. + 62333204s^2/135 - 17120272s/45\right)$
9	$\frac{1}{2^9 \cdot 9!}\left(s^9 - 60s^8 + 1554s^7 - 339752s^6/15 + 608195s^5/3 - 3413108s^4/3 \right.$ $\left. + 58415444s^3/15 - 36980592s^2/5 + 29621376s/5\right)$
10	$\frac{1}{2^{10} \cdot 10!}\left(s^{10} - 75s^9 + 2470s^8 - 140434s^7/3 + 1684795s^6/3 - 39716995s^5/9 \right.$ $\left. + 203835872s^4/9 - 656429348s^3/9 + 1197755152s^2/9 - 3435808384s/33\right)$
11	$\frac{1}{2^{11} \cdot 11!}\left(s^{11} - 275s^{10}/3 + 3740s^9 - 267982s^8/3 + 12429725s^7/9 - 129751721s^6/9 \right.$ $+ 924098054s^5/9 - 4421602196s^4/9 + 13554306392s^3/9 - 23924641952s^2/9$ $\left. + 6114845440s/3\right)$
12	$\frac{1}{2^{12} \cdot 12!}\left(s^{12} - 110s^{11} + 5445s^{10} - 479798s^9/3 + 9280205s^8/3 - 123929102s^7/3 \right.$ $+ 10475121877s^6/27 - 23012990638s^5/9 + 312088268404s^4/27 - 510208096792s^3/15$ $\left. + 55163536507232s^2/945 - 180122689115392s/4095\right)$
13	$\frac{1}{2^{13} \cdot 13!}\left(s^{13} - 130s^{12} + 23023s^{11}/3 - 815386s^{10}/3 + 19275685s^9/3 - 6720952810s^8/63 \right.$ $+ 171968412473s^7/135 - 55012846746s^6/5 + 1838220016204s^5/27$ $- 13210420846168s^4/45 + 788583485358656s^3/945 - 146776375250816s^2/105$ $\left. + 108917518148608s/105\right)$
14	$\frac{1}{2^{14} \cdot 14!}\left(s^{14} - 455s^{13}/3 + 31577s^{12}/3 - 2211209s^{11}/5 + 112857745s^{10}/9 \right.$ $- 2278256123s^9/9 + 504690193007s^8/135 - 16539447783859s^7/405$ $+ 44471816443054s^6/135 - 783270268643044s^5/405 + 360536160153832s^4/45$ $\left. - 8952938100310528s^3/405 + 1631507162446208s^2/45 - 1194301255328768s/45\right)$
15	$\frac{1}{2^{15} \cdot 15!}\left(s^{15} - 175s^{14} + 14105s^{13} - 6241963s^{12}/9 + 69674605s^{11}/3 - 5038346885s^{10}/9 \right.$ $+ 270176962913s^9/27 - 3633833813467s^8/27 + 36904046967788s^7/27$ $- 281151051200656s^6/27 + 1578507241259936s^5/27 - 6317482495400080s^4/27$ $\left. + 5661188276756608s^3/9 - 3037615468752640s^2/3 + 732188636928000s\right)$

are polynomials in s of degree k. Table 2.5 displays all those up to and including $k = 15$, while a general formula for them is given by

$$A_k(s) = (-1)^k \sum_{\substack{\lambda_1,\lambda_2,\lambda_3,\ldots,\lambda_k=0 \\ \sum_{i=1}^{k} i\lambda_i=k}}^{k,\lfloor k/2\rfloor,\lfloor k/3\rfloor,\ldots,1} (-1)^{N_k}(s)_{N_k} \prod_{i=1}^{k} \frac{1}{(i+1)^{\lambda_i}\,\lambda_i!}. \qquad (2.141)$$

Furthermore, (2.140) represents a small $|z|$-expansion with a finite disk of absolute convergence given by $|z| < 1$. For these values of z the equivalence symbol may be replaced by an equals sign.

Proof. From Appendix A we know that $1/(1 - z)$ represents the regularized value for the geometric series when $\Re z > 1$. If one replaces z in this result by $-x$ and integrates over x between zero and z, then one obtains

$$\sum_{k=0}^{\infty} \frac{(-1)^k}{k+1} z^{k+1} \begin{cases} = \ln(1+z), & \Re z > -1, \\ \equiv \ln(1+z), & \Re z \leq -1. \end{cases} \quad (2.142)$$

That is, $\ln(1 + z)$ represents the regularized value of the series on the lhs of (2.142) for $\Re z \leq -1$. For a more detailed discussion of this result, the reader should consult Refs. [43] and [44]. We can generalize the above result to all values of z by replacing the equals sign with an equivalence symbol when $\Re z > -1$. If the ensuing equivalence is introduced into $z^s / \ln^s(1 + z)$, then we arrive at

$$\frac{z^s}{\ln^s(1+z)} \equiv \frac{1}{(1 - z/2 + z^2/3 - z^3/4 + z^4/5 - \cdots)^s}, \quad (2.143)$$

which is valid for all z.

We are now in a position to apply the partition method for a power series expansion to (2.143). As in the previous section, we note that the rhs of (2.143) can be treated as the regularized value of the binomial series according to Lemma A.1 in Appendix A. For $\Re z < 1$, the equals sign can also be replaced by the equivalence symbol in which case we have an equivalence statement for all values of z. By using this result we find that (2.143) becomes

$$\frac{z^s}{\ln^s(1+z)} \equiv \sum_{k=0}^{\infty} \frac{\Gamma(k+s)}{\Gamma(s)\,k!} \left(\frac{z}{2} - \frac{z^2}{3} + \frac{z^3}{4} + \cdots \right)^k. \quad (2.144)$$

The series inside the large parentheses in the above result converges absolutely to $1 - \ln(1 + z)/z$ for $|z| < 1$. On the other hand, the series over k on the rhs of (2.144) is absolutely convergent for $|\ln(1+z)/z - 1| < 1$. The latter condition follows if the first condition is obeyed. Therefore, to ensure absolute convergence, we require that $|z| < 1$ for all values of s. A different approach is required to determine the other values of z for which the series on the rhs of (2.144) is convergent or divergent. For $|z| < 1$, however, the equivalence symbol can be replaced by an equals sign as indicated in

Theorem 2.3. Moreover, expanding (2.144) yields

$$\ln^{-s}(1+z) \equiv z^{-s}\left(1 + \frac{\Gamma(s+1)}{\Gamma(s)}\frac{z}{2} + \left(\frac{\Gamma(s+1)}{\Gamma(s)}\left(-\frac{1}{3}\right) + \frac{\Gamma(s+2)}{2!\,\Gamma(s)}\frac{1}{4}\right)z^2\right.$$
$$\left. + \left(\frac{\Gamma(s+3)}{3!\,\Gamma(s)}\frac{1}{8} - \frac{\Gamma(s+2)}{\Gamma(s)}\frac{1}{6} + \frac{\Gamma(s+1)}{\Gamma(s)}\frac{1}{4}\right)z^3 + \cdots\right) = \sum_{k=0}^{\infty} A_k(s)\, z^{k-1}.$$

(2.145)

Hence we observe that $A_0(s) = 1$, $A_1(s) = s/2$, $A_2(s) = s^2/8 - 5s/24$, while $A_3(s)$ is given by

$$A_3(s) = \frac{(s+2)(s+1)s}{48} - \frac{(s+1)s}{6} + \frac{s}{4} = \frac{1}{48}\left(s^3 - 5s^2 + 6s\right). \qquad (2.146)$$

All these results appear at the top of Table 2.5.

To evaluate the higher order coefficients by using the partition method for a power series expansion, we need to adapt the method in a similar manner as in the previous section. From all previous examples, we have seen that the values assigned to the parts in the partitions are dependent upon the coefficients of the inner series. Therefore, in the case of (2.144) the values assigned to the parts will be based on the coefficients of the series inside the large parentheses, which is the same series that appeared in the calculation of the reciprocal logarithm numbers in Chapter 1. Consequently, each part i is assigned a value of $(-1)^{i+1}/(i+1)$ once again.

As stated at the beginning of this chapter the examples in Chapter 1 dealt with cases where the outer series was the geometric series, which means in turn that the multinomial factor in the partition method for a power series expansion is $N_k!\prod_{i=1}^{k}1/\lambda_i!$. In the previous section, however, the outer series was the binomial series, which affected the multinomial factor by the introduction of the extra factor, $\Gamma(N_k + s)/\Gamma(s)\,N_k!$. In other words, the length of a specific partition or N_k takes the place of k when the partition method for a power series expansion is applied to the rhs of (2.144). As a result, the multinomial factor becomes $(s)_{N_k}\prod_{i=1}^{k}1/\lambda_i!$, which, as expected, reduces to the multinomial factor of Chapter 1 when $s = 1$.

To clarify the situation, let us calculate $A_5(s)$. First, there are seven partitions summing to 5, viz. $\{5\}$, $\{4,1\}$, $\{3,1_2\}$, $\{3,2\}$, $\{2,1_3\}$, $\{2_2,1\}$ and $\{1_5\}$. According to the previous paragraph we assign values of $1/2$, $-1/3$, $1/4$, $-1/5$ and $1/6$ respectively to parts: one, two, three, four and five. For the partition involving a single part, viz. $\{5\}$, the modified multinomial factor becomes $(s)_1$, which is a polynomial of degree 1. As there are two parts in the partition $\{4, 1\}$, the modified multinomial factor equals $(s)_2$ or a polynomial of degree 2, while for $\{3, 1_2\}$ it is $(s)_3/2!$ or a polynomial of degree 3. Consequently, the modified multinomial factor for a partition with N_k parts yields a polynomial of degree N_k in s to the generalized reciprocal logarithm numbers. Thus, the partition with the greatest number of parts, viz. k ones, contributes the highest order term or s^k. Generally, the contribution from each partition can be expressed as

$$C\left(\lambda_1, \lambda_2, \ldots, \lambda_k\right) = (s)_{N_k} \prod_{i=1}^{k} \frac{(-1)^{\lambda_i(i+1)}}{\lambda_i! \, (i+1)^{\lambda_i}} . \tag{2.147}$$

Summing all the contributions from the seven partitions summing to 5 yields

$$
\begin{aligned}
A_5(s) &= (s)_1\left(\frac{1}{6}\right) - (s)_2\left(\frac{1}{5}\right)\left(\frac{1}{2}\right) + (s)_3\left(\frac{1}{2}\right)^2\left(\frac{1}{4}\right)\left(\frac{1}{2!}\right) - (s)_2\left(\frac{1}{3}\right)\left(\frac{1}{4}\right) \\
&\quad - (s)_4\left(\frac{1}{3}\right)\left(\frac{1}{2}\right)^3\left(\frac{1}{3!}\right) + (s)_3\left(\frac{1}{3}\right)^2\left(\frac{1}{2}\right)\left(\frac{1}{2!}\right) + (s)_5\left(\frac{1}{2}\right)^5\left(\frac{1}{5!}\right) \\
&= \frac{1}{11520}\left(760s - 802s^2 + 305s^3 - 50s^4 + 3s^5\right) .
\end{aligned}
\tag{2.148}
$$

If we sum (2.147) over all partitions summing to k, then we arrive at (2.141). The other values of the $A_k(s)$ displayed in Table 2.5 have been obtained by introducing (2.141) into Mathematica [32]. As expected, for $s = 1$ these results reduce to the reciprocal logarithm numbers, A_k, discussed in Chapter 1 and Ref. [3]. Consequently, these coefficients are referred to as the generalized reciprocal logarithm numbers. In addition, for $s = -1$ they reduce to the coefficients of the standard Taylor series expansion for $\ln(1 + z)$, namely $A_k(-1) = (-1)^k/(k + 1)$. This completes the proof of Theorem 2.3.

We can use (2.141) in conjunction with (1.56) and (1.59) to determine general formulas for the highest order terms of the generalized reciprocal logarithm numbers. As for the generalized cosecant numbers we note that the highest order term is derived from the partition with k ones. Therefore, the contribution to the generalized reciprocal logarithm numbers from this partition is given by

$$A_k^{\{1_k\}}(s) = \frac{(s)_k}{2^k k!} = \frac{(-1)^k}{2^k k!} \sum_{j=0}^{k} (-1)^j S_k^{(j)} s^j . \tag{2.149}$$

Besides producing the highest order term for the reciprocal logarithm numbers, we observe that the above result provides a contribution to all the other powers of s. Consequently, introducing (1.59) into (2.148) yields

$$A_k^{\{1_k\}}(s) = \frac{1}{2^k}\left(\frac{s^k}{k!} + \frac{s^{k-1}}{2(k-2)!} + \frac{(3k-1)}{24}\frac{s^{k-2}}{(k-3)!}\right.$$
$$\left. + \frac{k(k-1)}{48}\frac{s^{k-3}}{(k-4)!} + O\!\left(s^{k-4}\right)\right). \tag{2.150}$$

There is also a term with s^{k-1} provided by the contribution from the partition with $k-2$ ones and one two, which we denote by $\{1_{k-2},2\}$. This contribution can be expressed as

$$A_k^{\{1_{k-2},2\}}(s) = -\frac{(s)_{k-1}}{3\cdot 2^{k-2}(k-2)!} = \frac{(-1)^k}{3\cdot 2^{k-2}(k-2)!}\sum_{j=0}^{k-1}(-1)^j S_{k-1}^{(j)}\, s^j. \tag{2.151}$$

By introducing the $k-1$ version of (1.59) into the above result, we find that it can be expressed as

$$A_k^{\{1_{k-2},2\}}(s) = -\frac{1}{3\cdot 2^{k-2}(k-2)!}\left(s^{k-1} + \frac{(k-1)(k-2)}{2}s^{k-2}\right.$$
$$\left. + \frac{(3k-4)}{24}(k-1)(k-2)(k-3)\,s^{k-3} + O\!\left(s^{k-4}\right)\right). \tag{2.152}$$

Both (2.149) and (2.151) yield terms of order of s^{k-2}. However, there are terms of this order due to the partitions with $k-2$ parts of which there are two types. The first is the partition with $k-3$ ones and one three, viz. $\{1_{k-3},3\}$, and the other is the partition with $k-4$ ones and two twos, denoted by $\{1_{k-4},2_2\}$. From (2.141) we obtain

$$A_k^{\{1_{k-3},3\}}(s) = \frac{(s)_{k-2}}{4\cdot 2^{k-3}(k-3)!} = \frac{(-1)^k}{4\cdot 2^{k-3}(k-3)!}\sum_{j=0}^{k-2}(-1)^j S_{k-2}^{(j)}\, s^j, \tag{2.153}$$

and

$$A_k^{\{1_{k-4},2_2\}}(s) = \frac{(s)_{k-2}}{2^2 2!\cdot 2^{k-4}(k-4)!} = \frac{(-1)^k}{8\cdot 2^{k-4}(k-4)!}\sum_{j=0}^{k-2}(-1)^j S_{k-2}^{(j)}\, s^j. \tag{2.154}$$

By combining (2.150), (2.152), (2.153) and (2.154), we find after some algebra that the reciprocal logarithm numbers can be expressed as

$$A_k(s) = \frac{1}{k!}\left(\frac{s}{2}\right)^k \Theta(k-1) - \frac{5}{12(k-2)!}\left(\frac{s}{2}\right)^{k-1}\Theta(k-2)$$
$$+ \frac{(25k-3)}{12\cdot 4!(k-3)!}\left(\frac{s}{2}\right)^{k-2}\Theta(k-3) - \frac{(625k^2 - 225k - 64)}{72\cdot 6!(k-4)!}$$
$$\times \left(\frac{s}{2}\right)^{k-3}\Theta(k-4) + O\!\left(\left(\frac{s}{2}\right)^{k-4}\right)\Theta(k-5), \tag{2.155}$$

where the Heaviside step-function $\Theta(x)$ is defined as

$$\Theta(x) = \begin{cases} 1, & x \geq 0, \\ 0, & x < 0. \end{cases} \tag{2.156}$$

The above result, which appears as (192) in Ref. [4], can be checked with the values of the various coefficients displayed in Table 2.5.

As a result of Theorem 2.3, we have

$$z^{-s} \ln^s(1+z) \equiv \frac{1}{z^s/\ln^s(1+z)} \equiv \frac{1}{1+\sum_{k=1}^{\infty} A_k(s) z^k}, \tag{2.157}$$

where the equivalence symbol can be replaced by an equals sign for $|z| < 1$. The lhs of (2.157) can also be replaced by the rhs of (2.140) with coefficients now expressed in terms of $A_k(-s)$. On the rhs, since the outer series is again the geometric series, we arrive at

$$\sum_{k=0}^{\infty} A_k(-s) z^k = \sum_{k=0}^{\infty} (-1)^k \left(A_1(s)z + A_2(s)z^2 + A_3(s)z^3 + \cdots \right)^k. \tag{2.158}$$

By expanding the rhs of this result, we find that $A_1(-s) = -A_1(s)$, $A_2(-s) = A_1(s)^2 - A_2(s)$, $A_3(-s) = 2A_1(s)A_2(s) - A_3(s) - A_1(s)^3$ and so on.

To calculate the higher order values, we apply the partition method for a power series in the same manner as in Chapter 1, where the outer series in the various examples was the geometric series. By comparing (2.158) with (1.39) in particular, we observe that the coefficients of the inner series are $-A_k(s)$ rather than $(-1)^{k+1}/(k+1)$. Moreover, the multinomial factor is simply $\prod_{i=1}^{k} N_k!/\lambda_i!$ again. Therefore, the contribution to $A_k(-s)$ from a partition is represented as

$$C(\lambda_1, \lambda_2, \ldots, \lambda_k) = (-1)^{N_k} N_k! \prod_{i=1}^{k} \frac{A_i(s)^{\lambda_i}}{\lambda_i!}. \tag{2.159}$$

To obtain $A_k(-s)$, we simply sum the contributions from all the partitions summing to k as in the proof of Theorem 2.3. Hence we arrive at

$$A_k(-s) = \sum_{\substack{\lambda_1, \lambda_2, \lambda_3, \ldots, \lambda_k=0 \\ \sum_{i=1}^{k} i\lambda_i = k}}^{k, \lfloor k/2 \rfloor, \lfloor k/3 \rfloor, \ldots, 1} (-1)^{N_k} N_k! \prod_{i=1}^{k} \frac{A_i(s)^{\lambda_i}}{\lambda_i!}. \tag{2.160}$$

Theorem 2 in Ref. [3] is concerned with determining values of $A_k(2)$, which are denoted by F_k. They are easily obtained by squaring (1.37), which yields $F_k = \sum_{j=0}^{k} A_{k-j} A_j$. Later in the theorem it is shown that these coefficients can be obtained from a recurrence relation involving only the reciprocal logarithm numbers, viz.

$$A_k(2) = -(k-2)A_{k-1} - (k-1)A_k .$$ (2.161)

Before this result can be used in (2.160), we require an expression for $A_k(-2)$, which is obtained by squaring the standard power/Taylor series expansion for $\ln(1+z)/z$. Hence we obtain

$$z^{-2}\ln(1+z)^2 = \sum_{k=0}^{\infty}(-z)^k\sum_{j=0}^{k}\frac{1}{(k-j+1)(j+1)} .$$ (2.162)

Equating the rhs of (2.162) with (2.140) written in terms of $A_k(-2)$ yields

$$A_k(-2) = \frac{2(-1)^k}{k+2}\sum_{j=0}^{k}\frac{1}{j+1} = \frac{2(-1)^k}{k+2}H_{k+1} ,$$ (2.163)

where H_{k+1} denotes the $k+1$-th harmonic number [45].

If we introduce the results for $A_k(2)$ and $A_k(-2)$ into (2.160), then we arrive at

$$(k-2)A_{k-1} + (k-1)A_k = (-1)^{k+1}\sum_{\substack{\lambda_1,\lambda_2,\lambda_3,...,\lambda_k=0 \\ \sum_{i=1}^{k} i\lambda_i=k}}^{k,\lfloor k/2\rfloor,\lfloor k/3\rfloor,...,1}(-1)^{N_k}N_k!\prod_{i=1}^{k}\left(\frac{H_{i+1}}{i/2+1}\right)^{\lambda_i}\frac{1}{\lambda_i!} ,$$

(2.164)

and

$$H_{k+1} = (-1)^k(k/2+1)\sum_{\substack{\lambda_1,\lambda_2,\lambda_3,...,\lambda_k=0 \\ \sum_{i=1}^{k} i\lambda_i=k}}^{k,\lfloor k/2\rfloor,\lfloor k/3\rfloor,...,1}N_k!\prod_{i=1}^{k}\left((i-2)A_{i-1} + (i-1)A_i\right)^{\lambda_i}/\lambda_i! .$$

(2.165)

Note that the corresponding results in Ref. [4] have the phase factor of $(-1)^k$ missing in them.

In Ref. [4] it is also shown that the generalized reciprocal logarithm numbers obey the finite sum of

$$A_k(s+t) = \sum_{j=0}^{k}A_{k-j}(s)A_j(t) .$$ (2.166)

If $s = -z$ and $t = z$ in (2.166), then the above result reduces to

$$\sum_{j=0}^{k}A_{k-j}(-z)A_j(z) = \begin{cases} 0, & k \geq 1 , \\ 1, & k = 0 . \end{cases}$$ (2.167)

On the other hand, if $s = 1 - z$ and $t = z$, then (2.166) becomes

$$\sum_{j=0}^{k} A_{k-j}(1-z)A_j(z) = A_k , \tag{2.168}$$

while for $s = -1 - z$ and $t = z$,

$$\sum_{j=0}^{k} A_{k-j}(-1-z)A_j(z) = A_k(-1) = \frac{(-1)^k}{k+1} . \tag{2.169}$$

Moreover, if we put $t = z$ and $s = 1$, then this yields

$$\sum_{j=0}^{k-1} A_{k-j}A_j(z) = A_k(z+1) - A_k(z) . \tag{2.170}$$

Interestingly, for $t = z$ and $s = -1$, (2.166) reduces to

$$\sum_{j=0}^{k-1} \frac{(-1)^j}{k-j+1} A_j(z) = (-1)^k(A_k(z-1) - A_k(z)) . \tag{2.171}$$

Note that by putting $z = 1$ in (2.171) we recover the recurrence relation for the reciprocal logarithm numbers given by (1.55). From these results, we see that (2.166) can be regarded as a master equation.

By (1) differentiating $\ln^{-s}(1+z)$ with respect to (w.r.t.) z, (2) multiplying both sides by z^{s+1}, and (3) introducing (2.140), we arrive at

$$z^{s+1} \frac{d}{dz} z^{-s} \sum_{k=0}^{\infty} A_k(s)z^k = -\frac{s}{1+z} \sum_{k=0}^{\infty} A_k(s+1)z^k . \tag{2.172}$$

Because (2.172) is both convergent and divergent for the same values of z on both sides irrespective of the value of s, the above result represents an equation. Furthermore, if we multiply both sides by $1 + z$ and equate like powers of z, then we find that

$$(k - s)A_k(s) + (k - s - 1)A_{k-1}(s) = -sA_k(s+1) . \tag{2.173}$$

Introducing (2.170) into the above result yields

$$kA_k(s) = -(k - s/2 - 1)A_{k-1}(s) - s\sum_{j=0}^{k-2} A_{k-j}A_j(s) , \tag{2.174}$$

which serves as the recurrence relation for the generalized reciprocal logarithm numbers with the same value of s throughout. The above result is also interesting because regardless of the value of s, one still requires the reciprocal logarithm numbers, $A_k(1)$. Hence the reciprocal logarithm numbers serve as a basis for determining their generalized versions.

The recurrence relation given by (2.174) can be implemented together with either (1.55) or (1.60) in Mathematica [32] to determine values of the generalized reciprocal logarithm numbers. This is a much faster method than using the still unoptimized partition method for a power series expansion. For example, by programming (2.174) and (1.55) as

A[k_, s_] := -(1 - s/(2 k) - 1/k) A[k - 1, s] - (s/k) Sum[A[k - j] A[j, s], {j, 0, k - 2}]

A[k_] := Sum[(-1)^(k - j + 1) A[j]/(k - j + 1), {j, 0, k - 1}]

A[0] := 1

A[0, s_] := 1

and then typing in the command Timing[Simplify[A[20,s]]] into a Venom laptop with 8 Gb RAM, one obtains the following output:

{85.056241, (s (-43193713057849290954030907392 + 62602841826891798235633152000 s
- 41678600533075858237063073280 s^2 + 170246245996231255706322222720 s^3
- 48008220348586562727240130208 s^4 + 99582834146411524759148760 s^5
- 15792752459459707153634520 s^6 + 1963118037544813585628940 s^7
- 19439676848751706309702500 s^8 + 154901338472603052676775 s^9
- 99841219273743147700500 s^10 + 521028628796733282225 s^11 - 21941915193000702900 s^12
+ 739685494622809350 s^13 - 19682345874658500 s^14 + 404199009002250 s^15
- 6180541740450 s^16 + 66257989875 s^17 - 444386250 s^18 + 1403325 s^19))
/357999806840777823014092800000}

Therefore, it has taken 85 CPU seconds to compute $A_{20}(s)$ without having to type in or evaluate the preceding 19 values of the generalized reciprocal logarithm numbers.

Since this section has been concerned with problems of a logarithmic nature, we shall conclude this section by studying the nested logarithm of $\ln \ln(1 + z)$. Before we consider this problem using the reciprocal logarithm numbers, the partition method for a power series expansion can be applied by introducing the Taylor/Maclaurin series expansion for $\log(1 + z)$ twice into this function. After the first introduction, we arrive at

$$
\ln \ln(1 + z) \equiv \ln\left(z - \frac{z^2}{2} + \frac{z^3}{3} - \frac{z^4}{4} + \cdots\right)
$$
$$
= \ln z + \ln\left(1 - \frac{z}{2} + \frac{z^2}{3} - \frac{z^3}{4} + \cdots\right). \tag{2.175}
$$

For $|z| < 1$, we may replace the equivalence symbol by an equals sign. In the above result the equals sign merely indicates that either form on the rhs is equivalent to the nested logarithm on the lhs. Now we introduce the Taylor series for logarithm into the second logarithm of the final form of (2.175). This gives

$$\ln\ln(1+z) \equiv \ln z - \sum_{k=1}^{\infty} \frac{1}{k}\left(\frac{z}{2} - \frac{z^2}{3} + \frac{z^3}{4} - \frac{z^4}{5} + \cdots\right)^k. \tag{2.176}$$

Now we can apply the partition method for a power series expansion to the infinite sum on the rhs. Each part i is assigned the coefficient of power of z^i, viz. $(-1)^i/(i+1)$, while the same multinomial factor in the examples of Chapter 1 still applies. The only modification to the partition method for a power series expansion occurs as a result of the $1/k$ factor in the summand, which has the effect of introducing a factor of $1/N_k$ into the sum over partitions in (1.38). Denoting the coefficients of the resulting power series expansion by $A_k^{(2)}$, we obtain

$$\ln\ln(1+z) \equiv \ln z + \sum_{k=1}^{\infty} A_k^{(2)} z^k, \tag{2.177}$$

where

$$A_k^{(2)} = (-1)^{k+1} \sum_{\substack{\lambda_1,\lambda_2,\lambda_3,\ldots,\lambda_k=0 \\ \sum_{i=1}^{k} i\lambda_i=k}}^{k,\lfloor k/2\rfloor,\lfloor k/3\rfloor,\ldots,1} (-1)^{N_k}(N_k-1)! \prod_{i=1}^{k} \frac{1}{\lambda_i!} \frac{1}{(i+1)^{\lambda_i}}. \tag{2.178}$$

The nested logarithm presented above can be expressed alternatively as

$$\ln\ln(1+z) = -\ln\left(\frac{z}{z\ln(1+z)}\right). \tag{2.179}$$

Consequently, we can introduce (1.37) into this result, which yields

$$\ln\ln(1+z) \equiv \ln z - \ln\left(1 + \sum_{k=1}^{\infty} A_k z^k\right). \tag{2.180}$$

Expanding the logarithm as in (2.175), we obtain

$$\ln\ln(1+z) \equiv \ln z + \sum_{k=1}^{\infty} \frac{(-1)^k}{k}\left(A_1 z + A_2 z^2 + A_3 z^3 + A_4 z^4 + \cdots\right)^k. \tag{2.181}$$

Hence we observe that the coefficients of the powers in the inner series are given by A_i. That is, each part i in a partition is assigned a value of A_i. Moreover, the multinomial factor used throughout Chapter 1 still applies except that it is now multiplied by $(-1)^{N_k}/N_k$. Therefore, the coefficients

Table 2.6 Doubly-nested and reciprocal logarithm numbers

k	$A_k^{(2)}$	A_k
0		1
1	$-1/2$	$1/2$
2	$5/24$	$-1/12$
3	$-1/8$	$1/24$
4	$251/2880$	$-19/720$
5	$-19/288$	$3/160$
6	$19807/362880$	$-863/60480$
7	$-751/17280$	$275/24192$
8	$1070017/29030400$	$-33953/3628800$
9	$-2857/89600$	$8183/1036800$
10	$26842253/958003200$	$-3250433/479001600$

$A_k^{(2)}$ can be written alternatively as

$$A_k^{(2)} = \sum_{\substack{\lambda_1,\lambda_2,\lambda_3,\dots,\lambda_k=0 \\ \sum_{i=1}^{k} i\lambda_i=k}}^{k,\lfloor k/2\rfloor,\lfloor k/3\rfloor,\dots,1} (-1)^{N_k}(N_k-1)! \prod_{i=1}^{k} \frac{A_i^{\lambda_i}}{\lambda_i!}. \qquad (2.182)$$

The second column of Table 2.6 displays the doubly-nested logarithm numbers determined from (2.182) up to $k = 10$ and also provides the corresponding reciprocal logarithm numbers in the adjacent column for comparative purposes. Besides oscillating in sign, both sets of numbers tend to zero in magnitude as $k \to \infty$, although the convergence is nowhere near as rapid as the cosecant and secant numbers presented in Chapter 1.

We can also derive a recurrence relation for the $A_k^{(2)}$ with the aid of Theorem 1.2. First, we differentiate (2.177) w.r.t. z and multiply both sides by $z + 1$. This gives

$$\frac{1}{\ln(1+z)} = \frac{1}{z} + 1 + \sum_{k=1}^{\infty} k A_k^{(2)} z^{k-1}. \qquad (2.183)$$

Next we introduce (1.37) on the lhs. We also assume that $|z| < 1$ so that we can replace the equivalence symbol by an equals sign. Consequently, we can equate like powers on both sides of the resulting equation. After some algebra we obtain

$$A_k^{(2)} = \frac{1}{k} A_k - \left(1 - \frac{1}{k}\right) A_{k-1}^{(2)}. \qquad (2.184)$$

Therefore, to determine the $A_k^{(2)}$ via this approach, one needs to have already stored the k-th reciprocal logarithm number, while all its predecessors are required in the evaluation of $A_{k-1}^{(2)}$.

The coefficients of the nested logarithm are also related to the coefficients $C_{2,k}$ in Ref. [26], where the latter are referred to as the Cauchy numbers of the second kind. Specifically, they are given by

$$C_{2,k} = (-1)^k k \cdot k! \, A_k^{(2)} \, . \tag{2.185}$$

In addition, Blagouchine points out that the $C_{2,k}$ are related to a finite series involving the Stirling numbers of the first kind much like the reciprocal logarithm numbers in (1.60). Consequently, we find that

$$A_k^{(2)} = \frac{1}{k \cdot k!} \sum_{j=1}^{k} \frac{|s_k^{(j)}|}{j+1} \, . \tag{2.186}$$

Furthermore, he derives an integral representation for the Cauchy numbers of the second kind, which is of a similar form to (1.45). In terms of the doubly-nested logarithm numbers, this becomes

$$A_k^{(2)} = \frac{(-1)^k}{k \cdot k!} \int_0^1 (\gamma)_k \, dy \, . \tag{2.187}$$

By using this result he is able to present the asymptotic behaviour of the Cauchy numbers of the second kind, which appears as (51) in the same reference. The derivation of this result, however, is only presented in Appendix B of the arXiv version of his paper [46]. There he also determines bounds for the $C_{2,k}$, which in terms of the doubly-nested logarithm numbers can be expressed as

$$\frac{1}{k \ln(k+1)} - \frac{1}{k \ln^2(k+1)} \leq (-1)^k A_k^{(2)} \leq \frac{1}{k \ln k} - \frac{\gamma}{k \ln^2 k} \, , \tag{2.188}$$

where $k \geq 3$ and γ represents Euler's constant.

It has also been mentioned that the power series in (2.177) is absolutely convergent for $|z| < 1$. This, however, is again too restrictive. In fact, from (2.180) it can be seen that the power series expansion is absolutely convergent when $|\sum_{k=1}^{\infty} A_k z^k| < 1$ or when $|z/\ln(1+z) - 1| < 1$. For real values this means that z can range between -1 and 2.51, which means in turn that the equivalence symbol in the power series expansion can be replaced by an equals sign. Consequently, one finds that

$$\sum_{k=1}^{\infty} A_k^{(2)} = \ln \ln 2 \, . \tag{2.189}$$

Furthermore, by differentiating (2.180) w.r.t. z, one obtains

$$1 + z + (1 + z) \sum_{k=1}^{\infty} k A_k^{(2)} z^k \equiv \frac{z}{\ln(1 + z)} . \qquad (2.190)$$

Substituting the rhs of the above result by (1.37) yields

$$\sum_{k=0}^{\infty} A_k z^k = 1 + z + \sum_{k=1}^{\infty} k A_k^{(2)} \left(z^k + z^{k+1} \right) . \qquad (2.191)$$

By equating like powers of z in (2.191), we observe that $A_1 = 1 + A_1^{(2)}$, while

$$A_k = k A_k^{(2)} + (k - 1) A_{k-1}^{(2)} , \quad k > 1 , \qquad (2.192)$$

which is another version of (2.184). Alternatively, we can divide (2.190) by $1 + z$ and replace the rhs by the product of (1.37) and the geometric series. Therefore, we arrive at

$$1 + \sum_{k=1}^{\infty} k A_k^{(2)} z^k \equiv \sum_{k=0}^{\infty} (-z)^k \sum_{k=0}^{\infty} A_k z^k . \qquad (2.193)$$

All the series in (2.193) are at least absolutely convergent for $|z| < 1$, which means that the equivalence symbol can be replaced by an equals sign for these values of z. Since z is still fairly arbitrary, we can equate like powers of z on both sides, thereby obtaining

$$A_k^{(2)} = \frac{(-1)^k}{k} \sum_{j=0}^{k} (-1)^j A_j . \qquad (2.194)$$

The above result, which shows that the doubly-nested logarithm numbers can be expressed purely in terms of the reciprocal logarithm numbers, can also be obtained by solving the recurrence relation given by (2.192).

To conclude this section, we note that in Ref. [3] Euler's constant is expressed as an infinite sum over the reciprocal logarithm numbers, viz.

$$\gamma = \sum_{k=1}^{\infty} \frac{(-1)^{k+1}}{k} A_k . \qquad (2.195)$$

If we introduce (2.192) into (2.195), then we obtain

$$\gamma = 1 + \sum_{k=1}^{\infty} \frac{(-1)^{k+1}}{k + 1} A_k^{(2)} . \qquad (2.196)$$

This alternative result for Euler's constant decreases monotonically from unity as k increases, whereas (2.195) increases monotonically from zero for increasing k.

2.4 GENERALIZATION OF ELLIPTIC INTEGRALS

In Chapter 1 we observed that the values assigned to the parts in the partition method for a power series need not necessarily be numeric. Consequently, the method was adapted to derive the Bernoulli and Euler polynomials. In this chapter we have seen that the method can be adapted to handle the introduction of an arbitrary power of ρ, which results in an extra step that modifies multinomial factor. In this section we aim to derive a power series expansion in situations, where both modifications occur simultaneously. Therefore, we shall consider the function $f(x, a, \rho) = (1 + a\cos x)^{-\rho}$, which as we shall see represents the generalization of the integrands appearing in elliptic integrals. Then by considering special values of a and ρ, we shall derive expansions for the three elliptic integrals, $F(x, \kappa)$, $E(x, \kappa)$ and $\Pi(n, x, \kappa)$. Finally, we shall apply the method to the cases where the cosine in $f(x, a, \rho)$ is replaced by sine and tangent.

Theorem 2.4. Via the partition method for a power series expansion the power series expansion for the reciprocal of $(1 + a\cos x)$ raised to an arbitrary power ρ is found to be

$$(1 + a\cos x)^{-\rho} \equiv (1 + a)^{-\rho} \left(1 + \sum_{k=1}^{\infty} q_{\rho,k}(a) x^{2k} \right), \qquad (2.197)$$

where the coefficients $q_{\rho,k}(a)$ are given by

$$q_{\rho,k}(a) = (-1)^k \sum_{\substack{\lambda_1,\lambda_2,\lambda_3,\dots,\lambda_k=0 \\ \sum_{i=1}^{k} i\lambda_i=k}}^{k,\lfloor k/2 \rfloor,\lfloor k/3 \rfloor,\dots,1} (\rho)_{N_k} \left(-\frac{a}{a+1} \right)^{N_k} \prod_{i=1}^{k} \frac{1}{\lambda_i!((2i)!)^{\lambda_i}}, \qquad (2.198)$$

and are polynomials of degree k in both ρ and a. Furthermore, the power series expansion is absolutely convergent for $|1 - \cos x| < |1 + 1/a|$ and conditionally convergent for $|1 - \cos x| > |1 + 1/a|$ and $\Re((1 - \cos x)/(1 + 1/a)) < 1$.

Proof. By introducing the power series for cosine into $f(x) = (1 + a\cos x)^{-\rho}$, one finds after a little algebra that

$$(1 + a\cos x)^{-\rho} = (1 + a)^{-\rho} \left(1 - \frac{a}{1+a}\frac{x^2}{2!} + \frac{a}{1+a}\frac{x^4}{4!} - \cdots \right)^{-\rho}. \qquad (2.199)$$

The second term on the rhs of (2.199) can be regarded as the regularized value of the binomial series. Therefore, (2.199) can be expressed as

$$\left(\frac{1+a}{1+a\cos x}\right)^{\rho} \equiv \sum_{k=0}^{\infty} \frac{(\rho)_k}{k!} \left(\frac{a}{a+1}\frac{x^2}{2!} - \frac{a}{a+1}\frac{x^4}{4!} + \frac{a}{a+1}\frac{x^6}{6!} + \cdots\right)^k.$$

(2.200)

From the previous examples in this chapter we have seen that the variable in the binomial or outer series determines the conditions under which the rhs of (2.200) converges. That is, the convergence of the rhs of (2.200) is not dependent upon ρ, but on the magnitude of the inner series, which is equal to $(1-\cos x)/(1+1/a)$. If $|1-\cos x| < |1+1/a|$, then the series on the rhs of (2.200) is absolutely convergent, whereas it is conditionally convergent for $|1-\cos x| > |1+1/a|$ and $\Re((1-\cos x)/(1+1/a)) < 1$. Under both sets of conditions the equivalence symbol can be replaced by an equals sign. For all other values of x it will be divergent, in which case only the equivalence symbol applies.

We have also observed that the value assigned to each part i in a partition is dependent upon the coefficient of x^{2i} in the series inside the large brackets on the rhs of (2.200). Therefore, to apply the partition method for a power series expansion to (2.200), each part i is assigned a value of $(-1)^{i+1}a/(a+1)(2i)!$. Because of the power ρ, the multinomial factor of $\prod_{i=1}^{k} N_k!/\lambda_i!$ is now multiplied by a factor $(\rho)_{N_k}/N_k!$. This is consistent with the fact that k in (2.200) plays the role of N_k in the partition method for a power series expansion. Consequently, the contribution due to each partition is given by

$$C(\lambda_1, \lambda_2, \ldots, \lambda_k) = (\rho)_{N_k} \prod_{i=1}^{k} \frac{1}{\lambda_i!} \left(\frac{(-1)^{i+1}a}{(a+1)(2i)!}\right)^{\lambda_i}.$$

(2.201)

The coefficients of the power series expansion on the rhs of (2.200) are obtained by adding the contributions due to all the partitions summing to k. Denoting the coefficients by $q_{\rho,k}(a)$, we find that they can be written as

$$q_{\rho,k}(a) = \sum_{\substack{\lambda_1,\lambda_2,\lambda_3,\ldots,\lambda_k=0 \\ \sum_{i=1}^{k} i\lambda_i=k}}^{k,\lfloor k/2\rfloor,\lfloor k/3\rfloor,\ldots,1} (\rho)_{N_k} \prod_{i=1}^{k} \frac{1}{\lambda_i!} \left(\frac{(-1)^{i+1}a}{(a+1)(2i)!}\right)^{\lambda_i}.$$

(2.202)

After a little algebra, we finally arrive at (2.198). This completes the proof of Theorem 2.4.

Since N_k ranges from unity to k, there will always be at least a factor of ρa in the numerator of (2.198), while the largest factors in the denominator are $(1+a)^k$ and $(2k)!$. Consequently, the coefficients in Theorem 2.4 can be expressed as

$$q_{\rho,k}(a) = \frac{\rho\,a}{(2k)!}\,(a+1)^{-k}\,C_k(\rho,a)\,, \tag{2.203}$$

where $C_k(\rho,a)$ are polynomials of degree $k-1$ in both ρ and a. By using (2.198) we find that the polynomials are given by

$$C_k(\rho,a) = (-1)^{k+1}(2k)!(a+1)^{k-1}\sum_{\substack{\lambda_1,\lambda_2,\lambda_3,\dots,\lambda_k=0 \\ \sum_{i=1}^{k} i\lambda_i = k}}^{k,\lfloor k/2\rfloor,\lfloor k/3\rfloor,\dots,1} (\rho+1)_{N_k-1}$$

$$\times \left(-\frac{a}{a+1}\right)^{N_k-1}\prod_{i=1}^{k}\frac{1}{\lambda_i!\,((2i)!)^{\lambda_i}}\,. \tag{2.204}$$

Table 2.7 displays the $C_k(\rho,a)$ up to and including $k=12$. They have been calculated using the general forms for the coefficients in the partition method for a power series expansion based on the programming methodology appearing in Chapter 5 and then using the Sum, Expand and Simplify routines in Mathematica. The highest order term in $a+1$ is determined by the partition with only one part, viz. $\{k\}$. In this case we have $N_k=1$ and $\lambda_k=1$, while all the other multiplicities vanish. This means that the leading term in powers of $a+1$ is given by $(-(a+1))^{k-1}$. The next highest order term is determined by all the partitions with two parts in them ($N_k=2$) such as $\{1,k-1\}$, $\{2,k-2\}$, and so on up to and including $\{\lfloor k/2\rfloor,k-\lfloor k/2\rfloor\}$, where $\lfloor x\rfloor$ again denotes the greatest integer less than or equal to x. For these partitions all the multiplicities are equal to unity except when k is an even integer. In this case there is only one non-zero multiplicity for the partition $\{\lfloor k/2\rfloor,k-\lfloor k/2\rfloor\}$, i.e. $\lambda_{k/2}=2$. Therefore, the sum over all 2-part partitions becomes

$$C_k^{(k-2)}(\rho,a) = (-1)^k a\,(a+1)^{k-2}\,(\rho+1)_1\left(\sum_{j=1}^{\lfloor k/2\rfloor}\binom{2k}{2j}\right.$$

$$\left. -\left(\frac{1+(-1)^k}{2}\right)\binom{2k-1}{k}\right), \tag{2.205}$$

where the upper subscript signifies the power of $a+1$. With the aid of No. 4.2.1.10 on p. 607 of Ref. [42] the first order coefficients reduce to

$$C_k^{(k-2)}(\rho,a) = (-1)^k a\,(a+1)^{k-2}\,(\rho+1)_1\left(2^{2k-2}-1\right). \tag{2.206}$$

Table 2.7 The terms $C_k(\rho, a)$ in the coefficients of the power series expansion for $(1 + a\cos x)^{-\rho}$

k	$C_k(\rho, a)$
1	1
2	$(3\rho + 2)a - 1$
3	$(15\rho^2 + 30\rho + 16)a^2 - (15\rho + 13)a + 1$
4	$(105\rho^3 + 420\rho^2 + 588\rho + 272)a^3 - 3(70\rho^2 + 168\rho + 99)a^2 + 3(21\rho + 20)a - 1$
5	$(a+1)^4 - 255a(a+1)^3(\rho+1)_1 + 2205a^2(a+1)^2(\rho+1)_2$
	$\quad - 3150a^3(a+1)(\rho+1)_3 + 945a^4(\rho+1)_4$
6	$-(a+1)^5 + 1023a(a+1)^4(\rho+1)_1 - 21120a^2(a+1)^3(\rho+1)_2$
	$\quad + 65835a^3(a+1)^2(\rho+1)_4 - 51975a^4(a+1)(\rho+1)_5 + 10395a^5(\rho+1)_6$
7	$(a+1)^6 - 4095a(a+1)^5(\rho+1)_1 + 195195a^2(a+1)^4(\rho+1)_2$
	$\quad - 1201200a^3(a+1)^3(\rho+1)_3 + 1891890a^4(a+1)^2(\rho+1)_5$
	$\quad - 945945a^5(a+1)(\rho+1)_6 + 135135a^6(\rho+1)_7$
8	$-(a+1)^7 + 16383a(a+1)^6(\rho+1)_1 - 1777230a^2(a+1)^5(\rho+1)_2$
	$\quad + 20585565a^3(a+1)^4(\rho+1)_3 - 58108050a^4(a+1)^3(\rho+1)_4$
	$\quad + 54864810a^5(a+1)^2(\rho+1)_5 - 18918900a^6(a+1)(\rho+1)_6$
	$\quad + 2027025a^7(\rho+1)_7$
9	$(a+1)^8 - 65535a(a+1)^7(\rho+1)_1 + 16076985a^2(a+1)^6(\rho+1)_2$
	$\quad - 341809650a^3(a+1)^5(\rho+1)_3 + 1637971335a^4(a+1)^4(\rho+1)_4$
	$\quad - 2614321710a^5(a+1)^3(\rho+1)_5 + 1640268630a^6(a+1)^2(\rho+1)_6$
	$\quad - 413513100a^7(a+1)(\rho+1)_7 + 34459425a^8(\rho+1)_8$
10	$-(a+1)^9 + 262143a(a+1)^8(\rho+1)_1 - 145020540a^2(a+1)^7(1+\rho)_2$
	$\quad + 5581493295a^3(a+1)^6(\rho+1)_3 - 44025570225a^4(a+1)^5(\rho+1)_4$
	$\quad + 112133266245a^5(a+1)^4(\rho+1)_5 - 114359345100a^6(a+1)^3(\rho+1)_6$
	$\quad + 51068867850a^7(a+1)^2(\rho+1)_7 - 9820936125a^8(a+1)(\rho+1)_8$
	$\quad + 654729075a^9(\rho+1)_9$
11	$(a+1)^{10} - 1048575a(a+1)^9(\rho+1)_1 + 1306495575a^2(a+1)^8(\rho+1)_2$
	$\quad - 90319036500a^3(1+\alpha)^7(\rho+1)_3 + 1150872695280a^4(a+1)^6)(\rho+1)_4$
	$\quad - 4521078857295a^5(a+1)^5(\rho+1)_5 + 7061340371085a^6(a+1)^4(\rho+1)_6$
	$\quad - 4983797718900a^7(a+1)^3(\rho+1)_7 + 1663666579575a^8(a+1)^2(\rho+1)_8$
	$\quad - 252070693875a^9(a+1)(\rho+1)_9 + 13749310575a^{10}(\rho+1)_{10}$
12	$-(a+1)^{11} + 4194303a(a+1)^{10}(\rho+1)_1 - 11763703050a^2(a+1)^9(\rho+1)_2$
	$\quad + 1454250053025a^3(a+1)^8(\rho+1)_3 - 29584688710500a^4(a+1)^7(\rho+1)_4$
	$\quad + 175418438510700a^5(a+1)^6(\rho+1)_5 - 404779703328000a^6(a+1)^5(\rho+1)_6$
	$\quad + 424883159575875a^7(a+1)^4(\rho+1)_7 - 219481554166875a^8(a+1)^3(\rho+1)_8$
	$\quad + 56816734399425a^9(a+1)^2(\rho+1) - 6957151150950a^{10}(a+1)(\rho+1)_{10}$
	$\quad + 316234143225a^{11}(\rho+1)_{11}$

To obtain the next highest coefficient, we need to determine all the contributions from the 3-part partitions summing to k. This is best done by fixing the first partition and then summing over all possible pairs as in the earlier calculation of $S(k, 3)$. For example, if we put the first part equal to unity, then all we need to do is sum all the contributions from the pairs of numbers that sum to $k - 1$, where the first part is less than or equal to $\lfloor (k-1)/2 \rfloor$. The sum of these contributions is represented by

$$S_1(k) = \frac{1}{2!} \left(\sum_{j=1}^{\lfloor (k-1)/2 \rfloor} \frac{1}{(2j)!(2k-2-2j)!} - \frac{1}{4(2k-4)!} \right.$$
$$\left. - \frac{1}{4((k-1)!)^2} \left(\frac{1+(-1)^{k-1}}{2} \right) \right). \tag{2.207}$$

The second term on the rhs of (2.207) corrects the first term, which double counts the contribution from the partition $\{1_2, k-2\}$, while the third term corrects for the double-counting of the partition $\{1, ((k-1)/2)_2\}$, which only arises when $k - 1$ is an even integer. Again, we apply No. 4.2.1.10 from Ref. [42] to (2.207), which gives

$$S_1(k) = \frac{1}{2!(2k-2)!} \left(2^{2k-4} - 1 - \frac{1}{2} \binom{2k-2}{2} \right). \tag{2.208}$$

The above result represents the sum of the contributions where one of the parts is equal to unity, although it does not include the $k = 3$ case where all the partitions are equal to unity, which will be discussed shortly. We turn our attention to the partitions with one two as the least part. In this instance we need to consider all the pairs of numbers summing to $k - 2$, but on this occasion the summation begins at $j = 2$, reflecting the fact that all the partitions with unity in them have already been considered. Therefore, the sum of all the contributions with one two as the least part is represented as

$$S_2(k) = \frac{1}{4!} \left(\sum_{j=1}^{\lfloor (k-2)/2 \rfloor} \frac{1}{(2j)!(2k-4-2j)!} - \frac{1}{2 \cdot 4!(2k-8)!} \right.$$
$$\left. - \frac{1}{2((k-2)!)^2} \left(\frac{1+(-1)^{k-1}}{2} \right) \right). \tag{2.209}$$

Once again, we apply No. 4.2.1.10 in Ref. [42], thereby obtaining

$$S_2(k) = \frac{1}{4!(2k-4)!} \left(2^{2k-6} - 1 - \binom{2k-4}{4} - \frac{1}{2} \binom{2k-4}{2} \right). \tag{2.210}$$

A pattern is developing here. If we wish to sum the contributions from all 3-part partitions where the smallest part is j, but excluding all three parts equal to one another, then we find that

$$S_j(k) = \frac{1}{(2j)!(2k-2j)!}\left(2^{2k-2j-2} - 1 - \sum_{i=1}^{j-1}\binom{2k-2j}{2i}\right.$$
$$\left. -\frac{1}{2}\binom{2k-2j}{2j}\right). \tag{2.211}$$

To obtain the contributions from all 3-part partitions excluding those where all the parts are identical, we just sum over all values of j ranging from unity to $\lfloor k/3 \rfloor$.

Now we turn our attention to the partitions where all the parts are identical. These only occur whenever k is a multiple of three. When this happens, there is only one extra contribution on top of the sum over $S_j(k)$. Because the extra contribution from the 3-part partitions with all the parts identical to each other occur for specific values of k, we require (2.31) again with $\ell = 3$. Then the contribution from a 3-part partition with all parts equal to one another for any value of k is found to be

$$S_0(k) = \frac{1}{3! \cdot ((2k/3)!)^3}\frac{1}{3}\left(1 + 2(-1)^k \cos(k\pi/3)\right), \tag{2.212}$$

where the final bracketed expression on the rhs emerges from using (2.31). The contribution to $C_k(\rho, a)$ from all 3-part partitions becomes the sum over j from unity to $\lfloor k/3 \rfloor$ with the summand equal to $S_j(k)$ plus (2.212) with the resulting quantity multiplied by the factor outside the sum over all partitions in (2.204). Finally, we arrive at

$$C_k^{(k-3)}(\rho, a) = (-1)^{k+1}a^2(a+1)^{k-3}(\rho+1)_2\left(\sum_{j=1}^{\lfloor k/3 \rfloor}\binom{2k}{2j}\left(2^{2k-2j-2} - 1\right.\right.$$
$$\left.\left. -\sum_{i=1}^{j-1}\binom{2k-2j}{2i} - \frac{1}{2}\binom{2k-2j}{2j}\right) + \frac{(2k)!}{18((2k/3)!)^3}\right.$$
$$\left. \times \left(1 + 2(-1)^k\cos(k\pi/3)\right)\right). \tag{2.213}$$

Thus, by introducing this result in Mathematica [32] and setting $k = 10$ in (2.213), we obtain the term with -145020540 in Table 2.7 in addition to the other results for their corresponding values of k.

The lowest order term in powers of $a + 1$, which is the zeroth power or final term of each result displayed in Table 2.7, is determined by calculating the contribution from the partition with the most parts, i.e., $\{1_k\}$. Since there is only one partition, where $\lambda_1 = k$ and the other multiplicities vanish, the sum over partitions reduces to $1/2^k\,k!$. Then according to (2.204), we observe that

$$C_k^{(0)}(\rho, a) = (2k - 1)!!\, a^{k-1}\, (\rho + 1)_{k-1} , \qquad (2.214)$$

where $(2k - 1)!! = (2k - 1)(2k - 3)\cdots 1$. The next lowest order term is the term with the linear power of $a + 1$ appearing in each of the results in Table 2.7. It is determined by calculating the contribution from the partitions with $k - 1$ parts. As in the zeroth order term, there is only one partition contributing to the sum over partitions, namely $\{1_{k-2}, 2\}$. Since $\prod_{i=1}^{k} 1/\lambda_i!((2i))^{\lambda_i} = 1/2^{k-2}(4!(k - 2)!$, (2.204) yields

$$C_k^{(1)}(\rho, a) = -\frac{2^{k-1}}{3}\, k(k - 1)(1/2)_k\, a^{k-2}(a + 1)\,(\rho + 1)_{k-2} . \qquad (2.215)$$

The term yielding the quadratic/square power of $(a + 1)$ in the $C_k(\rho, a)$ is determined by considering all the partitions with $k - 2$ parts. There are two such partitions, $\{1_{k-3}, 3\}$ and $\{1_{k-4}, 2_2\}$. In this instance the sum over the partitions in (2.204) reduces to

$$\sum_{\substack{\lambda_1,\lambda_2,\lambda_3,\ldots,\lambda_k=0 \\ \sum_{i=1}^{k} i\lambda_i=k,\ \sum_{i=1}^{k}\lambda_i=k-2}}^{k,\lfloor k/2\rfloor,\lfloor k/3\rfloor,\ldots,1} \prod_{i=1}^{k}\frac{1}{\lambda_i!((2i)!)^{\lambda_i}} = \Big(\frac{1}{6!(k-3)!(2!)^{k-3}}$$

$$+ \frac{1}{2!(4!)^2(k-4)!(2!)^{k-4}}\Big). \qquad (2.216)$$

By introducing the above result in (2.204) and carrying out a little algebra, we eventually arrive at

$$C_k^{(2)}(\rho, a) = \frac{(5k - 11)}{2880}\,\frac{(2k)!}{2^{k-3}(k - 3)!}\, a^{k-3}(a + 1)^2\,(\rho + 1)_{k-3} . \qquad (2.217)$$

By putting $k = 10$ and 12 in the above result after it is introduced into Mathematica [32], one obtains respectively the terms with 51068867850 and 56816734399425, which appear towards the bottom of Table 2.7.

Corollary to Theorem 2.4. The recurrence relation for the coefficients $C_k(\rho, a)$ as defined by (2.203) is

$$C_k(\rho, a) = (2k - 1)\, \rho\, a\, C_{k-1}(\rho, a) - a\sum_{j=1}^{k-1}\frac{(2k - 1)!}{(2j - 1)!}\,(a + 1)^{k-j-1}$$

$$\times\, C_j(\rho, a)\big(a^{-1}c_{k-j} + c_{k-j}(1)\big) , \qquad (2.218)$$

where the c_k are the cosecant numbers of Chapter 1, given by either (1.13), (1.15) or (1.16), and the $c_k(1)$ are the cosecant polynomials evaluated at unity as given in (1.86).

Proof. By differentiating $f(x, a, \rho)$ w.r.t. x, we obtain

$$\left(1 + a\cos x\right)^{-\rho} = \frac{(1 + a\cos x)}{\rho a\sin x} \frac{d}{dx}(1 + a\cos x)^{-\rho}. \tag{2.219}$$

We now introduce (2.197) into the above result, but we need to ensure that it is at least absolutely convergent in order to replace the equivalence symbol by an equals sign. From Theorem 2.4 we know that the series expansion for $f(x, a, \rho)$ is absolutely convergent for $|a(1 - \cos x)/(a+1)| < 1$. Therefore, it will be absolutely convergent at least for $\Re a > 0$ and $|x| < \pi/2$. In addition, we need to introduce (1.3) and (1.85) on the rhs. These expansions are absolutely convergent for $|x| < \pi$, which is covered by the restriction we have put on x for the series expansion for $f(x, a, \rho)$. Therefore, for the specified values of x and a we can replace the equivalence symbols in all the expansions by equals signs. Then (2.219) becomes

$$1 + \sum_{k=1}^{\infty} q_{\rho,k}(a)\, x^{2k} = \rho^{-1} \sum_{k=0}^{\infty}\left(a^{-1} c_k + c_k(1)\right)x^{2k-1} \sum_{k=0}^{\infty} 2k\, q_{\rho,k}(a)\, x^{2k-1}, \tag{2.220}$$

where the cosecant numbers are obtained from either (1.13), (1.15) or (1.16) and the $c_k(1)$ from (1.86). Multiplying both series on the rhs, we arrive at

$$1 + \sum_{k=1}^{\infty} q_{\rho,k}(a)\, x^{2k} = 2(\rho x^2)^{-1} \sum_{k=0}^{\infty} x^{2k} \sum_{j=1}^{k} j\, q_{\rho,j}(a)$$
$$\times \left(a^{-1} c_{k-j} + c_{k-j}(1)\right). \tag{2.221}$$

Since x is arbitrary, we can equate like powers on both sides. This gives

$$q_{\rho,k}(a)\, x^{2k} = 2\rho^{-1} \sum_{j=1}^{k+1} j\, q_{\rho,j}(a)\left(a^{-1} c_{k+1-j} + c_{k+1-j}(1)\right). \tag{2.222}$$

Next we isolate the $q_{\rho,k+1}$ term from the sum and put all the remaining terms together on the other side of the equation. Hence we find that

$$q_{\rho,k+1}(a)\, x^{2k} = \frac{\rho a}{2(k+1)(a+1)}\left(q_{\rho,k}(a) - \frac{2}{\rho} \sum_{j=1}^{k} j\, q_{\rho,j}(a)\right.$$
$$\left. \times \left(a^{-1} c_{k+1-j} + c_{k+1-j}(1)\right)\right). \tag{2.223}$$

Now we replace $q_{\rho,k}(a)$ by $C_k(\rho, a)$ by inserting (2.203). After a little manipulation, we obtain

$$C_{k+1}(\rho, a) = (2k+1)\, \rho\, a\, C_k(\rho, a) - a \sum_{j=1}^{k} \frac{(2k+1)!}{(2j-1)!}\, C_j(\rho, a)$$

$$\times \left(a^{-1} c_{k+1-j} + c_{k+1-j}(1) \right) (a+1)^{k-j}. \tag{2.224}$$

Finally, replacing $k+1$ by k in the above result gives (2.218). Moreover, since there are no singularities, (2.218) can be analytically continued to all values of a. This completes the proof of the corollary.

When (2.218) is implemented in Mathematica, it produces the same results as in Table 2.7 except that they are totally expanded out, which is of limited use. However, the output can be simplified drastically by using the Simplify routine, which has the capability of combining like terms. E.g., the output for $C_5(\rho, a)$ on introducing the recurrence relation and then typing in Simplify[C[5, ρ, a]] is

Out[10]:1 - a (251 + 255 ρ) + 3 a^2 (1217 + 1950 ρ + 735 ρ^2) - a^3 (10841 + 22185 ρ + 14490 ρ^2 + 3150 ρ^3) + a^4 (7936 + 18960 ρ + 16380 ρ^2 + 6300 ρ^3 + 945 ρ^4).

Whilst this is identical to the fifth result in Table 2.7, it does not provide the same structure as obtained from the partition method for a power series expansion. For example, we were able to derive general formulas for the three highest and lowest terms in the coefficients for all values of k, but the recurrence relation cannot provide such insight. So, although the recurrence relation is fast, it is unable to provide any information on how the terms comprising the coefficients are affected by the value of k. So, we see once again that the partition method for a power series expansion provides a totally different perspective to problems when they can be solved by other means.

It should also be mentioned that for $\rho = a = 1$, the coefficients, $q_{\rho,k}(a)$ given by (2.202), reduce to the secant-squared numbers, q_k, discussed earlier in this chapter, where they were given by (2.125) in terms of a sum over partitions and by (2.126) in terms of a recurrence relation.

One consequence of Theorem 2.4 is that we can evaluate the integral of $f(x, a, \rho)$ with the lower limit set equal to zero. Consequently, we obtain

$$\int_0^x (1 + a\cos x)^{-\rho}\, dx \equiv (1+a)^{-\rho} \left(1 + \sum_{k=1}^{\infty} \frac{q_{\rho,k}(a)}{2k+1}\, x^{2k+1} \right). \tag{2.225}$$

Furthermore, introducing (2.203) into the above result yields

$$\int_0^x (1 + a\cos x)^{-\rho}\, dx \equiv (1 + a)^{-\rho} \left(1 + \rho\, ax \sum_{k=1}^{\infty} \frac{C_k(\rho, a)}{(2k+1)!} \left(\frac{x^2}{a+1}\right)^k\right),$$

(2.226)

which resembles the power series of the hyperbolic sine function when the divergent $C_k(\rho, a)$ are replaced by unity.

If we put $a = -\kappa^2/(\kappa^2 - 2)$, then we can write

$$\frac{1}{(1 + a\cos x)^{\rho}} = \left(\frac{1 - \kappa^2/2}{1 - \kappa^2 \sin^2(x/2)}\right)^{\rho}.$$

(2.227)

Replacing the lhs of (2.197) with the above result and using (2.203), we arrive at

$$\frac{1}{(1 - \kappa^2 \sin^2 x)^{\rho}} \equiv 1 + \frac{\rho \kappa^2}{2} \sum_{k=1}^{\infty} \left(1 - \frac{\kappa^2}{2}\right)^{k-1} C_k\left(\rho, \frac{\kappa^2}{2 - \kappa^2}\right) \frac{(2x)^{2k}}{(2k)!},$$

(2.228)

where x has been replaced by $2x$. For $|\kappa| < 1$, the equivalence symbol can be replaced by an equals sign. Integrating the equation from 0 to x yields

$$\int_0^x \frac{dx}{(1 - \kappa^2 \sin^2 x)^{\rho}} = x + \frac{\rho \kappa^2}{4} \sum_{k=1}^{\infty} \left(1 - \frac{\kappa^2}{2}\right)^{k-1}$$
$$\times C_k\left(\rho, \frac{\kappa^2}{2 - \kappa^2}\right) \frac{(2x)^{2k+1}}{(2k+1)!}.$$

(2.229)

If $\rho = 1/2$ in (2.229), then the integral on the lhs becomes the elliptic integral of the first kind denoted by $F(x, \kappa)$, while κ is referred to as the elliptic modulus. Hence (2.229) yields

$$F(x, \kappa) = x + \frac{\kappa^2}{8} \sum_{k=1}^{\infty} \left(1 - \frac{\kappa^2}{2}\right)^{k-1} C_k\left(\frac{1}{2}, \frac{\kappa^2}{2 - \kappa^2}\right) \frac{(2x)^{2k+1}}{(2k+1)!}.$$

(2.230)

Table 2.8 lists the $C_k(1/2, \kappa^2/(2 - \kappa^2))$ appearing in the coefficients of the powers of x in (2.230) up to and including $k = 13$. As can be seen from the table, each term is preceded by a factor of $1/(2(2 - \kappa^2))^{k-1}$, which has the effect of cancelling the factor of $2^{2k}(1 - \kappa^2/2)^{k-1}$ in the summand of (2.229), followed by a polynomial in even powers of κ of degree $2k - 2$. The coefficient of the highest order term in these polynomials can be determined by introducing $a = \kappa^2/(2 - \kappa^2)$ and $\rho = 1/2$ into (2.214) and removing the powers of 2 and $2 - \kappa^2$. Hence we find that this coefficient

Table 2.8 The $C_k(1/2, \kappa^2/(2-\kappa^2))$ in the power series expansion for the elliptic integral of the first kind, $F(x, \kappa)$, given by (2.230)

k	$C_k(1/2, \kappa^2/(2-\kappa^2))$
1	1
2	$\frac{1}{2(2-\kappa^2)}\left(9\kappa^2 - 4\right)$
3	$\frac{1}{4(2-\kappa^2)^2}\left(225\kappa^4 - 180\kappa^2 + 16\right)$
4	$\frac{1}{8(2-\kappa^2)^3}\left(11025\kappa^6 - 12600\kappa^2 + 3024\kappa^2 - 64\right)$
5	$\frac{1}{16(2-\kappa^2)^4}\left(893025\kappa^8 - 1323000\kappa^6 + 529200\kappa^4 - 48960\kappa^2 + 256\right)$
6	$\frac{1}{32(2-\kappa^2)^5}\left(108056025\kappa^{10} - 196465500\kappa^8 + 110602800\kappa^6 - 20275200\kappa^4 \right.$ $\left. + 785664\kappa^2 - 1024\right)$
7	$\frac{1}{64(2-\kappa^2)^6}\left(18261468225\kappa^{12} - 39332393100\kappa^{10} + 28605376800\kappa^8 \right.$ $\left. - 8072064000\kappa^6 + 749548800\kappa^4 - 12579840\kappa^2 + 4096\right)$
8	$\frac{1}{128(2-\kappa^2)^7}\left(4108830350625\kappa^{14} - 10226422206000\kappa^{12} + 9125115199200\kappa^{10} \right.$ $- 3514374864000\kappa^8 + 553339987200\kappa^6 - 27298252800\kappa^4 + 201314304\kappa^2$ $\left. - 16384\right)$
9	$\frac{1}{256(2-\kappa^2)^8}\left(1187451971330625\kappa^{16} - 3352805566110000\kappa^{14} \right.$ $+ 3546523221040800\kappa^{12} - 1739255947228800\kappa^{10} + 396258025363200\kappa^8$ $\left. - 36751373568000\kappa^6 + 987769958400\kappa^4 - 3221176320\kappa^2 + 65536\right)$
10	$\frac{1}{512(2-\kappa^2)^9}\left(428670161650355625\kappa^{18} - 1353695247316912500\kappa^{16} \right.$ $+ 1656285949658340000\kappa^{14} - 989052806405664000\kappa^{12}$ $+ 298400077469894400\kappa^{10} - 42602663795328000\kappa^8$ $\left. + 2400488636313600\kappa^6 - 35640247910400\kappa^4 + 51539410944\kappa^2 - 262144\right)$
11	$\frac{1}{1024(2-\kappa^2)^{10}}\left(189043541287806830625\kappa^{20} - 660152048941547662500\kappa^{18} \right.$ $+ 917263899581939910000\kappa^{16} - 646546084553809440000\kappa^{14}$ $+ 244283963147922297600\kappa^{12} - 48124533474899481600\kappa^{10}$ $+ 4454705959074201600\kappa^8 - 155377644871680000\kappa^6$ $\left. + 1284337410048000\kappa^4 - 824632934400\kappa^2 + 1048576\right)$
12	$\frac{1}{2048(2-\kappa^2)^{11}}\left(10000403334124981340062\overline{5}\kappa^{22} - 38262412756652102518500\overline{0}\kappa^{20} \right.$ $+ 59519308732569937251000\kappa^{18} - 48404532190059945900000\kappa^{16}$ $+ 22047968935405773360000\kappa^{14} - 5601270293425078272000\kappa^{12}$ $+ 746895224143350374400\kappa^{10} - 45805618514202624000\kappa^8$ $+ 100071018048798720000\kappa^6 - 46256762585088000\kappa^4 + 13194136387584\kappa^2$ $\left. - 4194304\right)$
13	$\frac{1}{4096(2-\kappa^2)^{12}}\left(625025208382811333753906256250\overline{2}5\kappa^{24} \right.$ $- 2600104866872495148416250000\kappa^{22} + 4476702292528295994664500000\kappa^{20}$ $- 4128175961363961537030000000\kappa^{18} + 2205493145191169097264000000\kappa^{16}$ $- 690456586348452059136000000\kappa^{14} + 122407427039154155458560000\kappa^{12}$ $- 11309539211686097141760000\kappa^{10} + 466161937603978936320000\kappa^8$ $- 64272109687898112000\kappa^6 + 1665573306472857600\kappa^4$ $\left. - 211106219950080\kappa^2 + 16777216\right)$

is equal to $((2k-1)!!)^2$. On the other hand, the constant in these polynomials can be determined by introducing $a = \kappa^2/(2-\kappa^2)$ into $(-(a+1))^{k-1}$, which represents the highest order term in powers of $a+1$ in the $C_k(\rho, a)$. As a consequence, we find that the constant in the polynomials displayed in Table 2.8 is equal to $(-4)^{k-1}$.

The results in Table 2.8 can also be verified in mathematical software packages such as Mathematica and Maple. First, the results in Table 2.8 must be programmed into the software. Typically, the third result would be programmed in Mathematica [32] as

FCoeff[3, κ_]:= (16 - 180 κ^2 + 225 κ^4)/(4 (κ^2 - 2)^2) .

Then the Sum command can be invoked as follows:

Felliptic[κ_, x_, n_] := x + (κ^2/8) Sum[(1 - κ^2/2)^(k - 1)FCoeff[k, κ] (2 x)^(2 k + 1)/(2 k + 1)!, {k, 1, n}] .

This has been programmed for the first n results in Table 2.8. Entering the command Simplify[Felliptic[κ,x,4]], i.e., only considering the first four results from the table, yields

Out[7]:= x + x^3 κ^2/6 + x^5 κ^2 (-4 + 9 κ^2) /120 + x^7 κ^2 (16 - 180 κ^2 + 225 κ^4)/5040 + x^9 (-κ^2/5670 + κ^4/120 - 5 κ^6/144 + 35 κ^8/1152) .

The elliptic integral of the first kind is a built-in function called EllipticF[ϕ,m] in Mathematica. It also has a similar name in Maple. The main difference between the integral in (2.229) and the version in Mathematica is that in the case of the latter the modulus m is not squared. That is, in order to get the built-in form to agree with the above results, the modulus must be squared in Mathematica. Furthermore, in order to obtain the Taylor series, we require the Series[f,x,n] routine. This yields the terms up to and including the n-th order term. Since the highest order term in the result for Felliptic[κ,x,4] is x^9, we need to put n=10 in Series[f,x,n]. Consequently, entering Series[EllipticF[x, κ^2],{x,0,10}] into Mathematica produces the following output:

Out[9]:= x + (κ^2 x^3)/6 + 1/120 (-4 κ^2 + 9 κ^4) x^5 + (16 κ^2 - 180 κ^4 + 225 κ^6) x^7/5040 + (-64 κ^2 + 3024 κ^4 - 12600 κ^6 + 11025 κ^8) x^9/362880 + O[x^{11}] .

This is identical to the output for Felliptic[κ,x,4] given above.

If we put $\rho = -1/2$ in (2.229), then we obtain a power series expansion for the elliptic integral of the second kind which is denoted by $E(x, k)$.

Hence we arrive at

$$E(x, \kappa) = x - \frac{\kappa^2}{8} \sum_{k=1}^{\infty} \left(1 - \frac{\kappa^2}{2}\right)^{k-1} C_k\left(-\frac{1}{2}, \frac{\kappa^2}{2 - \kappa^2}\right) \frac{(2x)^{2k+1}}{(2k+1)!}. \qquad (2.231)$$

Table 2.9 presents the values of $C_k(-1/2, \kappa^2/(2 - \kappa^2))$ for k up to and including 13. Again, we see that the denominator for each value of k is equal to $1/(2(2 - \kappa^2))^{k-1}$, which cancels the factors outside in the summand of (2.231). From (2.214) we find that the coefficient of the highest order in κ is $(2k - 1)!!(2k - 3)!!$, while the constant term obtained from $(-(a + 1))^{k-1}$ remains the same as in $C(1/2, \kappa^2/(2 - \kappa^2))$, namely $(-4)^{k-1}$.

If we enter the results in Table 2.9 as FCoeff2[k, κ] into Mathematica and apply the Sum command to (2.231) so that

Eelliptic[κ_, x_, n_] := x - (κ^2/8) Sum[(1 - κ^2/2)^(k - 1) FCoeff2[k, κ] (2 x)^(2 k + 1)/(2 k + 1)!, {k, 1, n}] ,

then for only four results (n=4) the following output appears

Out[31]:= x - x³ κ²/6 - x⁵ κ² (-4 + 3 κ²)/120 - x⁷ κ² (16 - 60 κ² + 45 κ⁴)/5040 + x⁹ (κ²/5670 - κ⁴/360 + κ⁶/144 - 5 κ⁸/1152) .

On the other hand, we can use the built-in function EllipticE and the Series command in Mathematica with n=10. Hence following output is generated

Out[33]:= x - κ² x³/6 + (1/120) (4 κ² - 3 κ⁴) x⁵ + (1/5040)(-16 κ² + 60 κ⁴ - 45 κ⁶) x⁷ + (1/362880) (64 κ² - 1008 κ⁴ + 2520 κ⁶ - 1575 κ⁸) x⁹) + O[x¹¹] .

Thus, both forms for the elliptic integral of the second kind, $E(x, \kappa)$, yield identical results.

The elliptic integral of the third kind is defined as

$$\Pi(n, x, \kappa) = \int_0^x \frac{dx}{(1 - n\sin^2 x)\sqrt{1 - \kappa^2 \sin^2 x}}, \qquad (2.232)$$

where n is known as the elliptic characteristic. Obviously, this elliptic integral is considerably more complex than the two kinds studied above, but it can still be handled using the material in Theorem 2.4. In this instance the best approach is to re-cast both terms in the denominator into power series expansions using (2.228). Thus, the first term is expressed as

$$\frac{1}{1 - n\sin^2 x} = \sum_{j=0}^{\infty} v_j x^{2j}, \qquad (2.233)$$

Table 2.9 Coefficients $C_k(-1/2, \kappa^2/(2 - \kappa^2))$ in the power series expansion for the elliptic integral of the second kind, $E(x, \kappa)$

k	$C_k(-1/2, \kappa^2/(2 - \kappa^2))$
1	1
2	$\frac{1}{2(2-\kappa^2)}\left(3\kappa^2 - 4\right)$
3	$\frac{1}{4(2-\kappa^2)^2}\left(45\kappa^4 - 60\kappa^2 + 16\right)$
4	$\frac{1}{8(2-\kappa^2)^3}\left(1575\kappa^6 - 2520\kappa^4 + 1008\kappa^2 - 64\right)$
5	$\frac{1}{16(2-\kappa^2)^4}\left(99225\kappa^8 - 189000\kappa^6 + 105840\kappa^4 - 16320\kappa^2 + 256\right)$
6	$\frac{1}{32(2-\kappa^2)^5}\left(9823275\kappa^{10} - 21829500\kappa^8 + 15800400\kappa^6 - 4055040\kappa^4 \right.$ $\left. + 261888\kappa^2 - 1024\right)$
7	$\frac{1}{64(2-\kappa^2)^6}\left(1404728325\kappa^{12} - 3575672100\kappa^{10} + 3178375200\kappa^8 \right.$ $\left. - 1153152000\kappa^6 + 149909760\kappa^4 - 4193280\kappa^2 + 4096\right)$
8	$\frac{1}{128(2-\kappa^2)^7}\left(273922023375\kappa^{14} - 786647862000\kappa^{12} + 829555927200\kappa^{10} \right.$ $\left. - 390486096000\kappa^8 + 79048569600\kappa^6 - 5459650560\kappa^4 + 67104768\kappa^2 \right.$ $\left. - 16384\right)$
9	$\frac{1}{256(2-\kappa^2)^8}\left(69850115960625\kappa^{16} - 223520371074000\kappa^{14} \right.$ $\left. + 272809478541600\kappa^{12} - 158114177020800\kappa^{10} + 44028669484800\kappa^8 \right.$ $\left. - 5250196224000\kappa^6 + 197553991680\kappa^4 - 1073725440\kappa^2 + 65536\right)$
10	$\frac{1}{512(2-\kappa^2)^9}\left(22561587455281875\kappa^{18} - 79629132195112500\kappa^{16} \right.$ $\left. + 110419063310556000\kappa^{14} - 76080985108128000\kappa^{12} \right.$ $\left. + 27127279769990400\kappa^{10} - 4733629310592000\kappa^8 + 342926948044800\kappa^6 \right.$ $\left. - 7128049582080\kappa^4 + 17179803648\kappa^2 - 262144\right)$
11	$\frac{1}{1024(2-\kappa^2)^{10}}\left(9002073394657468125\kappa^{20} - 34744844681134087500\kappa^{18} \right.$ $\left. + 53956699975408230000\kappa^{16} - 43103072303587296000\kappa^{14} \right.$ $\left. + 18791074088301715200\kappa^{12} - 4374957588627225600\kappa^{10} \right.$ $\left. + 494967328786022400\kappa^8 - 22196806410240000\kappa^6 \right.$ $\left. + 256867482009600\kappa^4 - 274877644800\kappa^2 + 1048576\right)$
12	$\frac{1}{2048(2-\kappa^2)^{11}}\left(4348001449619557104375\kappa^{22} - 18220196550786715485000\kappa^{20} \right.$ $\left. + 31325951964510493290000\kappa^{18} - 28473254229447027000000\kappa^{16} \right.$ $\left. + 14698645956937182240000\kappa^{14} - 4308669456480829440000\kappa^{12} \right.$ $\left. + 678995658312136704000\kappa^{10} - 50895131682447360000\kappa^8 \right.$ $\left. + 1429585972125696000\kappa^6 - 9251352517017600\kappa^4 \right.$ $\left. + 4398045462528\kappa^2 - 4194304\right)$
13	$\frac{1}{4096(2-\kappa^2)^{12}}\left(2500100833531245335015625\kappa^{24} \right.$ $\left. - 11304803769010848471375000\kappa^{22} + 21317629964420457117450000\kappa^{20} \right.$ $\left. - 21727241901915587037000000\kappa^{18} + 12973489089359818219200000\kappa^{16} \right.$ $\left. - 4603043908989680394240000\kappa^{14} + 941595592608878118912000\kappa^{12} \right.$ $\left. - 102813992833509974016000\kappa^{10} + 5179577084488654848000\kappa^8 \right.$ $\left. - 9181729955414016000\kappa^6 + 333114661294571520\kappa^4 \right.$ $\left. - 70368739983360\kappa^2 + 16777216\right)$

where $v_0 = 1$ and for $j \geq 1$,

$$v_j = \frac{2^{2j}}{(2j)!} \left(\frac{n}{2}\right)\left(1 - \frac{n}{2}\right)^{j-1} C_j\left(1, \frac{n}{2-n}\right). \tag{2.234}$$

The v_j are polynomials of degree j in n with the first few given by $v_1 = n$, $v_2 = n^2 - n/3$, $v_3 = n^3 - 2n^2/3 + 2n/45$, etc. On the other hand, the second term in the integrand for $\Pi(n, x, \kappa)$ can be written as

$$\frac{1}{(1 - \kappa^2 \sin^2 x)^{1/2}} = \sum_{j=0}^{\infty} w_j x^{2j}, \tag{2.235}$$

where $w_0 = 1$ and for $j \geq 1$,

$$w_j = \frac{2^{2j}}{(2j)!} \left(\frac{\kappa^2}{4}\right)\left(1 - \frac{\kappa^2}{2}\right)^{j-1} C_j\left(\frac{1}{2}, \frac{\kappa^2}{2-\kappa^2}\right). \tag{2.236}$$

Multiplying both series gives

$$\frac{1}{(1 - n\sin^2 x)\sqrt{1 - \kappa^2 \sin^2 x}} = \sum_{k=0}^{\infty} P(k, n)\, x^{2k}, \tag{2.237}$$

where

$$P(k, n, \kappa) = \sum_{j=0}^{k} v_j w_{k-j}. \tag{2.238}$$

By introducing the above result into (2.232), we obtain

$$\Pi(n, x, \kappa) = \sum_{k=0}^{\infty} P(k, n, \kappa)\, \frac{x^{2k+1}}{2k+1}. \tag{2.239}$$

The $P(k, n, \kappa)$ can be evaluated quickly by using the Sum and Simplify routines in Mathematica. They are displayed in Table 2.9 up to and including $k = 9$. From the table we see that they are polynomials of degree k in the elliptic characteristic. Moreover, the coefficients of the powers of n are polynomials in powers of κ^2, whose degree is twice the difference between k and the power of n. Hence the polynomial is of degree 12 in κ when n is first order or linear in $P(7, n, \kappa)$. In addition, the polynomial coefficients oscillate in sign, while the highest power of n has unity as its coefficient. The second highest power of n has $\kappa^2/2 - (k - 1)/3$ as its coefficient. In fact, for each power n^{k-l}, the highest power of κ is $2l$ and its coefficient is equal to $\Gamma(l + 1/2)/l!\Gamma(1/2)$. For example, for $k = 5$ and $l = 5$, the highest power of κ is κ^{10} and its coefficient is equal to $\Gamma(11/2)/5!\Gamma(1/2) = 63/256$, which is indeed the case on inspection of Table 2.10.

Table 2.10 The $P(k, n, \kappa)$ as given by (2.238)

k	$P(k, n, \kappa)$
0	1
1	$n + \kappa^2/2$
2	$n^2 + n(\kappa^2/2 - 1/3) + 3\kappa^4/8 - \kappa^2/6$
3	$n^3 + n^2(\kappa^2/2 - 2/3) + n(3\kappa^4/8 - \kappa^2/3 + 2/45) + 5\kappa^6/16 - \kappa^4/4 + \kappa^2/45$
4	$n^4 + n^3(\kappa^2/2 - 1) + n^2(3\kappa^4/8 - \kappa^2/2 + 1/5) + n(5\kappa^6/16 - 3\kappa^4/8 + \kappa^2/10$ $- 1/315) + 35\kappa^8/128 - 5\kappa^6/16 + 3\kappa^4/40 - \kappa^2/630$
5	$n^5 + n^4(\kappa^2/2 - 4/3) + n^3(3\kappa^4/8 - 2\kappa^2/3 + 7/15) + n^2(5\kappa^6/16 - \kappa^4/2$ $+ 7\kappa^2/30 - 34/945) + n(35\kappa^8/128 - 5\kappa^6/12 + 7\kappa^4/40 - 17\kappa^2/945 + 2/14175)$ $+ 63\kappa^{10}/256 - 35\kappa^8/96 + 7\kappa^6/48 - 17\kappa^4/1260 + \kappa^2/14175$
6	$n^6 + n^5(\kappa^2/2 - 5/3) + n^4(3\kappa^4/8 - 5\kappa^2/6 + 38/45) + n^3(5\kappa^6/16 - 5\kappa^4/8$ $+ 19\kappa^2/45 - 128/945) + n^2(35\kappa^8/128 - 25\kappa^6/48 + 19\kappa^4/60 - 64\kappa^2/945$ $+ 62/14175) + n(63\kappa^{10}/256 - 175\kappa^8/384 + 19\kappa^6/72 - 16\kappa^4/315$ $+ 31\kappa^2/14175 - 2/467775) + 231\kappa^{12}/1024 - 105\kappa^{10}/256 + 133\kappa^6/576$ $- 8\kappa^6/189 + 31\kappa^4/18900 - \kappa^2/467775$
7	$n^7 + n^6(\kappa^2/2 - 2) + n^5(3\kappa^4/8 - \kappa^2 + 4/3) + n^4(5\kappa^6/16 - 3\kappa^4/4 + 2\kappa^2/3$ $- 64/189) + n^3(35\kappa^8/128 - 5\kappa^6/8 + \kappa^4/2 - 32\kappa^2/189 + 26/945)$ $+ n^2(63\kappa^{10}/256 - 35\kappa^8/64 + 5\kappa^6/12 - 8\kappa^4/63 + 13\kappa^2/945 - 4/10395)$ $+ n(231\kappa^{12}/1024 - 63\kappa^{10}/128 + 35\kappa^8/96 - 20\kappa^6/189 + 13\kappa^4/1260$ $- 2\kappa^2/10395 + 4/42567525) + 429\kappa^{14}/2048 - 231\kappa^{12}/512 + 21\kappa^{10}/64$ $- 5\kappa^8/54 + 13\kappa^6/1512 - \kappa^4/6930 + 2\kappa^2/42567525$
8	$n^8 + n^7(\kappa^2/2 - 7/3) + n^6(3\kappa^4/8 - 7\kappa^2/6 + 29/15) + n^5(5\kappa^6/16 - 7\kappa^4/8$ $+ 29\kappa^2/30 - 43/63) + n^4(35\kappa^8/128 - 35\kappa^6/48 + 29\kappa^4/40 - 43\kappa^2/126$ $+ 457/4725) + n^3(63\kappa^{10}/256 - 245\kappa^8/384 + 29\kappa^6/48 - 43\kappa^4/168$ $+ 457\kappa^2/9450 - 31/7425) + n^2(231\kappa^{12}/1024 - 147\kappa^{10}/256 + 203\kappa^8/384$ $- 215\kappa^6/1008 + 457\kappa^4/12600 - 31\kappa^2/14850 + 5461/212837625)$ $+ n(429\kappa^{14}/2048 - 539\kappa^{12}/1024 + 609\kappa^{10}/1280 - 215\kappa^8/1152$ $+ 457\kappa^6/15120 - 31\kappa^4/19800 + 5461\kappa^2/425675250 - 1/638512875)$ $+ 6435\kappa^{16}/32768 - 1001\kappa^{14}/2048 + 2233\kappa^{12}/5120 - 43\kappa^{10}/256$ $+ 457\kappa^8/17280 - 31\kappa^6/23760 + 5461\kappa^4/567567000 - \kappa^2/1277025750$
9	$n^9 + n^8(\kappa^2/2 - 8/3) + n^7(3\kappa^4/8 - 4\kappa^2/3 + 119/45) + n^6(5\kappa^6/16 - \kappa^4$ $+ 119\kappa^2/90 - 1138/945) + n^5(35\kappa^8/128 - 5\kappa^6/6 + 119\kappa^4/120 - 569\kappa^2/945$ $+ 713/2835) + n^4(63\kappa^{10}/256 - 35\kappa^8/48 + 119\kappa^6/144 - 569\kappa^4/1260$ $+ 713\kappa^2/5670 - 1964/93555) + n^3(231\kappa^{12}/1024 - 21\kappa^{10}/32 + 833\kappa^8/1152$ $- 569\kappa^6/1512 + 713\kappa^4/7560 - 982\kappa^2/93555 + 63047/127702575)$ $+ n^2(429\kappa^{14}/2048 - 77\kappa^{12}/128 + 833\kappa^{10}/1280 - 569\kappa^8/1728 + 713\kappa^6/9072$ $- 491\kappa^6/62370 + 63047\kappa^4/255405150 - 514/383107725) + n(6435\kappa^{16}/32768$ $- 143\kappa^{14}/256 + 9163\kappa^{12}/15360 - 569\kappa^{10}/1920 + 713\kappa^8/10368$ $- 491\kappa^6/74844 + 63047\kappa^4/340540200 - 257\kappa^2/383107725 + 2/97692469875)$ $+ 12155\kappa^{18}/65536 - 2145\kappa^{16}/4096 + 17017\kappa^{14}/30720 - 6259\kappa^{12}/23040$ $+ 713\kappa^{10}/11520 - 491\kappa^8/85536 + 63047\kappa^6/408648240 - 257\kappa^4/510810300$ $+ \kappa^2/97692469875$

We can also derive interesting integrals involving the coefficients of the elliptic integrals. For example, according to No. 6.141(1) in Ref. [14], Catalan's constant G is defined as

$$G = \frac{1}{2} \int_0^1 K(\kappa) \, d\kappa \,, \tag{2.240}$$

where $K(\kappa)$, the complete elliptic integral of the first kind, is given by

$$K(\kappa) = \int_0^{\pi/2} \frac{dx}{\sqrt{1 - \kappa^2 \sin^2 x}} \,. \tag{2.241}$$

Hence $K(\kappa)$ reduces to (2.230) with ρ and x equal to $1/2$ and $\pi/2$ respectively. By introducing this result into (2.240), we arrive at

$$G = \frac{\pi}{4} + \frac{1}{16} \sum_{k=1}^{\infty} \frac{\pi^{2k+1}}{(2k+1)!} \int_0^1 d\kappa \, \kappa^2 \left(1 - \frac{\kappa^2}{2}\right)^{k-1} C_k\left(\frac{1}{2}, \frac{\kappa^2}{2 - \kappa^2}\right). \tag{2.242}$$

In Section 6 of Ref. [5] it is shown that Catalan's constant can also be written as

$$G = \frac{1}{2} \sum_{k=0}^{\infty} \frac{c_k}{2k+1} \left(\frac{\pi}{2}\right)^{2k+1}. \tag{2.243}$$

The series in (2.243) is not only monotonically increasing or positive definite, but also indicates that Catalan's constant is irrational because the sum is over powers of $\pi/2$, which are all transcendental. In fact, since (2.243) is a convergent series in the same powers of $\pi/2$ as (2.242) and each power of π is unique, we can equate both results for G at each power of π. Hence for $k \geq 1$ we obtain

$$\int_0^1 \kappa^2 \left(1 - \frac{\kappa^2}{2}\right)^{k-1} C_k\left(\frac{1}{2}, \frac{\kappa^2}{2 - \kappa^2}\right) d\kappa = 2^{2-2k} \Gamma(2k+1) \, c_k. \tag{2.244}$$

This is a remarkable result relating the coefficients, $C_k(1/2, \kappa^2/(2 - \kappa^2))$, with the cosecant numbers and is easily checked by using the results in Tables 1.1 and 2.8. For example, for $k = 3$ in Table 2.8, we have $C_3(1/2, \kappa^2/(2 - \kappa^2)) = (225\kappa^4 - 180\kappa^2 + 16)/4(2 - \kappa^2)^2$. Introducing this result into the lhs of (2.244) yields

$$\frac{1}{16} \int_0^1 (225\kappa^6 - 180\kappa^4 + 16\kappa^2) d\kappa = 31/336. \tag{2.245}$$

From Table 1.1, we have $c_3 = 31/15120$. When this is multiplied by $2^{-4}\Gamma(7)$ or $6!/16$, we obtain $31/336$ as in the integral.

We conclude this chapter by replacing the cosine function in $f(x, \rho, a)$ with different trigonometric functions such as $\sin x$ and $\tan x$. In the case of the sine function the analogue of (2.200) becomes

$$(1 + a \sin x)^{-\rho} \equiv \sum_{k=0}^{\infty} \frac{(\rho)_k}{k!} \left(ax - \frac{ax^3}{3!} + \frac{ax^5}{5!} - \frac{ax^7}{7!} + \cdots \right)^k. \tag{2.246}$$

From Appendix A the expansion on the rhs will be absolutely convergent for $|a \sin x| < 1$ and conditionally convergent for $|\Re a \sin x| < 1$ and $|a \sin x| > 1$. Then the equivalence symbol can be replaced by an equals sign. For all other values of a and x, the expansion is divergent and so the equivalence symbol applies indicating that the lhs represents the regularized value of the expansion on the rhs. Comparing (2.246) with (2.200), we see that there is no factor of $a+1$ in the denominator of each power on the rhs and the powers of x are odd instead of even. This means that the partition method for a power series will now involve partitions composed of parts that are only odd numbers. We can still apply the method but since the coefficients of the even powers equal zero, they can be discarded. That is, we assign a value of $(-1)^i/(2i+1)!$ to each part $2i+1$ and zero to each part $2i$. As a consequence, the contribution of each partition is given by

$$C(\lambda_1, \lambda_2, \ldots, \lambda_k) = (\rho)_{N_k} \prod_{i=1}^{k} \frac{1}{\lambda_i!} p_i^{\lambda_i}, \tag{2.247}$$

where for j, a non-negative integer,

$$p_i = \begin{cases} (-1)^j/(2j+1)!, & i = 2j+1, \\ 0 & i = 2j. \end{cases} \tag{2.248}$$

Unlike (2.197), which is a power series expansion in even powers of x, the expansion emanating from (2.246) will be in all powers of x despite the fact that partitions with only odd parts contribute to the coefficients. Therefore following Theorem 2.4, we have

$$(1 + a \sin x)^{-\rho} \equiv 1 + \sum_{k=1}^{\infty} r_k(\rho, a) x^k, \tag{2.249}$$

where the coefficients $r_k(\rho, a)$ are given by

$$r_k(\rho, a) = \sum_{\substack{\lambda_1, \lambda_3, \lambda_5 \ldots, \lambda_\ell = 0 \\ \sum_{i=1}^{k} i \lambda_i = k}}^{k, \lfloor k/3 \rfloor, \lfloor k/5 \rfloor, \ldots, 1} (-1)^{N_k} (\rho)_{N_k} a^{N_k} \prod_{i=1,3,5}^{\ell} \frac{(-1)^{(i-1)/2}}{\lambda_i! i!^{\lambda_i}}, \tag{2.250}$$

Table 2.11 Coefficients $r_k(\rho, a)$ in the power series expansion of $(1 + a\sin x)^{-\rho}$

k	$r_k(\rho, a)$
1	$-a\rho$
2	$\frac{1}{2!} a^2(\rho)_2$
3	$\frac{1}{3!} \left(a(\rho)_1 - a^3(\rho)_3 \right)$
4	$\frac{1}{4!} \left(-4a^2(\rho)_2 + a^4(\rho)_4 \right)$
5	$\frac{1}{5!} \left(-a(\rho)_1 + 10a^3(\rho)_3 - a^5(\rho)_5 \right)$
6	$\frac{1}{6!} \left(16a^2(\rho)_2 - 20a^4(\rho)_4 + a^6(\rho)_6 \right)$
7	$\frac{1}{7!} \left(a(\rho)_1 - 91a^3(\rho)_3 + 35a^5(\rho)_5 - a^7(\rho)_7 \right)$
8	$\frac{1}{8!} \left(-64a^2(\rho)_2 + 336a^4(\rho)_4 - 56a^6(\rho)_6 + a^8(\rho)_8 \right)$
9	$\frac{1}{9!} \left(-a(\rho)_1 + 820a^3(\rho)_3 - 966a^5(\rho)_5 + 84a^7(\rho)_7 - a^9(\rho)_9 \right)$
10	$\frac{1}{10!} \left(256a^2(\rho)_2 - 5440a^4(\rho)_4 + 2352a^6(\rho)_6 - 120a^8(\rho)_8 + a^{10}(\rho)_{10} \right)$

and $\ell = k + ((-1)^{k+1} - 1)/2$. Note that the sum has been restricted to the partitions with odd parts. As a consequence, the upper limit ℓ has been introduced to take into account that k may be even in which case the highest part is $k - 1$. In addition, the product is only taken over odd values of i. Integrating (2.249) yields

$$\int_0^x \frac{dx}{(1 + a\sin x)^\rho} \equiv 1 + \sum_{k=1}^\infty \frac{r_k(\rho, a)}{k+1} x^{k+1}. \tag{2.251}$$

The equivalence symbol can be replaced by an equals sign for $|a| < 1$. Therefore, to obtain the power series expansion for the above integral, all we need to do is divide each $r_k(\rho, a)$ by $k + 1$ and multiply each power of x by x in (2.249).

Table 2.11 displays the first ten values of the $r_k(\rho, a)$ in terms of Pochhammer polynomials for compactness. If we apply the Expand routine in Mathematica [32] to these results, then the $r_k(\rho, a)$ are found to be composed of varying powers of a and ρ up to k. For example, for $k = 7$, one obtains

$$r_7(\rho, a) = \frac{a\rho}{5040} - \frac{13a^3\rho}{360} + \frac{a^5\rho}{6} - \frac{a^7\rho}{7} - \frac{13a^3\rho^2}{240} + \frac{25a^5\rho^2}{72} - \frac{7a^7\rho^2}{20}$$
$$- \frac{13a^3\rho^3}{720} + \frac{35a^5\rho^3}{144} - \frac{29a^7\rho^3}{90} + \frac{5a^5\rho^4}{72} - \frac{7a^7\rho^4}{48} + \frac{a^5\rho^5}{144}$$
$$- \frac{5a^7\rho^5}{144} - \frac{a^7\rho^6}{240} - \frac{a^7\rho^7}{5040}. \tag{2.252}$$

Furthermore, by substituting (1.56) into the tabulated results, we are able to observe how the coefficients in the $r_k(a, \rho)$ are affected by the Stirling numbers of the first kind.

From the table it can be seen that for odd values of k, the Pochhammer symbols possess odd subscript arguments, while for even values of k the subscript arguments are even. This occurs because only an odd number of odd integers can yield an odd number, while only an even number of odd integers can yield an even number. As a consequence, the number of partitions contributing to each $r_k(\rho, a)$ is drastically reduced compared with previous examples. For odd values of k, i.e. $k = 2m + 1$, the first term in the table is given by $(-1)^{m+1} a(\rho)_1/k!$ corresponding to the single part partition $\{k\}$. For even values of k, the first term is obtained by summing all pairs of odd parts summing to k, which yields $(-1)^{k/2+1} 2^{k-2} a^2(\rho)_2/k!$. On the other hand, for each value of k in the table the highest order term is determined by the partition with k ones or $\{1_k\}$. Hence the final term in each $r_k(\rho, a)$ is given by $(-a)^k (\rho)_k/k!$.

As in the case of the generalized elliptic integrals, we can also derive a recurrence relation for the $r_k(\rho, a)$. By differentiating the lhs of (2.249), we find that

$$\left(\frac{1}{\cos x} + a \tan x\right) \frac{d}{dx} (1 + a \sin x)^{-\rho} = -\rho a (1 + a \sin x)^{-\rho}. \qquad (2.253)$$

We now introduce (1.25) and (1.101) to replace respectively $1/\cos x$ and $\tan x$ in the above result. In addition, we assume that $|x| < \pi/2$, so that the equivalence symbols can be replaced by equals signs. Hence (2.253) becomes

$$\left(\sum_{k=0}^{\infty} d_k x^{2k} + a \sum_{k=0}^{\infty} d_k(1) x^{2k+1}\right) \sum_{k=0}^{\infty} k \, r_k(\rho, a) x^{k-1} = -\rho a (1 + a \sin x)^{-\rho}. \qquad (2.254)$$

Since the first two series on the lhs sum over even and odd powers of x separately, they can be combined into one series whose coefficients can be represented by

$$d_k^* = \frac{(1 + (-1)^k)}{2} d_{\lfloor k/2 \rfloor} + \frac{(1 - (-1)^k)}{2} a d_{\lfloor k/2 \rfloor}(1) , \qquad (2.255)$$

where $\lfloor x \rfloor$ once again denotes the floor function. By using the above definition we can multiply both series on the lhs to obtain

$$\sum_{k=0}^{\infty} x^k \sum_{j=0}^{k} j \, r_j(\rho, a) \, d_{k-j}^* = \rho a \sum_{k=0}^{\infty} r_k(\rho, a) x^{k+1} . \qquad (2.256)$$

Since x is still fairly arbitrary, we can equate like powers of x on both sides, which after some algebra yields

$$r_k(\rho, a) = -\frac{\rho a}{k} r_{k-1}(\rho, a) - \frac{1}{k} \sum_{j=1}^{k-1} j r_j(\rho, a) d^*_{k-j}. \qquad (2.257)$$

When this recurrence relation is programmed in Mathematica with (1.32) and (1.100), it produces the same results displayed in Table 2.11 although in a more expanded format.

We now turn our attention to the situation where the sine function is replaced by $\tan x$. In this case we need to replace the sine power series by (1.101). Therefore, we have

$$\left(1 + a \tan x\right)^{-\rho} \equiv \sum_{k=0}^{\infty} (-a)^k \frac{(\rho)_k}{k!} \left(d_0(1)x + d_1(1)x^3 + d_2(1)x^5 + \cdots\right)^k. \qquad (2.258)$$

Comparing this result with (2.246), we see that the only difference occurs with the coefficients of the inner series. Instead of $(-1)^{k+1}/(2k+1)!$, we have $d_k(1)$ for each power x^{2k+1}. Therefore, the contribution from each partition is again given by (2.247) except p_i becomes

$$p_i = \begin{cases} -d_j(1), & i = 2j+1, \\ 0, & i = 2j, \end{cases} \qquad (2.259)$$

where the $d_j(1)$ are given in (1.102). In this case we shall denote the coefficients of the power series in (2.247) as $u_k(\rho, a)$. That is,

$$\left(1 + a \tan x\right)^{-\rho} \equiv 1 + \sum_{k=1}^{\infty} u_k(\rho, a) x^k. \qquad (2.260)$$

By introducing (2.260) into the contribution from each partition, viz. (2.247), and summing over all the partitions with odd parts, we find that the coefficients for the above function can be expressed as

$$u_k(\rho, a) = \sum_{\substack{\lambda_1, \lambda_3, \lambda_5 \ldots, \lambda_\ell = 0 \\ \sum_{i=1}^{k} i\lambda_i = k}}^{k, \lfloor k/3 \rfloor, \lfloor k/5 \rfloor, \ldots, 1} (-1)^{N_k} (\rho)_{N_k} a^{N_k} \prod_{i=1,3,5}^{\ell} \frac{d_{\lfloor i/2 \rfloor}(1)}{\lambda_i!}, \qquad (2.261)$$

where ℓ is given as before below (2.250).

As in the case of the $r_k(\rho, a)$, the $u_k(\rho, a)$ can also be evaluated via a recurrence relation. Once again, we differentiate w.r.t. x, which yields

$$\left(\cos^2 x + a \sin x \cos x\right) \frac{d}{dx} \left(1 + a \tan x\right)^{-\rho} = -\rho a \left(1 + a \tan x\right)^{-\rho}. \qquad (2.262)$$

We can express the first term on the lhs in terms of the double angle formulas for both cosine and sine, while we replace the remaining terms by (2.260). This is valid as long as $|a \tan x| < 1$. Then (2.262) becomes

$$\frac{1}{2}\left(\cos(2x) + a\sin(2x)\right) \sum_{k=0}^{\infty} k\, u_k(\rho, a)\, x^{k-1} = -\rho a \sum_{k=0}^{\infty} u_k(\rho, a)\, x^k. \qquad (2.263)$$

Next we expand the bracketed expression on the lhs as a power series by writing it as

$$\sum_{k=0}^{\infty} e_k\, x^k = \frac{1}{2}\left(\cos(2x) + a\sin(2x)\right), \qquad (2.264)$$

where

$$e_k = \begin{cases} 1, & k = 0, \\[2mm] \dfrac{(-4)^k\, a}{(2k+1)!}, & k \text{ odd}, \\[3mm] \dfrac{(-4)^{2k}}{2 \cdot (2k)!}, & k \text{ even}. \end{cases} \qquad (2.265)$$

Therefore, multiplying the series on the lhs of (2.264) with the other series appearing on the lhs of (2.263) yields

$$\sum_{k=0}^{\infty} x^k \sum_{j=0}^{k} j\, u_j(\rho, a)\, e_{k-j} = -\rho a \sum_{k=0}^{\infty} u_k(\rho, a)\, x^{k+1}. \qquad (2.266)$$

By equating like powers on both sides and carrying out a little algebra, we finally arrive at for $k \geq 1$,

$$u_k(\rho, a) = -\frac{\rho a}{k}\, u_{k-1}(\rho, a) - \frac{1}{k} \sum_{j=1}^{k-1} j\, u_j(\rho, a)\, e_{k-j}, \qquad (2.267)$$

while $u_0(\rho, a) = 1$.

Table 2.12 presents the first ten $u_k(\rho, a)$ aside from $u_0(\rho, a)$. They have been evaluated by using the same general forms derived from the programming methodology presented in later chapters and have also been verified by introducing the recurrence relation given above into Mathematica [32]. As expected, they have a similar structure to the results in Table 2.11 in that for odd values of k, the Pochhammer terms possess odd subscript arguments corresponding to the fact that only odd numbers of odd parts can sum to k, while for even values of k, the Pochhammer terms possess even subscript arguments since only an even number of odd parts can sum to k. In addition, the highest order term in ρ is identical to the highest order term

Table 2.12 Coefficients $u_k(\rho, a)$ in the power series expansion of $(1 + a\tan x)^{-\rho}$

k	$u_k(\rho, a)$
1	$-a\rho$
2	$\frac{1}{2!}\, a^2(\rho)_2$
3	$-\frac{1}{3!}\left(2a(\rho)_1 + a^3(\rho)_3\right)$
4	$\frac{1}{4!}\left(8a^2(\rho)_2 + a^4(\rho)_4\right)$
5	$-\frac{1}{5!}\left(16a(\rho)_1 + 20a^3(\rho)_3 + a^5(\rho)_5\right)$
6	$\frac{1}{6!}\left(136a^2(\rho)_2 + 40a^4(\rho)_4 + a^6(\rho)_6\right)$
7	$-\frac{1}{7!}\left(272a(\rho)_1 + 616a^3(\rho)_3 + 70a^5(\rho)_5 + a^7(\rho)_7\right)$
8	$\frac{1}{8!}\left(3968a^2(\rho)_2 + 2016a^4(\rho)_4 + 112a^6(\rho)_6 + a^8(\rho)_8\right)$
9	$-\frac{1}{9!}\left(7936a(\rho)_1 + 28160a^3(\rho)_3 + 5376a^5(\rho)_5 + 168a^7(\rho)_7 + a^9(\rho)_9\right)$
10	$\frac{1}{10!}\left(176896a^2(\rho)_2 + 135680a^4(\rho)_4 + 12432a^6(\rho)_6 + 240a^8(\rho)_8 + a^{10}(\rho)_{10}\right)$

in the $r_k(\rho, a)$, viz. $(-1)^k(\rho)_k a^k/k!$. However, there are differences too. For example, the coefficients of the $r_k(\rho, a)$ oscillate in sign, whereas the coefficients in the $u_k(\rho, a)$ are all uniform in sign, although they are governed by a factor of $(-1)^k$ outside. For odd values of k the lowest order term in a is determined by the single part partition, $\{k\}$, which according to (2.259) contributes a value of $d_{\lfloor k/2 \rfloor}(1)$. When k is even, the lowest order term in a is second order with its coefficient being the sum of the contributions from all 2-part partitions with odd parts. Using (2.247) and (2.259), we find that this sum can be expressed as

$$\sum_{\substack{\lambda_1,\lambda_3,\dots,\lambda_{\ell_2}=0 \\ \sum_{i=1,3,5}^{\ell_2} i\lambda_i=k,\ \sum_{i=1,3,5}^{\ell_2}\lambda_i=2}}^{k,\lfloor k/2\rfloor,\lfloor k/3\rfloor,\dots,2} a^2\,(\rho)_2 \prod_{i=1,3,5}^{\ell_2} \frac{p_i^{\lambda_i}}{\lambda_i!} = a^2\,(\rho)_2 \sum_{i=1,3,5}^{\ell_2} d_{\lfloor i/2\rfloor}(1) d_{\lfloor(k-i)/2\rfloor}(1)$$

$$-\frac{\left(1-(-1)^{k/2}\right)}{4}\, a^2\,(\rho)_2\, d_{\lfloor k/4\rfloor}(1)^2 ,\tag{2.268}$$

where $\ell_2 = k/2 - (1 + (-1)^{k/2})/2$. The last term accounts for the situation when $k/2$ is odd since the multiplicity $\lambda_{k/2}$ is then equal to 2, whereas for all the other partitions the multiplicities are equal to unity. When implemented in Mathematica, the above result gives the coefficients of the a^2 terms in Table 2.12 such as $176896/10!$ or $691/14175$ for $k = 10$.

The final case considered in this section is the replacement of the trigonometric function in $f(x, a, \rho)$ by $\cot x$. Then we introduce (1.85) into this modified form of $f(x, a, \rho)$, which is only valid for those values of x, where (1.85) can be represented as an equation, viz. $|\Re x| < \pi$. Thus, for these values of x we obtain

$$\left(1 + a\cot x\right)^{-\rho} = \left(\frac{x}{a}\right)^{\rho} \left(1 + \frac{x}{a} + \sum_{k=1}^{\infty} c_k(1) x^{2k}\right)^{-\rho}. \tag{2.269}$$

Now we can treat the rhs of the above result as the regularized value of the binomial series. Consequently, we find that

$$\left(1 + a\cot x\right)^{-\rho} \equiv \left(\frac{x}{a}\right)^{\rho} \sum_{k=0}^{\infty} \frac{(\rho)_k}{k!} \left(-\frac{x}{a} - \sum_{k=1}^{\infty} c_k(1) x^{2k}\right)^{k}, \tag{2.270}$$

provided $|\Re x| < \pi$. Moreover, from the study of the binomial theorem in Appendix A the equivalence symbol can be replaced by an equals sign for $\Re\, a\cot x > -1$, while the rhs is absolutely convergent for $|a\cot x| < 1$. So, we apply the partition method for a power series expansion for these values of a and x. Since the coefficients of the odd powers of x greater than unity vanish, the partitions that we need to consider are those consisting of only even numbers and unity. That is, there are no contributions from partitions involving odd parts greater than or equal to 3. Hence (2.248) becomes

$$C\left(\lambda_1, \lambda_2, \lambda_4, \ldots, \lambda_{2\lfloor k/2\rfloor}\right) = (\rho)_{N_k} \prod_{i=1,2,4,\ldots}^{2\lfloor k/2\rfloor} \frac{1}{\lambda_i!} p_i^{\lambda_i}, \tag{2.271}$$

where for j, a non-negative integer,

$$p_i = \begin{cases} -c_j(1), & i = 2j, \\ -1/a, & i = 1, \\ 0, & i = 2j+1, \quad j \neq 0. \end{cases} \tag{2.272}$$

The third case in (2.272) has been provided for completeness even though it is redundant as far as (2.271) is concerned. If we denote the coefficients in the power series expansion of the rhs of (2.270) by $v_k(\rho, a)$ and assume that $|a\cot x| < 1$, then we arrive at

$$\left(1 + a\cot x\right)^{-\rho} = \left(\frac{x}{a}\right)^{\rho} \left(1 + \sum_{k=1}^{\infty} v_k(\rho, a) x^k\right), \tag{2.273}$$

where

$$v_k(\rho, a) = \sum_{\substack{\lambda_1, \lambda_2, \lambda_4, \ldots, \lambda_{2\lfloor k/2\rfloor}=0 \\ \sum_{i=1,2,4,\ldots}^{2\lfloor k/2\rfloor} i\lambda_i = k}}^{k, \lfloor k/2\rfloor, \lfloor k/4\rfloor, \ldots, 1} (\rho)_{N_k} \prod_{i=1,2,4,\ldots}^{2\lfloor k/2\rfloor} \frac{p_i^{\lambda_i}}{\lambda_i!}. \tag{2.274}$$

Table 2.13 Coefficients $v_k(\rho, a)$ in the power series expansion of $(1 + a \cot x)^{-\rho}$

k	$v_k(\rho, a)$
1	$-\rho\, a^{-1}$
2	$\frac{1}{3!}\left(2(\rho)_1 + 3\,(\rho)_2\, a^{-2}\right)$
3	$-\frac{1}{3!}\left(2(\rho)_2\, a^{-1} + (\rho)_3\, a^{-3}\right)$
4	$\frac{1}{3\cdot5!}\left(8(\rho)_1 + 20(\rho)_2 + 60(\rho)_3\, a^{-2} + 15(\rho)_4\, a^{-4}\right)$
5	$-\frac{1}{3\cdot5!}\left(8a(\rho)_2\, a^{-1} + 20(\rho)_3\, a^{-1} + 20(\rho)_4\, a^{-3} + 3(\rho)_5\, a^{-5}\right)$
6	$\frac{1}{9\cdot7!}\left(96(\rho)_1 + 336(\rho)_2 + 280(\rho)_3 + 504(\rho)_3\, a^{-2} + 1260(\rho)_4\, a^{-2}\right.$ $\left. + 630(\rho)_5\, a^{-4} + 63(\rho)_6\, a^{-6}\right)$
7	$-\frac{1}{9\cdot7!}\left(96(\rho)_2\, a^{-1} + 336(\rho)_3\, a^{-1} + 280(\rho)_4\, a^{-1} + 168(\rho)_4\, a^{-3}\right.$ $\left. + 420(\rho)_5\, a^{-3} + 126(\rho)_6\, a^{-5} + 9(\rho)_7\, a^{-7}\right)$
8	$\frac{1}{15\cdot9!}\left(1152(\rho)_1 + 5184(\rho)_2 + 6720(\rho)_3 + 5760(\rho)_3\, a^{-2} + 2800(\rho)_4\right.$ $+ 20160(\rho)_4\, a^{-2} + 5040(\rho)_4\, a^{-4} + 16800(\rho)_5\, a^{-2} + 12600(\rho)_6\, a^{-4}$ $\left. + 2520(\rho)_7\, a^{-6} + 135(\rho)_8\, a^{-8}\right)$
9	$-\frac{1}{15\cdot9!}\left((1152(\rho)_2 + 5184(\rho)_3 + 6720(\rho)_4 + 2800(\rho)_5)a^{-1} + (1920(\rho)_4\right.$ $+ 6720(\rho)_5 + 5600(\rho)_6)\, a^{-3} + (1008(\rho)_6 + 2520(\rho)_7)\, a^{-5} + 360(\rho)_8\, a^{-7}$ $\left. + 15(\rho)_9\, a^{-9}\right)$
10	$\frac{1}{9\cdot11!}\left(7680(\rho)_1 + 42240(\rho)_2 + 71808(\rho)_3 + 49280(\rho)_4 + 12320(\rho)_5\right.$ $+ (38016(\rho)_3 + 171072(\rho)_4 + 221760(\rho)_5 + 92400(\rho)_6)\, a^{-2}$ $+ (110800(\rho)_6 + 92400(\rho)_7)\, a^{-4} + (11088(\rho)_7 + 27720(\rho)_8)\, a^{-6}$ $\left. + 2970(\rho)_9\, a^{-8} + 99(\rho)_{10} a^{-10}\right)$

The above result can be analytically continued to all values of x in the complex plane, but then the equals sign should be replaced by an equivalence symbol for those values where it is divergent.

Table 2.13 presents the first ten $v_k(\rho, a)$ calculated from (2.274). In these results the Pochhammer subscript indicates the total number of parts in the partitions, while the power of $1/a$ indicates the number of ones that were counted in the partitions. Therefore, the lowest power of a or highest power of $1/a$ in these results is determined by the partition with the most ones, viz. $\{1_k\}$. Its contribution to the $v_k(\rho, a)$ is simply $(\rho)_k\, a^{-k}/k!$. The second lowest power in a is determined from the partition $\{1_{k-2}, 2\}$, whose contribution to the coefficients is $(-1)^k(\rho)_{k-2}\, a^{-(k-2)}/(3 \cdot (k - 2)!)$ since $c_1(1) = -1/3$. The lowest order term in ρ represents the contribution from the partition $\{k\}$, when k is even and $\{k - 1, 1\}$ when k is odd. Hence for even values of k the lowest order term in ρ is $-(\rho)_1\, c_{k/2}(1)$, whereas for odd values of k, it is $(\rho)_2\, c_{(k-1)/2}(1)\, a^{-1}$.

As has been the case for the preceding examples in this section, we can also derive a recurrence relation for the $v_k(\rho, a)$. Again, we begin by differentiating $f(x) = (1 + a\cot x)^{-\rho}$ w.r.t. to x, which yields

$$\left(\sin^2 x + a\sin x \cos x\right)\frac{d}{dx}f(x) = \rho a\left(1 + a\cot x\right)^{-\rho}. \tag{2.275}$$

By introducing (2.274) into the above equation, we obtain

$$\left(\frac{1 - \cos(2x)}{2x} + \frac{a}{2x}\sin(2x)\right)\left(\rho + \rho\sum_{k=1}^{\infty}v_k(\rho, a)\,z^k + \sum_{k=1}^{\infty}k\,v_k(\rho, a)\,z^{k-1}\right)$$

$$= \rho a\sum_{k=0}^{\infty}v_k(\rho, a)\,z^k. \tag{2.276}$$

Next we expand the first term on the lhs in terms of the power series expansions for sine and cosine. The expansion of the term with cosine inside the large brackets produces an infinite series in odd powers of x, while that for the sine term produces an infinite series in even powers of x. Both series can then be combined into one series involving powers of x with the coefficients given by

$$g_k = \begin{cases} (-1)^{\lfloor k/2 \rfloor}\,2^k/(k+1)!, & k \text{ odd}, \\ (-1)^{\lfloor k/2 \rfloor}\,2^k\,a/(k+1)!, & k \text{ even}. \end{cases} \tag{2.277}$$

As a consequence, we can multiply the series with the g_k coefficients with both series involving the $v_k(\rho, a)$ on the lhs of (2.276). This yields

$$\sum_{k=0}^{\infty}x^k\sum_{j=0}^{k}(j+\rho)\,g_{k-j}\,v_j(\rho, a) = \rho a\sum_{k=0}^{\infty}v_k(\rho, a)\,z^k. \tag{2.278}$$

Even though x has been restricted, it is still sufficiently arbitrary to allow like powers on both sides of the above equation to be equated to one another. Therefore we obtain the following recurrence relation:

$$v_k(\rho, a) = -\frac{1}{ka}\sum_{j=0}^{k-1}(j+\rho)\,g_{k-j}\,v_j(\rho, a), \tag{2.279}$$

where $v_0(\rho, a) = 1$. As expected, this recurrence relation produces the same results displayed in Table 2.13, although in a different format.

CHAPTER 3

Generating Partitions

So far, this book has been concerned with introducing the partition method for a power series expansion and demonstrating its versatility by applying it to increasingly more advanced problems culminating in the generalization of the elliptic integrals discussed in the final section of previous chapter. Consequently, we are now in a better position to create a general theory behind the method so that a programming methodology can be developed. Before this methodology can be developed, however, we need to consider the problem of generating partitions, which represents the subject of this chapter.

As has been observed in the previous chapters, the partition method for a power series expansion is, broadly speaking, composed of two separate steps. The first step involves determining all the partitions summing to an integer k, which represents the order of the variable in the resulting power series expansion. The second and more complicated step is to calculate the contribution made by each partition to the coefficient of the k-th order term in the series expansion. It is this step upon which the theoretical framework resulting in the programming methodology will be based in later chapters. Nevertheless, before the contribution due to each partition can be evaluated, we need to know all the parts i and their multiplicities λ_i in each partition. Furthermore, the first step can in turn be broken down into two separate tasks or processes: the first dealing with the generation of partitions, and the second involving the processing of the partitions. It should also be mentioned here that once the first step has been performed, the results can be stored for each value of k so that it becomes the input for the different cases/problems in the second step. Ultimately, one will be limited by the memory capacity of the computing system because the number of partitions experiences combinatorial explosion, e.g., the number of partitions summing to 100 is 190 569 292, whereby it becomes impractical to store multiplicities of each partition in an array despite the fact that many of them are equal to zero as we have seen in Table 1.2 of Chapter 1.

Although both tasks in the first step are required in the development of a programming methodology for the partition method for a power series expansion, it should be pointed out that the first task of generating

The Partition Method for a Power Series Expansion.
DOI: http://dx.doi.org/10.1016/B978-0-12-804466-7.00003-6

partitions represents an interesting problem in its own right since it continues to be the source for new algorithms as evidenced by Refs. [47–50]. Whilst none of these references presents an algorithm based on the partition trees appearing in Figs. 1.1 and 1.2, we shall discuss some of them to indicate their strengths and weaknesses both in generating partitions and for implementation into the partition method for a power series expansion. Ultimately, we shall find that the algorithm based on partition trees, which was first sketched out in Ref. [3], will prove to be the most suitable for the partition method for a power series expansion, but when it comes to scanning the partitions only and not processing them, other algorithms based on a lexicographic approach are faster. It will also be seen especially in Chapter 7 that specific types or classes of partitions can be determined by making simple modifications to the algorithm based on partition trees, whereas these are often difficult to determine, if not impossible, with the other partition-generating algorithms. In such cases a completely different algorithm is usually required. Hence the algorithm based on partition trees is far more versatile than the other more well-known algorithms. Furthermore, as a result of the material in Chapter 5, we shall re-formulate the partition method for a power series expansion in terms of an operator approach. Thus, the code presented here will come to represent the partition operator denoted by $L_{P,k}[\cdot]$ in Chapter 6, while the modified codes yielding specific classes of partitions will represent different operators. This will lead to many new and fascinating results when we study the various generating functions in the theory of partitions in the remaining chapters.

As discussed in Refs. [3–5], when applying the partition method for a power series expansion there is actually no need to generate all the partitions at low orders. In these cases one can write down all the partitions on a sheet of paper and then proceed to the second step of determining the contributions due to each partition. The problem occurs when we wish to evaluate the higher order terms, especially if our aim is to derive an extremely accurate approximation to the original function or problem by including as many orders as possible. Determining higher orders increases the complexity dramatically due to the exponential increase in the number of partitions. For greater than the tenth order, it is no longer feasible to write down all the partitions and then to calculate their contributions to the coefficient. Hence a programming methodology is required despite the fact that this too will eventually become very slow at very high orders of the series expansion. For very high orders, i.e., much higher than

100, the availability of memory in the computing system becomes an issue. Consequently, the total number of partitions will need to be divided into subsets. As each subset is used to determine its contribution to the coefficient, the value would be stored and the subset discarded so that the calculation of the next subset could proceed. The coefficient becomes the sum of all the stored values obtained from the subsets. In addition, as a by-product of developing a programming methodology for the partition method for a power series expansion, we shall discover very interesting results in the theory of partitions. E.g., it has already been observed that the series expansion obtained via the method is identical to a Taylor/Maclaurin series when the latter method can be applied. The development of a programming methodology for the partition method means that the higher order derivatives in a Taylor/Maclaurin series can be expressed in terms of a sum of the contributions from all the partitions. Therefore, we now have a means of linking the continuous/differentiable property of a function with the discrete mathematics of partitions from enumerative combinatorics.

When it was mentioned above that there was no need to employ an algorithm to generate all the partitions in the partition method for a power series expansion, it was meant that there was no need to write them down for each value of k, since only the parts and their multiplicities are required as input for the second step. Representing a partition in this form or manner is known as the multiplicity representation, whereas the representation where each part is written down or printed out is known as the standard representation. Since the number of partitions summing to k, represented by the partition-number function $p(k)$, increases exponentially, for $k > 10$, it is no longer practical to write down the partitions in the standard representation for use in the partition method for a power series expansion. Nevertheless, whilst the multiplicity representation is ideal in the partition method for a power series expansion, one still needs to examine other algorithms for generating partitions because it could turn out that an algorithm generating partitions in the standard representation may only require minor modification to generate them in the multiplicity representation. Moreover, the generation of partitions in either representation continues to attract interest. Therefore, we shall review the existing algorithms, but ultimately our aim will be to determine the most appropriate for the partition method for a power series expansion. On the other hand, for those readers who are only interested in the problem of generating partitions, they are urged to consult the list of references just in case the other algorithms are more suited to their needs.

Having justified the need for generating all the integer partitions summing to an arbitrary integer, we now turn to the issue of finding an appropriate algorithm that will expedite the process, but will also do so in an appropriate form for the second step in the partition method for a power series expansion. For a time there seemed to be only one useful algorithm for generating partitions. This was McKay's algorithm [11], which was basically a succession rule whereby partitions were generated in linear time. It was developed further by Knuth [13], who used the fact that if the last part greater than unity is a two, then the next partition can be determined very quickly. This modification meant that each partition takes almost a constant amount of time to be generated. The Knuth/McKay algorithm, which is implemented in C/C++ below, generates the partitions summed to a global integer n in a particular form of the standard representation known as the reverse lexicographic order. Consequently, the parts in a partition, say $a[1]$ up to $a[k]$, are printed out according to $a[1] \geq a[2] \geq \cdots \geq a[k]$, while the first part of each new partition is less than or equal to the first part of the preceding partition. For example, the partitions summing to 5 in reverse lexicographic order are:

5|
4|1|
3|2|
3|1|1|
2|2|1|
2|1|1|1|
1|1|1|1|1|

The rules for generating partitions in reverse lexicographic order can be obtained from Refs. [13] and [51]. Briefly, if the partition is not composed entirely of ones, then it ends with a value of $x+1$ followed by zero or more ones. The next smallest partition in lexicographic order is obtained by replacing the segment of the partition $\{\ldots, x+1, 1, \ldots, 1\}$ by $\{\cdots, x, \ldots, x, r\}$, where the remainder r is less than or equal to x.

```
/* This C/C++ program generates partitions in reverse
   lexicographic order following the McKay/Knuth
   algorithm discussed on p. 38 of Fascicle 3
   of Vol. 4 of D.E. Knuth's The Art of Computer
   Programming. */

#include <stdio.h>
#include <memory.h>
```

```
#include <stdlib.h>

int main(int argc, char *argv[])
{
int *a, i, m, n, q, x;
if(argc != 2) printf("execution: ./knuth <partition#>\n");
else{
                n=atoi(argv[1]);
                a=(int *) malloc((n+1)*sizeof(int));
P1:        a[0]=0;
                m=1;
P2:        a[m]=n;
                q=m-(n==1);
P3:        for(i=1; i<=m; i++) printf("%i|", a[i]);
                printf(" \n" );
                if(a[q] != 2) goto P5;
P4:        a[q--]=1;
                a[++m]=1;
                goto P3;
P5:        if(q==0) goto end;
                x= a[q]-1;
                a[q]=x;
                n= m-q+1;
                m= q+1;
P6:        if (n<= x) goto P2;
                a[m++]=x;
                n -=x;
                goto P6;
end: ;
        }
printf("\n");
free(a);
return (0);
} .
```

By current standards the above code is considered to be slow for generating the partitions at each order due to the excessive unconditional branching. From a computational point of view it is also very non-structured. Hence it is not accordance with modern programming practice. A significantly faster algorithm for generating partitions in reverse lexicographic order has been developed by Zoghbi and Stojmenovic (ZS) in Ref. [47]. Actually, these authors present two algorithms in their paper, but the second, which generates the partitions in lexicographic order, is slower than the first. Nevertheless, if one runs the above code against a C/C++-coded version of the first algorithm, then one finds that it takes 1362 CPU seconds to print out the partitions summing to 80 on the screen of a Sony

VAIO laptop with 2 GB RAM compared with 1399 CPU seconds using the Knuth/McKay code given above. On the other hand, if the partitions are directed to an output file, then it takes only 30 CPU seconds with the Knuth/McKay code compared with 28 CPU seconds using the ZS code.

```
/* This program determines partitions in ascending order
following the algorithm devised by J. Kelleher.*/

#include <stdio.h>
#include <memory.h>
#include <stdlib.h>

int main(int argc, char *argv[])
{
int *a, i, m, n, ydummy, xdummy, count;
if(argc!= 2) printf("execution: ./kelleher <partition#>\n");
else{
    n=atoi(argv[1]);
    a=(int *) malloc((n+2)*sizeof(int));

    for(i=0; i<=(n+1);i++) a[i]=0;
    a[1]=n;
    count=1;
    while (count != 0){
            xdummy=a[count-1]+1;
            ydummy=a[count]-1;
            count=count-1;
            while (xdummy <= ydummy){
                    a[count]=xdummy;
                    ydummy=ydummy-xdummy;
                    count=count+1;
                                        }
            a[count]=xdummy+ydummy;
            for (i=0;i<= count;i++){
                    printf("%i",a[i]);
                    if (i<count) printf("|");
                                        }
                    printf("\n");
                    }
    }
printf("\n");
free(a);
return (0);
}
```

The lexicographic approach is not the only method of ordering partitions. In Refs. [48] and [49] Kelleher and O'Sullivan develop algorithms for generating partitions in ascending order, but not in lexicographic order. A C/C++-code based on one of the algorithms is listed above. When this code is run for the partitions summing to 5, it prints out

```
1|1|1|1|1
1|1|1|2
1|1|3
1|2|2
1|4
2|3
5 .
```

By inverting the reverse lexicographic order of the partitions summing to 5 given earlier, we see that the partition $\{1_2, 3\}$ in the Kelleher/O'Sullivan code appears before $\{1, 2_2\}$, while in the McKay/Knuth code it appears after the latter partition. A similar situation occurs with the partitions, $\{1, 4\}$ and $\{2, 3\}$. Nevertheless, the partitions generated by both codes are arranged in ascending order. However, Kelleher and O'Sullivan are able to take advantage of the different order to develop an even more efficient version of the code listed above. As a result, they state that their optimized version is far more superior to the ZS code. Specifically, if the above code is run to yield the $15\,796\,476$ ($= p(80)$) partitions summing to 80 on the same Sony laptop as the previous code, then it takes 1342 CPU seconds to print out the partitions. Yet, if one does the same with the more efficient version of their code, then it takes the same amount of time. On the other hand, if the partitions are directed to an output file, then it is found that the above code takes 29 CPU seconds, while the more efficient version takes 28 CPU seconds, the same time taken as the ZS code.

According to Ref. [52], Kelleher has found on his computing system that partitions summing to 80 are generated at a rate of 1.30×10^8 per second using the first algorithm, while with the second algorithm the rate is 2.87×10^8 per second. For the ZS code the rate is 1.26×10^8 per second, while with the Knuth/McKay code the partitions are generated at a rate of 1.73×10^8 per second. The reason why there was no marked difference in performance in the C/C++ versions of the Kelleher codes is attributed to the manner in which the partitions were printed out.

Unfortunately, the above codes do not utilize the recursive nature of partitions, which can be observed by realizing that the partitions summing to $k + 1$ include all the partitions summing to k with an extra part of unity in them in addition to other partitions possessing parts greater than unity.

In fact, according to p. 45 of Ref. [13], the number of partitions summing to k with m parts, which is denoted by $\begin{vmatrix} k \\ m \end{vmatrix}$, obeys the following recurrence relation:

$$\begin{vmatrix} k \\ m \end{vmatrix} = \begin{vmatrix} k-1 \\ m-1 \end{vmatrix} + \begin{vmatrix} k-m \\ m \end{vmatrix}. \tag{3.1}$$

Furthermore, $\begin{vmatrix} k+m \\ m \end{vmatrix}$ represents the number of partitions summing to k with at most m parts. If $P(k, m)$ represents the partitions summing to k with at most m parts, then using (3.1) we find that

$$P(k, m) = P(k-1, m-1) + P(k-m, m), \tag{3.2}$$

which is derived on p. 96 of Ref. [53].

To incorporate recursion into the process of generating the partitions, we need to construct an algorithm that utilizes the partition trees displayed in Figs. 1.1 and 1.2 of Chapter 1. It should be noted that these trees are different from the binary tree approach in the work of Yamanaka et al. [50], which seeks to generate each partition in the standard representation in a constant time rather than an average time. This work also represents a further development on Fenner and Loizou [54], who appear to have been first to develop a binary tree representation for partitions generated in lexicographic order.

Since the construction and recursive properties of partition trees have been described extensively in Chapter 1, only a summary of the main points will be presented here to assist in understanding the algorithm that is to follow. Partition trees are constructed by drawing branch lines to all pairs of numbers that can be summed to the seed number k, where the first number in the tuple is an integer less than or equal to $\lfloor k/2 \rfloor$. Thus, branches are drawn initially to $(0, 6)$, $(1, 5)$, $(2, 4)$ and $(3, 3)$ in Fig. 1.2. Whenever a zero appears in the first member of a tuple, the path terminates as exemplified by $(0, 6)$. For the other pairs branch lines are drawn to all pairs with integers that sum to the second number according to the same rule. Thus, for $(1, 5)$ paths branch out to $(0, 5)$, $(1, 4)$ and $(2, 3)$, but not to $(3, 2)$ or $(4, 1)$. This recursive approach continues until all paths are terminated by terminating tuples as in Fig. 1.2. Moreover, to avoid repeating or duplicated partitions, when creating branches from a tuple, the first member of the following tuple must be greater than or equal to the first member of the previous tuple unless it is a terminating tuple. This means that if we consider the branches from the tuple $(2, 4)$ in Fig. 1.2, then there are no branches where unity appears as the first member in a tuple. Similarly, for the tuple $(3, 3)$, there will be no tuples branching out from it with either unity or two as the first member of a tuple.

The first member of each tuple plus the second member of the terminating tuple in each path constitute a partition summing to the seed number k. E.g., the path in Fig. 1.2 consisting of $(1, 5)$, $(1, 4)$, $(2, 2)$ and $(0, 2)$ represents the partition $\{1_2, 2_2\}$, while that consisting of $(1, 5)$, $(1, 4)$, $(1, 3)$ and $(0, 3)$ represents the partition $\{1_3, 3\}$ In addition, the number of branches along each path gives the number of parts or length of the partition, whilst those tuples with zeros in vertical columns represent the partitions with the same number of parts in them. For $k > 3$ the last path in the figure consists of $([k/2] + 1, [k/2])$ and $(0, [k/2])$ when k is odd and $(k/2, k/2)$ and $(0, k/2)$ when k is even. Therefore, in both cases the partition tree terminates at the central partition, which is given by $\{\lfloor k/2 \rfloor + 1, \lfloor k/2 \rfloor\}$ and $\{k/2, k/2\}$ for odd and even values of k, respectively. To determine $\left| \begin{matrix} k \\ m \end{matrix} \right|$, we simply count all the tuples with zero m branches away from the seed number. Hence from Fig. 1.2, we see that $\left| \begin{matrix} 6 \\ 3 \end{matrix} \right|$ equals three. In addition, it can be seen that the number of partitions with exactly m parts equals the number of partitions whose largest part is m. This will become more apparent when we discuss conjugate partitions in Chapter 6.

Excluding terminating tuples in Fig. 1.2 we see that the second member in a tuple decrements with each rightward horizontal movement or right branch along the same path, while the first member in each tuple increments with each downward vertical movement for branches from the same tuple or the seed number. Thus partition trees are two-dimensional in nature. This is particularly interesting as it means that we are describing the rare instance of bivariate recursion where one of the variables increments, while the other decrements. Because of this, there is no need to introduce a search algorithm to remove duplicated or repeated partitions. Consequently, only one algorithm will be required to generate partitions in the same order as those in a partition tree. In addition, since it has already been mentioned that partition trees terminate at the central partition, we shall refer to this algorithm as the bivariate recursive central partition or BRCP algorithm.

Let us now examine the recursive nature of partition trees more extensively so that we can gain the necessary insight for creating an elementary code of the BRCP algorithm. From Fig. 1.2 it can be seen that the total number of partitions summing to k, viz. $p(k)$, can be evaluated by summing the partitions that can be separated into m parts, where m ranges from 1 to the seed number k or 6 in the figure. Since there is only one partition with a single part and one with k parts, by scanning over the columns in a partition tree we obtain the following trivial result for $k \geq 3$:

$$p(k) = 2 + \sum_{m=2}^{k-1} \left| \begin{matrix} k \\ m \end{matrix} \right| . \tag{3.3}$$

On the other hand, if we scan the rows of a partition tree, then we find that the total number of partitions can also be obtained by denoting the number of partitions whose parts are greater than or equal to m by $p(k, m)$. Hence we arrive at

$$p(k) = 1 + \sum_{m=1}^{\lfloor k/2 \rfloor} p(k - m, m) , \tag{3.4}$$

where, again, $\lfloor x \rfloor$ is the greatest integer less than or equal to x. This result appears in Ref. [55] except that the variables inside $p(k, m)$ are interchanged. It will also be used in Chapter 6 when introducing the partition operator.

It has already been stated that $\left| \begin{matrix} k+m \\ m \end{matrix} \right|$ represents the number of partitions summing to k with at most m parts, i.e. $P(k, m)$. If we put $m = 2$ and $k = 4$, then we observe in Fig. 1.2 that there are three tuples with zeros in the column two branches away from the seed number. These represent the partitions, $\{1, 5\}$, $\{2, 4\}$ and $\{3_2\}$. Hence $P(4, 2) = 3$. From (3.1) we note that $P(4, 2)$ is also equal to the sum of $\left| \begin{matrix} 5 \\ 1 \end{matrix} \right|$ and $\left| \begin{matrix} 4 \\ 2 \end{matrix} \right|$. If we treat the five in the tuple $\{1, 5\}$ of the partition tree as a seed number, then $\left| \begin{matrix} 5 \\ 1 \end{matrix} \right|$ is equal to one corresponding to the tuple $\{0, 5\}$. Furthermore, if we now treat the four in the tuple $\{1, 4\}$ from $\{1, 5\}$ as a seed number, then we find that two branches further to the right $\left| \begin{matrix} 4 \\ 2 \end{matrix} \right|$ equal 2 corresponding to the terminating tuples $\{0, 3\}$ and $\{0, 2\}$. Note that we could not have used the four in the tuple $\{2, 4\}$ in the figure as the seed number because the partition tree here gives all the partitions summing to 4, whose parts are greater than unity. That is, the terminating tuple $\{0, 3\}$ is excluded in this instance. Now including the first branches emanating from the selected seed number of 4, we see that the two terminating tuples correspond to the partitions, $\{1, 3\}$ and $\{2_2\}$. Therefore, by summing the $\begin{vmatrix} 5 \\ 1 \end{vmatrix}$ and $\begin{vmatrix} 4 \\ 2 \end{vmatrix}$, we find once again that $P(4, 2)$ is equal to 3. Thus we see that the recursive nature of (3.1) is built into the structure of partition trees.

According to Knuth [13], $\begin{vmatrix} k \\ m \end{vmatrix}$ also represents the number of partitions summing to k, whose largest part is m. This connection can be observed by using Ferrers diagrams, which will be introduced later in this book. As an example, let us consider $\begin{vmatrix} 6 \\ 3 \end{vmatrix}$, which can be determined by summing all those tuples with a zero in the vertical column three branches away from the seed number in Fig. 1.2. This is found to equal 3. On the other hand, the largest part of a partition always appears in the final tuple ending a path in the partition tree. Hence the number of partitions whose largest part is 3 can be determined by summing all the paths ending with the terminating tuple of $(0, 3)$. There are three of these in Fig. 1.2 with the first occurring at the top of the fourth column from the seed number, the second at the bottom of the third column and the third at the bottom of the second column.

In discussing the ordering schemes of the preceding algorithms, we used the seven partitions where the seed number is equal to 5, i.e. $p(5) = 7$. We shall do likewise for partitions generated from a partition tree. Then the order in which they are generated is

5
1|4
1|1|3
1|1|1|2
1|1|1|1|1
1|2|2
2|3 .

Although each partition is in ascending order, it is evident that the partitions are generated in a very different order compared with the Kelleher/O'Sullivan and ZS codes. In fact, the reverse lexicographic approach of McKay and Knuth is closer to the above order despite the fact that each partition in the latter code is generated in descending order.

An elementary version of the BRCP algorithm in C/C++, which first appeared in Ref. [3], is

```
void brcp(int k,int j) {
  printf("%d",k);
  k=k-j;
  while (k > = j){
    printf(",%d(",j);
    brcp(k--, j++);
    printf(")");
              }
        }.
```

Fig. 1.2 lists the specific recursive call to the function at each non-terminating tuple. Note that the order of the variables in the above code has been interchanged compared with the tuples in a partition tree. Consequently, the first variable is decremented and the second variable incremented in the recursive call inside the function. For $k = 4$ the output from this code is

$$4, 1(3, 1(2, 1(1))), 2(2) .$$

By processing the commas and parentheses, we obtain the partitions in the order they appear in the partition tree, viz. $\{4\}$, $\{1, 3\}$, $\{1_2, 2\}$, $\{1_4\}$ and $\{2_2\}$. Although the output is very compact, the above code is not suitable for the implementation into the second step of the partition method for a power series expansion. In addition, it needs to be adapted in order to solve the various problems in the theory of partitions presented in Chapter 6, although we shall see that these modifications are not major. Even if we wish to generate the partitions on separate lines as in the other codes discussed in this chapter, then we have to make changes. Despite these shortcomings, the above code represents the simplest implementation of the BRCP algorithm. It is not only more structured and hence, more elegant than the reverse lexicographic algorithm of McKay and Knuth, but it is also more powerful or versatile. For example, one single call to **brcp(6,1)** results in all the other calls to the routine as shown in Fig. 1.2. In fact, the total number of calls to **brcp** yields the total number of partitions $p(k)$. By introducing a counter for the number of calls to **brcp**, we obtain the total number of partitions summing to the seed number without having to create new function.

Let us consider generating the partitions in a similar manner to the other codes discussed in this chapter. That is, each partition should appear on a separate line, but since our aim is to use the partitions as input for the second step of the partition method for a power series expansion, we shall generate them in the multiplicity representation. As a consequence, another simple function called **termgen** needs to be introduced into the BRCP code, which generates non-zero multiplicities via the variable *freq*. This variable is set equal to the various components in the array called *part*, which contains all the multiplicities determined by counting the same parts in a partition. However, the value of *freq* is only printed out if the multiplicity is non-zero as can be seen by the if condition accompanying the printf statement in the for loop of **termgen**. Because of the call to the new function **termgen**, the original **brcp** code has to be modified slightly so that **partgen** becomes

```c
#include <stdio.h>
#include <memory.h>
#include <stdlib.h>

int tot,*part;
long unsigned int term=1;

void termgen()
{
int freq,i;
printf("%ld: ",term++);
for(i=0;i<tot;i++){
                freq=part[i];
                if(freq) printf("%i(%i)",freq,i+1);
                }
printf("\n");
}

void brcp(int p,int q)
{
part[p-1]++;
termgen();
part[p-1]--;
p -= q;
while(p >= q){
        part[q-1]++;
        brcp(p--, q);
        part[q++ -1]--;
                }
}

int main(int argc,char *argv[])
{
int i;
```

```
if(argc !=2) printf("partgen <sum of the partitions >\n");
else{
    tot=atoi(argv[1]);
    part=(int *) malloc(tot*sizeof(int));
    if(part==NULL) printf("unable to allocate array\n\n");
    else{
        for(i=0;i<tot;i++) part[i]=0;
        idx(tot,1);
        free(part);
        }
    }
printf("\n");
return(0);
}
```

In this code the user enters the seed number, which is represented by the variable *tot* as before, while the variable *term* represents a running count of the partitions as they are being generated by **brcp**. Hence the final value of *term* is $p(k)$. Running the code with *tot* set equal to 5 yields the following output:

1: 1(5)
2: 1(1) 1(4)
3: 2(1) 1(3)
4: 3(1) 1(2)
5: 5(1)
6: 1(1) 2(2)
7: 1(2) 1(3) .

Thus, we obtain the seven partitions summing to 5 in the same order as they appear in a partition tree. The first value printed out on each line is the current value of *term* followed by a colon. As mentioned previously, each line of output only provides the nonzero multiplicities of the parts. These values are then accompanied by the specific values of the parts in parentheses. For example, 1(5) denotes the partition {5} where $\lambda_5 = 1$ and all the other λ_i vanish, while 3(1) 1(2) represents $\{1_3, 2\}$, where $\lambda_1 = 3$ and $\lambda_2 = 1$. As expected, the final partition is the central partition for $k = 5$, which is $\{2, 3\}$.

From the above output we see that the partitions summing to *tot* are printed out in ascending order. If the user wishes to have them generated in descending order, then all that is required is to alter the for loop in **termgen** to

```
for(i=tot-1;i>=0;i--){ etc.
```

For comparative purposes, **partgen** was run on the same Sony laptop as the other codes, where it was found that it took 1561 CPU seconds to compute all the partitions summing to 80 in the above format. Thus the execution time compared with the other previously mentioned codes has increased, primarily due to the processing of the partitions. However, if the output is directed to a file, thereby allowing it to be used as input to the second step of the partition method for a power series expansion, then it only takes 26 CPU seconds to execute, which makes it the best performing code discussed in this chapter.

Whilst the other algorithms/codes prove to be faster than a fully-optimized version of the BRCP algorithm in certain situations, they, however, do not possess the versatility or flexibility of the latter. This versatility will be exploited when we embark on creating a programming methodology of the partition method for a power series expansion in the following chapters. It should be noted that Refs. [13] and [50] present different algorithms for generating partitions with a specific number of parts, whilst the latter reference presents yet another algorithm for doubly-restricted partitions or where the parts lie in a specified range. In Chapter 6 we shall see that only minor modifications to the above code are required to solve these problems. Moreover, we can solve more complex problems such as determining conjugate partitions via Ferrers diagrams by employing the above code.

In order to give the reader a foretaste of this flexibility, let us modify **partgen** so that as it prints out each partition, it also prints out the rank of the partition. There are at least two definitions for the rank of a partition [56]. The first one, with which we shall be concerned here, states that the rank is the number obtained by subtracting the number of parts in the partition from the largest part in the partition, while the second one states that it is given by the size of the Durfee square of the partition [57].

According to the first definition of the rank, we require the total number parts of each partition in addition to the largest part of each partition. This means that at least two new integer variables are required, which we shall call *num_parts* and *largest_part*. Since the processing of partitions occurs in **termgen**, they only need to be declared as local variables in that function. To calculate the number of parts in a partition or its length, we sum all the components or values in the array *part* in a new for loop, which is inserted before the for loop where the processing of the partitions occurred in the original **termgen**. To determine the largest part in a partition, we set *largest_part* initially equal to one since this is the lowest possible value it can be. Then as **termgen** prints out the multiplicities of the parts, we put *largest_part* equal to the highest index in *part* that has a non-zero multiplicity. Finally, we introduce another printf statement that prints out the rank by calculating the difference between *largest_part* and *num_parts*. Thus, the modified version of **termgen** becomes

```
void termgen ()
{
int freq, i, num_parts, largest_part;
/*num_parts is the number of parts in a partition */
printf("Partition No: %ld is ",term++);
num_parts=0;
for(i=0;i<tot;i++){
      num_parts= num_parts+part[i];
          }
largest_part=1;
  for(i=0;i<tot;i++){
                  freq=part[i];
                  if(freq)
                     {
                      printf("%i(%i) ",freq, i+1);
                      largest_part=i+1;
                      }
            }
printf(" whose rank is %i \n",largest_part−num_parts);
}
```

If we run **partgen** with the new **termgen**, then the code produces the following output:

PARTITIONS SUMMING TO 5
Partition No: 1 is 1(5) whose rank is 4
Partition No: 2 is 1(1) 1(4) whose rank is 2
Partition No: 3 is 2(1) 1(3) whose rank is 0
Partition No: 4 is 3(1) 1(2) whose rank is –2
Partition No: 5 is 5(1) whose rank is –4
Partition No: 6 is 1(1) 2(2) whose rank is –1
Partition No: 7 is 1(2) 1(3) whose rank is 1 .

The header is printed out by inserting the statement

```
printf("PARTITIONS SUMMING TO %i \n",tot);
```

after the malloc routine in **main**.

Partition trees can also be interpreted differently when it is realized that the first branch from the seed number is the only partition with a single part, viz. {6} in Fig. 1.2. The partitions summing to $n-1$ or 5 then appear along the second branch to $\{1, 5\}$. If the partitions summing to $n-1$ have already been stored in an array, then all we need to do is increment the number of ones by one in all these partitions to get the partitions summing to n. The next branch emanating from the seed number represents the partition tree for $n-2$, but now all the parts in the partitions are greater than unity. So, if the partitions summing to $n-2$ have been stored previously, then we disregard those partitions where the number of ones is non-zero and increment the number of twos in the remaining partitions by one to get the partitions summing to n. The next branch from the seed number represents the partition tree for $n-3$ except that those partitions, where the number of ones and twos are non-zero, are now neglected. To obtain the partitions summing to n, we increment the number of threes by one in the remaining partitions. This process is continued until $[n/2]$ partition trees have been processed in this manner. Although the new interpretation may lead to less processing of the partitions, it comes at the expense of having to store all previous partitions in memory. Nevertheless, we shall return to this interpretation later in this book.

CHAPTER 4

General Theory

The previous chapter presented the BRCP algorithm, which generates partitions in the same order as they appear in partition trees such as those displayed in Figs. 1.1 and 1.2. Although it is not necessary to generate partitions via partition trees, the main advantage of the BRCP algorithm is that it can generate partitions naturally in the multiplicity representation, which is the required format for the second and more important step of the partition method for a power series expansion, i.e., where partitions are coded to yield specific contributions to the coefficients in the resulting power series. Therefore, as far as this chapter is concerned, it will be assumed that an adequate algorithm, not necessarily the BRCP algorithm, exists for generating partitions in the multiplicity representation so that we can concentrate on developing a general theory behind the second step of the partition method for a power series expansion. This is just in case a better algorithm should appear in the future.

In the early development of the partition method for a power series expansions as described in Chapter 1, each distinct part occurring in a partition was set equal to l_i, while λ_i represented the non-zero multiplicities. Thus, if there were j distinct parts in a partition, then the length of the partition was given by $\sum_{i=1}^{j} \lambda_i = N_k$, while $\sum_{i=1}^{j} \lambda_i l_i = k$, where k represented the order of the term under consideration in the power series expansion. This approach was later dropped in favour of a second convention where the subscript in the multiplicity represents the part in the partition regardless of whether the part appears in the partition or not. According to this convention, λ_1 represents the number of ones, λ_2 represents the number of twos, etc. Consequently, the length is always given by a sum between unity and k, viz. $\sum_{i=1}^{k} \lambda_i = N_k$. The former approach has the advantage that it eliminates the redundancy caused by the fact that most of the multiplicities are equal to zero in the partitions as can be seen from Tables 1.2 and 1.3. However, the problem with the first convention is that the total number of parts in a partition varies, which means that writing down general formulas or expressions involving all the partitions summing to a specific value becomes awkward. Consequently, we shall adopt the second convention when discussing partitions, which seems to be generally accepted convention [13]. Moreover, since $0! = 1$, a multiplicity of zero only

The Partition Method for a Power Series Expansion.
DOI: http://dx.doi.org/10.1016/B978-0-12-804466-7.00004-8

produces a harmless factor of unity, which has no bearing on the calculation of the coefficients. However, steps will need to be taken to avoid the zero multiplicities from affecting the CPU time, especially for very high orders greater than 100.

Before we can present a theorem involving the partition method for a power series expansion, we need to introduce a new definition. In the literature a composite function $(g \circ f)(x)$ is defined as being equal to $g(f(x))$. Here this definition is extended to a pseudo-composite function as follows.

Definition 4.1. The pseudo-composite function $(g_a \circ f)(x)$ is defined by

$$(g_a \circ f)(x) := g(af(x)) ,$$

where a need not necessarily be a number or a parameter.

Now we are in a position to introduce the main theorem in this chapter.

Theorem 4.1. Given (1) the function $f(z)$ can be expressed in terms of a power series referred to here as the inner power series, i.e., $f(z) \equiv \sum_{k=0}^{\infty} p_k y^k$, where $y = z^{\mu}$, and (2) the function $g(z)$ can be expressed in terms of another or outer power series, viz. $g(z) \equiv h(z) \sum_{k=0}^{\infty} q_k z^k$, where $h(z)$ is an arbitrary function or number, then for non-zero values of p_0 there exists a power series expansion for the quotient of the pseudo-composite functions, $(g_a \circ f)(z)$ and $(h_a \circ f)(z)$, which can be expressed as

$$\frac{(g_a \circ f)(z)}{(h_a \circ f)(z)} \equiv \sum_{k=0}^{\infty} D_k(a) \, y^k . \tag{4.1}$$

The first few coefficients in (4.1) are

$$D_0(a) = F(ap_0) , \quad D_1(a) = aF^{(1)}(ap_0) \, p_1 , \tag{4.2}$$

and

$$D_2(a) = \frac{a^2}{2} F^{(2)}(ap_0) \, p_1^2 + aF^{(1)}(ap_0) \, p_2 , \tag{4.3}$$

while the general formula for the coefficients of the k-th order term is

$$D_k(a) = \sum_{\substack{\lambda_1, \lambda_2, \lambda_3, \dots, \lambda_k = 0 \\ \sum_{i=1}^{k} i \lambda_i = k}}^{k, \lfloor k/2 \rfloor, \lfloor k/3 \rfloor, \dots, 1} a^{N_k} F^{(N_k)}(ap_0) \prod_{i=1}^{k} \frac{p_i^{\lambda_i}}{\lambda_i!} . \tag{4.4}$$

In these results $N_k = \sum_{i=1}^{k} \lambda_i$ and $F^{(N)}(ap_0)$ denotes the N-th derivative of the function $F(ap_0)$, which, in turn, represents the regularized value of the power series expansion $\sum_{j=0}^{\infty} q_j (ap_0)^j$. That is, $F(ap_0) \equiv \sum_{j=0}^{\infty} q_j (ap_0)^j$.

For the special case, where $p_0 = 0$, the coefficients reduce to $D_0(a) = q_0$, $D_1(a) = q_1 p_1 a$, $D_2(a) = p_2 q_1 a + p_1^2 q_2 a^2$, and $D_3(a) = p_2 q_1 a + 2 p_1 p_2 q_2 a^2 + p_1^3 q_3 a^3$. More generally, the coefficients are given by

$$D_k(a) = \sum_{\substack{\lambda_1, \lambda_2, \lambda_3, \ldots, \lambda_k = 0 \\ \sum_{i=1}^{k} i\lambda_i = k}}^{k, \lfloor k/2 \rfloor, \lfloor k/3 \rfloor, \ldots, 1} q_{N_k} a^{N_k} N_k! \prod_{i=1}^{k} \frac{p_i^{\lambda_i}}{\lambda_i!}. \tag{4.5}$$

In addition, when $D_0(a) \neq 0$, the inverted quotient of the pseudo-composite functions can be expressed in terms of another power series, which is

$$\frac{(h_a \circ f)(y)}{(g_a \circ f)(y)} \equiv \frac{1}{D_0(a)} \sum_{k=0}^{\infty} E_k(a) y^k, \tag{4.6}$$

with the coefficients $E_k(a)$ given by

$$E_k(a) = \sum_{\substack{\lambda_1, \lambda_2, \lambda_3, \ldots, \lambda_k = 0 \\ \sum_{i=1}^{k} = k}}^{k, \lfloor k/2 \rfloor, \lfloor k/3 \rfloor, \ldots, 1} (-1)^{N_k} N_k! D_0(a)^{-N_k} \prod_{i=1}^{k} \frac{D_i(a)^{\lambda_i}}{\lambda_i!}. \tag{4.7}$$

Finally, the coefficients $D_k(a)$ and $E_k(a)$ satisfy the following recurrence relation:

$$\sum_{j=1}^{k} D_j(a) E_{k-j}(a) = 0. \tag{4.8}$$

Remark. As in the case of a Taylor/Maclaurin series the power series given by (4.1) and (4.6) can be either (1) convergent for all values of the variable, (2) absolutely convergent within a finite radius of convergence or (3) asymptotic, where an asymptotic power series is defined here as a power series expansion whose radius of absolute convergence is zero. In the last case, however, the asymptotic series must be complete, which means that it will yield the values of the original function when it is regularized and hence does not require jump discontinuous terms as a result of the Stokes phenomenon [19,43,44].

Proof. Since $g(z)$ can be expressed in terms of a power series multiplied by the function $h(z)$, we have

$$\frac{(g_a \circ f)(z)}{(h_a \circ f)(z)} \equiv \left(q_0 + \sum_{k=1}^{\infty} q_k a^k f(z)^k \right). \tag{4.9}$$

Introducing the power series expansion for $f(z)$ into the above result yields

$$\frac{(g_a \circ f)(y)}{(h_a \circ f)(y)} \equiv \left(q_0 + q_1\, a \sum_{k=0}^{\infty} p_k\, y^k + q_2\, a^2 \left(\sum_{k=0}^{\infty} p_k\, y^k \right)^2 \right.$$
$$\left. + q_3\, a^3 \left(\sum_{k=0}^{\infty} p_k\, y^k \right)^3 + q_4\, a^4 \left(\sum_{k=0}^{\infty} p_k\, y^k \right)^4 + \cdots \right). \tag{4.10}$$

By isolating the zeroth order term of the power series expansion for $f(z)$ in (4.10), one obtains

$$\frac{(g_a \circ f)(y)}{(h_a \circ f)(y)} \equiv \left(q_0 + q_1\, a \left(p_0 + \sum_{k=1}^{\infty} p_k\, y^k \right) + q_2\, a^2 \left(p_0 + \sum_{k=1}^{\infty} p_k\, y^k \right)^2 \right.$$
$$\left. + q_3\, a^3 \left(p_0 + \sum_{k=1}^{\infty} p_k\, y^k \right)^3 + q_4\, a^4 \left(p_0 + \sum_{k=1}^{\infty} p_k\, y^k \right)^4 + \cdots \right). \tag{4.11}$$

If we expand the above result in descending powers of p_0, then we find that

$$\frac{(g_a \circ f)(y)}{(h_a \circ f)(y)} \equiv \sum_{j=0}^{\infty} q_j\, (a\,p_0)^j + \sum_{j=1}^{\infty} j\, q_j\, a^j\, p_0^{j-1} \sum_{k=1}^{\infty} p_k\, y^k + \sum_{j=2}^{\infty} \binom{j}{2} q_j\, a^j\, p_0^{j-2}$$
$$\times \left(\sum_{k=1}^{\infty} p_k\, y^k \right)^2 + \sum_{j=3}^{\infty} \binom{j}{3} q_j\, a^j\, p_0^{j-3} \left(\sum_{k=1}^{\infty} p_k\, y^k \right)^3 + \cdots . \tag{4.12}$$

We now represent the regularized value of the series $\sum_{k=0}^{\infty} q_k (a p_0)^k$ by $F(a p_0)$. That is,

$$\sum_{k=0}^{\infty} q_k (a p_0)^k \equiv F(a p_0) . \tag{4.13}$$

Consequently, the sums on the rhs of (4.12) can be simplified, but before doing so, we note that

$$\sum_{j=i}^{\infty} \binom{j}{i} q_j\, a^j\, p_0^{j-i} = \frac{a^i}{i!} \sum_{j=i}^{\infty} q_j (a p_0)^{j-i} \prod_{l=0}^{i-1} (j-l) = \frac{a^i}{i!} \frac{d^i}{dz^i} \sum_{j=0}^{\infty} q_j z^j \Bigg|_{z=a p_0} . \tag{4.14}$$

Introducing (4.13) into the above result gives

$$\sum_{j=i}^{\infty} \binom{j}{i} q_j\, a^j\, p_0^{j-i} \equiv \frac{a^i}{i!} \frac{d^i}{dz^i} F(z) \Bigg|_{z=a p_0} = \frac{a^i}{i!} F^{(i)} (a p_0) . \tag{4.15}$$

Therefore, (4.12) becomes

$$\frac{(g_a \circ f)(y)}{(h_a \circ f)(y)} \equiv \sum_{k=0}^{\infty} \frac{a^k}{k!} F^{(k)} (a p_0) \left(\sum_{j=1}^{\infty} p_j\, y^j \right)^k . \tag{4.16}$$

If (4.16) is expanded in powers of y, then we obtain

$$\frac{(g_a \circ f)(y)}{(h_a \circ f)(y)} \equiv F(ap_0) + a F^{(1)}(ap_0)p_1 y + \left(\frac{a^2}{2} F^{(2)}(ap_0) p_1^2 + a F^{(1)}(ap_0)p_2\right) y^2$$

$$+ \left(\frac{a^3}{3!} F^{(3)}(ap_0) + a^2 F^{(2)}(ap_0) p_1 p_2 + a F^{(1)}(ap_0) p_3\right) y^3 + O(y^4). \quad (4.17)$$

From this result we see that the coefficients of the zeroth, first and second order terms in y correspond to the results for $D_0(a)$, $D_1(a)$ and $D_2(a)$ given in the theorem. Furthermore, we observe that

$$D_3(a) = \frac{a^3}{3!} F^{(3)}(ap_0) p_1^3 + a^2 F^{(2)}(ap_0) p_1 p_2 + a F^{(1)}(ap_0) p_3. \quad (4.18)$$

The first few coefficients are relatively easy to write down, but beyond that it becomes progressively more difficult. It is at this stage that we need to introduce partitions in order to derive a general expression for the coefficients of all powers of y in (4.17).

If we look closely at the result for D_3, then we see that it is the sum of three separate contributions, which is due to the fact that there are only three partitions summing to 3, namely $\{1_3\}$, $\{1, 2\}$ and $\{3\}$. As stated in Refs. [3–5], in order to calculate the contribution due to each partition, a value must be assigned to each part in the partitions. In particular, Ref. [3] states that these assigned values depend upon the coefficients of the inner series, i.e. the power series expansion for $f(z)$, which in turn, becomes the variable in the outer power series for $g(z)$. Therefore, we assign the value of p_i to each part i. Moreover, there is a multinomial factor associated with each partition, which is not only dependent upon the multiplicities λ_i, but also on their sum, $N_k = \sum_{i=1}^{k} \lambda_i$. This factor arises from the fact that the value of k on the rhs of (4.16) is used to render each partition. That is, k in (4.16) corresponds to N_k in the partition method for a power series expansion. Often the multinomial factor simply becomes $N_k!/\lambda_1! \lambda_2! \cdots \lambda_k!$, but in (4.16) there is an extra factor of $a^k F^{(k)}(ap_0)/k!$ in the terms. As indicated in the first two chapters, since k in (4.16) plays the role of N_k in the partition method for a power series expansion, it means that the standard multinomial factor must be multiplied by a factor of $a^{N_k} F^{(N_k)}(ap_0)/N_k!$ for each partition. Therefore, the overall contribution from a partition to the coefficient of the resulting power series for the quotient of the pseudo-composite functions is given by

$$C\left(\lambda_1, \lambda_2, \ldots, \lambda_k\right) = a^{N_k} F^{(N_k)}(ap_0) \prod_{i=1}^{k} \frac{p_i^{\lambda_i}}{\lambda_i!}. \quad (4.19)$$

From this result it can be seen that we do not actually require the partitions themselves, only their multiplicities. Nevertheless, it should be noted that each set of multiplicities represents a specific partition. Hence each contribution given by (4.19) is unique.

To make the preceding material clearer, let us evaluate the fourth order term in y in (4.16) or D_4. This is determined by considering all the partitions summing to 4 of which there are five: $\{1_4\}$, $\{1_2, 2\}$, $\{2_2\}$, $\{1, 3\}$ and $\{4\}$. For the first partition, $\lambda_1 = 4$, while the other λ_i are zero. Therefore, from (4.19) we have

$$C(4, 0, 0, 0) = a^4 F^{(4)}(ap_0) p_1^4/4! . \tag{4.20}$$

In the case of the second partition $\lambda_1 = 2$, $\lambda_2 = 1$ and the other λ_i vanish, while for the third partition, only $\lambda_2 (= 2)$ does not vanish. According to (4.19), the contributions due to these partitions are

$$C(2, 1, 0, 0) = a^3 F^{(3)}(ap_0) p_1^2 p_2/3! , \tag{4.21}$$

and

$$C(0, 2, 0, 0) = a^2 F^{(2)}(ap_0) p_2^2/2! . \tag{4.22}$$

In the case of the fourth partition $\lambda_1 = \lambda_3 = 1$ with λ_2 and λ_4 equal to zero, whereas in the final partition only $\lambda_4 (= 1)$ is non-zero. Hence these partitions yield

$$C(1, 0, 1, 0) = a^2 F^{(2)}(ap_0) p_1 p_3/2! , \tag{4.23}$$

and

$$C(0, 0, 0, 1) = a F^{(1)}(ap_0) p_4 . \tag{4.24}$$

Finally, $D_4(a)$ is given by the sum of the five or $p(4)$ contributions given above.

To derive the general formula for the coefficients $D_k(a)$, we now sum over all the partitions summing to k. This entails summing over all values that the multiplicities of the parts can take. Each multiplicity, λ_i, ranges from 0 to its maximal value of $\lfloor k/i \rfloor$. Furthermore, all the partitions are constrained by the condition that

$$\sum_{i=1}^{k} \lambda_i i = \lambda_1 + 2\lambda_2 + 3\lambda_3 + \cdots + k\lambda_k = k . \tag{4.25}$$

From this result we see that λ_k equals zero for all partitions except $\{k\}$, in which case it will equal unity with the other λ_i equal to zero. That is, the multiplicities are more often zero than non-zero as can be seen from the sparseness in Tables 1.2 and 1.3. This means that much redundancy occurs when summing over the allowed values of λ_i. More succinctly, the sum over all partitions summing to k can be expressed as

$$D_k(a) = \sum_{\substack{\lambda_1,\lambda_2,\lambda_3,\ldots,\lambda_k=0 \\ \sum_{i=1}^{k} i\lambda_i=k}}^{k,\lfloor k/2\rfloor,\lfloor k/3\rfloor,\ldots,1} C(\lambda_1,\lambda_2,\ldots,\lambda_k) . \tag{4.26}$$

If (4.19) is introduced into the above equation, then we obtain the general formula given by (4.4).

Now we turn our attention to the important case of $p_0 = 0$ upon which nearly all the examples in Chapters 1 and 2 are based. This means that (4.11) reduces to

$$\frac{(g_a \circ f)(y)}{(h_a \circ f)(y)} \equiv q_0 + \sum_{k=1}^{\infty} q_k \, a^k \left(\sum_{j=1}^{\infty} p_j \, y^j\right)^k . \tag{4.27}$$

Expanding the first few powers in y yields

$$\frac{(g_a \circ f)(y)}{(h_a \circ f)(y)} \equiv q_0 + q_1 \, ap_0 \, y + \left(q_1 ap_2 + q_2 a^2 p_1^2\right) y^2 + \left(q_1 ap_3\right.$$
$$\left. + 2q_2 a^2 p_1 p_2 + q_3 a^3 p_1^3\right) y^3 + O\left(y^4\right) . \tag{4.28}$$

That is, we obtain a power series expansion in y except now the coefficients $D_k(a)$ take on a different form. In particular, we find that the first few coefficients, again denoted by D_k, become those listed above (4.5) in the theorem. Since (4.16) and (4.27) are isomorphic, we can, once again, apply the method for a power series expansion in order to derive the general formula for the coefficients of the above expansion. In fact, the only difference between the equivalence statements is that the terms for the outer series in (4.27) are multiplied by $q_k a^k$ as opposed to $a^k F^{(k)}(ap_0)/k!$ in (4.16). Consequently, the only change to the partition method occurs with the multinomial factor, which now becomes $q_{N_k} a^{N_k} N_k! \prod_{i=1}^{k} 1/\lambda_i!$. Thus, the contribution to the coefficients due to each partition is given by

$$C\left(\lambda_1,\lambda_2,\ldots,\lambda_k\right) = q^{N_k} a^{N_k} N_k! \prod_{i=1}^{k} \frac{p_i^{\lambda_i}}{\lambda_i!} . \tag{4.29}$$

Introducing this result into (4.26) gives (4.5).

Although it has been stated that the expansion given by (4.1) can be asymptotic, this does not mean that it will be divergent for all values of y. If it were, then inverting the equivalence implies that the inverted quotient of the pseudo-composite functions vanishes for all values of y. Therefore, there must be a region in the complex plane where the equivalence symbol can be replaced by an equals sign. In Ref. [3] it was found that an asymptotic series possesses zero radius of absolute convergence, but it can be either conditionally convergent or divergent depending upon which sector in the complex plane the variable is situated. For those values of y where (4.1) is convergent, we can invert the quotient of the pseudo-composite functions $(g_a \circ f)(y)$ and $(h_a \circ f)(y)$. Provided $D_0(a) \neq 0$, we find that

$$\frac{(h_a \circ f)(y)}{(g_a \circ f)(y)} = \frac{1}{D_0(a)} \frac{1}{\left(1 + \sum_{k=1}^{\infty} (D_k(a)/D_0(a)) y^k\right)} . \tag{4.30}$$

The rhs of this result can be regarded as the regularized value of the geometric series with the variable equal to $-\sum_{k=1}^{\infty} (D_k(a)/D_0(a)) y^k$. According to Appendix A or Refs. [18,19,43,44], the geometric series, which is given by $\sum_{k=0}^{\infty} z^k$, is either conditionally or absolutely convergent for $\Re z < 1$. For all other values of z in the principal branch of the complex plane, it is either divergent or undefined and thus, must be regularized to yield meaningful values. Treating the rhs of (4.30) as the limit of the geometric series means that $\left(\Re \sum_{k=1}^{\infty} (D_k(a)/D_0(a)) y^k\right) > -1$, while for all other values of y, the series will be either divergent or undefined with the latter case occurring when $\left(\Re \sum_{k=1}^{\infty} (D_k(a)/D_0(a)) y^k\right) = -1$. Therefore, (4.30) can be expressed as

$$\frac{(h_a \circ f)(y)}{(g_a \circ f)(y)} \equiv \frac{1}{D_0(a)} \sum_{k=0}^{\infty} \left(-\sum_{j=1}^{\infty} (D_j(a)/D_0(a)) y^j\right)^k . \tag{4.31}$$

By comparing (4.31) with (4.16), we see that they are isomorphic, which means in turn that the partition method for a power series expansion can be applied again. In this instance the coefficients of y^j in the inner series or the values assigned to the parts j in the partitions are $-D_j(a)/D_0(a)$ instead of p_j. Furthermore, the multinomial factor becomes the standard value of $N! \prod_{i=1}^{k} 1/\lambda_i!$ for partitions summing to k. Hence the contribution from each partition is represented by

$$C\left(\lambda_1, \lambda_2, \ldots, \lambda_k\right) = (-1)^{N_k} N_k! D_0(a)^{-N_k} \prod_{i=1}^{k} \frac{D_i(a)^{\lambda_i}}{\lambda_i!} . \tag{4.32}$$

Introducing the above result into (4.26) with the $D_k(a)$ on the lhs replaced by $E_k(a)$ yields (4.7).

To derive the final result in the theorem, we multiply (4.1) by (4.6), which yields

$$1 = \frac{(g_a \circ f)(\gamma)}{(h_a \circ f)(\gamma)} \frac{(h_a \circ f)(\gamma)}{(g_a \circ f)(\gamma)} \equiv \sum_{k=0}^{\infty} D_k(a)\gamma^k \left(\frac{1}{D_0(a)} \sum_{k=0}^{\infty} E_k(a)\,\gamma^k \right). \qquad (4.33)$$

Since $E_0 = 1$ from (4.7), we can separate the zeroth order term in γ, which cancels the term of unity on the lhs of the equivalence. Then the above result reduces to

$$\sum_{k=1}^{\infty} \gamma^k \sum_{j=0}^{k} D_j(a)E_{k-j}(a) \equiv 0. \qquad (4.34)$$

As indicated earlier in the proof, there will be some values of γ, actually a region in the complex plane, where the equivalence symbol can be replaced by an equals sign. Otherwise, either (4.1) or (4.6) would be zero for all values of γ. Given that there will be an infinite number of values of γ where an equals sign applies, the lhs of the above result will also vanish for these values of γ. For this to occur, it means that the inner series, which is independent of γ, must also vanish. Hence we arrive at the recurrence relation given by (4.8). This completes the proof of Theorem 4.1.

It should be stressed that Theorem 4.1 has avoided the issue of determining the radius of absolute convergence for the power series on the rhs of (4.1). This is because although one can determine a value for the radius of absolute convergence when deriving the resulting power series, it is often only an estimate, not the supremum, which we observed in Chapter 1. That is, one often needs to examine the resulting power series expansion by independent means to determine the radius of absolute convergence, if it exists. Moreover, one can introduce a divergent power series as the inner series and despite the fact that the outer series may also possess a finite radius of absolute convergence, the resulting power series arising from (4.1) may be convergent over the entire complex plane. E.g., the power series expansion for cosecant in Chapter 1 was found to possess a radius of absolute convergence equal to π. It was then used as the inner series in an inversion, where it became the variable for the outer series. The outer series in this case was the geometric series, whose radius of absolute convergence is unity. Thus, both series were restricted by radii of absolute convergence. However, the resulting power series arising from (4.1) was merely another form of the standard Taylor/Maclaurin series for sine, which is convergent for all values of the power variable.

Table 4.1 displays the coefficients $D_k(a)$ up to $k = 10$ for the important case where p_0 vanishes. In other words, the coefficients have been evaluated via (4.5), while the various terms appear in the order as they are generated by partition trees. The reason why the $p_0 = 0$ case is important is that in all the examples presented in Chapters 1 and 2 the coefficients have been determined by programming the symbolic results in Table 4.1 into Mathematica. In addition, a was equal to unity in all these problems. Thus, the results in Table 4.1 are quite general.

Because the general forms for the $D_k(a)$ in Theorem 4.1 involve summations over the partitions summing to k, it could be argued that the theorem is simply a re-statement of Faà di Bruno's formula [40,58]. According to page 137 of Ref. [53], the theorem for this formula states that if the f and g are two formal (Taylor/Maclaurin) series defined by

$$f := \sum_{k \geq 0} f_k \frac{u^k}{k!}, \quad \text{and} \quad g := \sum_{m \geq 0} g_m \frac{t^m}{m!},$$

with $f_0 = 0$, then the formal (Taylor/Maclaurin) series for the composite of g with f, i.e. $h := g \circ f$, is given by

$$h := \sum_{n \geq 0} h_n \frac{t^n}{n!}, \tag{4.35}$$

where $h_0 = g_0$ and

$$h_n = \sum_{1 \leq k \leq n} g_k \, \boldsymbol{B}_{n,k}(f_1, f_2, \ldots f_n). \tag{4.36}$$

In this result the $\boldsymbol{B}_{n,k}$ are referred to as the exponential Bell polynomials or Bell polynomials of the second kind [35]. In terms of the notation in this book they are written as

$$\boldsymbol{B}_{n,k}(x_1, x_2, \ldots, x_n) = \sum_{\substack{\lambda_1, \lambda_2, \lambda_3, \ldots, \lambda_n = 0 \\ \sum_{i=1}^{n} i\lambda_i = k, \ \sum_{i=1}^{n} \lambda_i = k}}^{n, \lfloor n/2 \rfloor, \lfloor n/3 \rfloor, \ldots, 1} n! \prod_{i=1}^{k} \frac{x_i^{\lambda_i}}{\lambda_i!(i!)^{\lambda_i}}, \tag{4.37}$$

where n is referred to as the weight in Ref. [53]. In addition, if we let $x_i = x$ in the above result, then we obtain the Bell polynomials of the first kind, which are given by (2.4) in Chapter 2. Because the above consists of $\left|\begin{matrix} n \\ k \end{matrix}\right|$ monomials rather than all the terms obtained by considering all the partitions summing to n, viz. $p(n)$, Comtet refers to these polynomials as the partial Bell polynomials. However, the sum in (4.36) removes the restriction caused by the condition, $\sum_{i=1}^{N} \lambda_i = k$ in (4.37).

Table 4.1 Coefficients $D_k(a)$ in the power series expansion given by (4.5)

k	$D_k(a)$
0	q_0
1	$p_1 q_1 a$
2	$p_2 q_1 a + p_1^2 q_2 a^2$
3	$p_3 q_1 a + 2 p_1 p_2 q_2 a^2 + p_1^3 q_3 a^3$
4	$p_4 q_1 a + 2 p_1 p_3 q_2 a^2 + 3 p_1^2 p_2 q_3 a^3 + p_1^4 q_4 a^4 + p_2^2 q_2 a^2$
5	$p_5 q_1 a + 2 p_1 p_4 q_2 a^2 + 3 p_1^2 p_3 q_3 a^3 + 4 p_1^3 p_2 q_4 a^4 + p_1^5 q_5 a^5 + 3 p_1 p_2^2 q_3 a^3$ $+ 2 p_2 p_3 q_2 a^2$
6	$p_6 q_1 a + 2 p_1 p_5 q_2 a^2 + 3 p_1^2 p_4 q_3 a^3 + 4 p_1^3 p_3 q_4 a^4 + 5 p_1^4 p_2 q_5 a^5 + p_1^6 q_6 a^6$ $+ 6 p_1^2 p_2^2 q_4 a^4 + 6 p_1 p_2 p_3 q_3 a^3 + 2 p_2 p_4 q_2 a^2 + p_2^3 q_3 a^3 + p_3^2 q_2 a^2$
7	$p_7 q_1 a + 2 p_1 p_6 q_2 a^2 + 3 p_1^2 p_5 q_3 a^3 + 4 p_1^3 p_4 q_4 a^4 + 5 p_1^4 p_3 q_5 a^5 + 6 p_1^5 p_2 q_6 a^6$ $+ p_1^7 q_7 a^7 + 10 p_1^3 p_2^2 q_5 a^5 + 12 p_1^2 p_2 p_3 q_4 a^4 + 6 p_1 p_2 p_4 q_3 a^3 + 4 p_1 p_2^3 q_4 a^4$ $+ 3 p_1 p_3^2 q_3 a^3 + 2 p_2 p_5 q_2 a^2 + 3 p_2^2 p_3 q_3 a^3 + 2 p_3 p_4 q_2 a^2$
8	$p_8 q_1 a + 2 p_1 p_7 q_2 a^2 + 3 p_1^2 p_6 q_3 a^3 + 4 p_1^3 p_5 q_4 a^4 + 5 p_1^4 p_4 q_5 a^5 + 6 p_1^5 p_3 q^6 a^6$ $+ 7 p_1^6 p_2 q_7 a^7 + p_1^8 q_8 a^8 + 15 p_1^4 p_2^2 q_6 a^6 + 20 p_1^3 p_2 p_3 q_5 a^5 + 12 p_1^2 p_2 p_4 q_4 a^4$ $+ 10 p_1^2 p_2^3 q_5 a^5 + 6 p_1^2 p_3^2 q_4 a^4 + 6 p_1 p_2 p_5 q_3 a^3 + 12 p_1 p_2^2 p_3 q_4 a^4$ $+ 6 p_1 p_3 p_4 q_3 a^3 + 2 p_2 p_6 q_2 a^2 + 3 p_2^2 p_4 q_3 a^3 + p_2^4 q_4 a^4 + 3 p_2 p_3^2 q_3 a^3 + 2 p_3 p_5 q_2 a^2$ $+ p_4^2 q_2 a^2$
9	$p_9 q_1 a + 2 p_1 p_8 q_2 a^2 + 3 p_1^2 p_7 q_3 a^3 + 4 p_1^3 p_6 q_4 a^4 + 5 p_1^4 p_5 q_5 a^5 + 6 p_1^5 p_4 q_6 a^6$ $+ 7 p_1^6 p_3 q_7 a^7 + 8 p_1^7 p_2 q_8 a^8 + p_1^9 q_9 a^9 + 21 p_1^5 p_2^2 q_7 a^7 + 30 p_1^4 p_2 p_3 q_6 a^6$ $+ 20 p_1^3 p_2 p_4 q_5 a^5 + 20 p_1^3 p_2^3 q_6 a^6 + 10 p_1^3 p_3^2 q_5 a^5 + 12 p_1^2 p_2 p_5 q_4 a^4$ $+ 30 p_1^2 p_2^2 p_3 q_5 a^5 + 12 p_1^2 p_3 p_4 q_4 a^4 + 6 p_1 p_2 p_6 q_3 a^3 + 12 p_1 p_2^2 p_4 q_4 a^4$ $+ 5 p_1 p_2^4 q_5 a^5 + 12 p_1 p_2 p_3^2 q_4 a^4 + 6 p_1 p_3 p_5 q_3 a^3 + 3 p_1 p_4^2 q_3 a^3 + 2 p_2 p_7 q_2 a^2$ $+ 3 p_2^2 p_5 q_3 a^3 + 4 p_2^3 p_3 q_4 a^4 + 6 p_2 p_3 p_4 q_3 a^3 + 2 p_3 p_6 q_2 a^2 + p_3^3 q_3 a^3 + 2 p_4 p_5 q_2 a^2$
10	$p_{10} q_1 a + 2 p_1 p_9 q_2 a^2 + 3 p_1^2 p_8 q_3 a^3 + 4 p_1^3 p_7 q_4 a^4 + 5 p_1^4 p_6 q_5 a^5 + 6 p_1^5 p_5 q_6 a^6$ $+ 7 p_1^6 p_4 q_7 a^7 + 8 p_1^7 p_3 q_8 a^8 + 9 p_1^8 p_2 q_9 a^9 + p_1^{10} q_{10} a^{10} + 28 p_1^6 p_2^2 q_8 a^8$ $+ 42 p_1^5 p_2 p_3 q_7 a^7 + 30 p_1^4 p_2 p_4 q_6 a^6 + 35 p_1^4 p_2^3 q_7 a^7 + 15 p_1^4 p_3^2 q_6 a^6 + 20 p_1^3 p_2 p_5 q_5 a^5$ $+ 60 p_1^3 p_2^2 p_3 q_6 a^6 + 20 p_1^3 p_3 p_4 q_5 a^5 + 12 p_1^2 p_2 p_6 q_4 a^4 + 30 p_1^2 p_2^2 p_4 q_5 a^5$ $+ 15 p_1^2 p_2^4 q_6 a^6 + 30 p_1^2 p_2 p_3^2 q_5 a^5 + 12 p_1^2 p_3 p_5 q_4 a^4 + 6 p_1^2 p_4^2 q_4 a^4 + 6 p_1 p_2 p_7 q_3 a^3$ $+ 12 p_1 p_2^2 p_5 q_4 a^4 + 20 p_1 p_2^3 p_3 q_5 a^5 + 24 p_1 p_2 p_3 p_4 q_4 a^4 + 6 p_1 p_3 p_6 q_3 a^3 + 4 p_1 p_3^3 q_4 a^4$ $+ 6 p_1 p_4 p_5 q_3 a^3 + 2 p_2 p_8 q_2 a^2 + 3 p_2^2 p_6 q_3 a^3 + 4 p_2^3 p_4 q_4 a^4 + p_2^5 q_5 a^5 + 6 p_2^2 p_3^2 q_4 a^4$ $+ 6 p_2 p_3 p_5 q_3 a^3 + 3 p_2 p_4^2 q_3 a^3 + 2 p_3 p_7 q_2 a^2 + 3 p_3^2 p_4 q_3 a^3 + 2 p_4 p_6 q_2 a^2 + p_5^2 q_2 a^2$

Table 4.2 Coefficients h_k in Faà di Bruno's formula or Riordan's Bell polynomials $Y_k(fg_1, \ldots, fg_k)$

k	h_k or $Y_k(gf_1, \ldots, gf_k)$
1	$g_1 f_1$
2	$g_1 f_2 + g_2 f_1^2$
3	$g_1 f_3 + 3g_2 f_1 f_2 + 3g_3 f_1^3$
4	$g_1 f_4 + 4g_2 f_1 f_3 + 3g_2 f_2^2 + 6g_3 f_1^2 f_2 + g_4 f_1^4$
5	$g_1 f_5 + 5g_2 f_1 f_4 + 10g_2 f_2 f_3 + 10g_3 f_1^2 f_3 + 15g_3 f_1 f_2^2 + 10g_4 f_1^3 f_2 + g_5 f_1^5$
6	$g_1 f_6 + 6g_2 f_1 f_5 + 15g_2 f_2 f_4 + 10g_2 f_3^2 + 15g_3 f_1^2 f_4 + 60g_3 f_1 f_2 f_3 + 15g_3 f_2^3$ $+ 20g_4 f_1^3 f_3 + 45g_4 f_1^2 f_2^2 + 15g_5 f_1^4 f_2 + g_6 f_1^6$

In Chapter 5 of Ref. [36] Riordan refers to the h_n in (4.37) as Bell polynomials without mentioning the seminal work of Faà di Bruno [59], [60]. Consequently, his definition incorporates all the partitions summing to n. In particular, he defines the Bell polynomials as

$$Y_n(gf_1, \ldots, gf_n) = \sum_{\substack{\lambda_1, \lambda_2, \lambda_3, \ldots, \lambda_n = 0 \\ \sum_{i=1}^{n} i\lambda_i = n,}}^{n, \lfloor n/2 \rfloor, \lfloor n/3 \rfloor, \ldots, 1} N_n! g_{N_n} \prod_{i=1}^{n} \frac{f_i^{\lambda_i}}{\lambda_i! (i!)^{\lambda_i}}, \qquad (4.38)$$

where $g_n = d^n g(t)/dt^n|_{t=0}$ and $f_n = d^n f(u)/du^n|_{u=0}$. As in Comtet's version of Faà di Bruno's formula, f and g are again differentiable functions. Moreover, despite the different interpretations in the Bell polynomials, the final forms for h_n and $Y_n(gf_1, \ldots, gf_n)$ are identical to each another.

The first six h_k or $Y_k(gf_1, \ldots, gf_k)$ are displayed in Table 4.2. We can compare them with the first six D_k in Table 4.1, once it is realized that the f_k and g_k play the roles of the inner and outer series in Theorem 4.1. Therefore, besides setting $a = 1$ in Table 4.1, we set $p_k = f_k$ and $q_k = g_k$. Although the results in both tables possess the same number of terms for each value of k, viz. $p(k)$, the coefficients of many of the terms are different. This is because the results in Table 4.2 require that both $f(x)$ and $g(x)$ be infinitely differentiable or smooth, whereas this is not the case with Theorem 4.1. However, if we put $a = 1$, $p_k = f_k/k!$ and $q_k = g_k/k!$ in Table 4.1, then for $k = 4$ we find that

$$p_4 q_1 + 2p_1 p_3 q_2 + 3p_1^2 p_2 q_3 + p_1^4 q_4 + p_2^2 q_2 = \frac{f_4 g_1}{4!} + 2\frac{f_1 f_3 g_2}{3! \cdot 2!} + 3\frac{f_1^2 f_2 g_3}{2! \cdot 3!}$$

$$+ \frac{f_1^4 g_4}{4!} + \frac{f_2^2 g_2}{(2!)^3} = \frac{1}{4!}\left(f_4 g_1 + 4f_1 f_3 g_2 + 6f_1^2 f_2 g_3 + f_1^4 g_1^4 + 3g_2 f_2^2\right). \quad (4.39)$$

This is, of course, $h_4/4!$ or $Y(gf_1, \ldots, gf_4)$. Hence we see that Faà di Bruno's formula is an example of the more general Theorem 4.1, which does not require that the functions be smooth. In addition, Theorem 4.1 considers: (1) pseudo-composite functions, which are also more general than the composite function $g \circ f$ used above, and (2) the case when $g_0 \neq 0$.

To re-inforce the preceding material, let us consider the application of Theorem 4.1 to the following function:

$$f(z) = \exp\left(az^\nu \sin^\rho z\right), \qquad (4.40)$$

where ν, ρ and z can be real or complex. This represents an unusual expansion in that the power variable z will also appear together as part of a in Theorem 4.1. In addition, when a is purely imaginary, we are effectively obtaining power series expansion for cosine and sine of the argument in (4.40). Before Theorem 4.1 can be applied, however, we require a power series expansion for $\sin^\rho z$. This has already been done in Theorem 2.1 of Chapter 2, where the coefficients were referred to as the generalized cosecant numbers. Hence the exponential in (4.40) can be expressed as

$$\exp\left(az^\nu \sin^\rho z\right) \equiv \exp\left(az^{\rho+\nu} \sum_{k=0}^{\infty} c_{-\rho,k} z^{2k}\right), \qquad (4.41)$$

where the equivalence symbol can be replaced by an equals sign for $|z| < \pi$. In fact, this will apply to the final result obtained via (4.1), although as indicated earlier, this may be too restrictive on z.

Since $c_{-\rho,0} = 1$ for all values of ρ, we obtain an isolated exponential factor of $\exp\left(az^{\rho+\nu}\right)$. Therefore, in order to apply Theorem 4.1, we set this term equal to $h_a \circ f$. Then (4.41) becomes

$$\frac{\exp\left(az^{\mu-\rho} \sin^\rho z\right)}{\exp\left(az^\mu\right)} \equiv \exp\left(az^\mu \sum_{k=1}^{\infty} c_{-\rho,k} z^{2k}\right), \qquad (4.42)$$

where $\mu = \rho + \nu$. Consequently, we see that the inner power series begins at $k = 1$, i.e., $p_0 = 0$, while the coefficients of the outer series are those for the exponential power series, viz. $q_k = 1/k!$. Since $p_0 = 0$, the coefficients of the resulting power series expansion can be obtained from (4.5) or directly from the results in Table 4.1 by setting $p_k = (1+(-1)^k)c_{-\rho,k/2}/2$ and $a = az^\mu$. On the other hand, when using Mathematica to calculate the coefficients up to z^{2k} in the resulting power series expansion, one must set each p_{2i} and p_{2i-1} equal to $c_{-\rho,i}$ and zero, respectively up to k. This is actually a clumsy approach for two reasons. The first is that there is much redundant calculation because half of the coefficients of the inner series vanish. The second is that we have to use the form for D_{2k} to obtain the coefficient of z^{2k}, but it would be much easier if we could use D_k instead since it possesses far less terms. For example, D_{20} has $p(20)$ or 627 terms, whereas D_{10} has only 42 terms. To circumvent this problem, we replace z^2 by y in (4.42) so that (4.42) can be expressed as

$$\frac{\exp\left(az^{\mu-\rho}\sin^\rho z\right)}{\exp\left(az^\mu\right)} \equiv \exp\left(ay^{\mu/2}\sum_{k=1}^{\infty}c_{-\rho,k}y^k\right)\bigg|_{y=z^2}. \tag{4.43}$$

Now we can carry out the calculations with $p_k = c_{-\rho,k}$, $a = ay^{\mu/2}$ and $q_k = 1/k!$. Finally, we replace y by z^2 to obtain the power series expansion in terms of z with the coefficients denoted by $Z_{\rho,\mu,k}(z)$. Therefore, we arrive at

$$\frac{\exp\left(az^{\mu-\rho}\sin^\rho z\right)}{\exp\left(az^\mu\right)} \equiv \sum_{k=0}^{\infty} Z_{\rho,\mu,k}(a,z)\, z^{2k}. \tag{4.44}$$

The rhs of the above result becomes a standard power series expansion only when $\mu = 0$ or $\rho = -\nu$. Otherwise, it represents an unusual expansion in which the coefficients are also dependent upon the power variable.

Table 4.3 presents the coefficients $Z_{\rho,\mu,k}(a,z)$ on the rhs of (4.44) up to $k = 10$. Oddly enough, but not recommended, they have been determined by using the "clumsy" approach mentioned above. The reason for doing this was to see if a Venom laptop with 8 Gb RAM was capable of handling $D_{20}(a)$ in a time expedient manner, which turned out to be the case. This provides the impetus for developing a programming methodology in the next chapter because here we have an instance where it is possible to obtain a power series expansion to thirty or forty orders with little difficulty. From the table we also see that the az^μ-dependence in the exponential on the lhs of (4.44) appears in Bell polynomials of the first kind, which have been studied in Chapter 2. There it was stated that the coefficients of these polynomials were the Stirling numbers of the second kind, i.e., $B_k(x) = \sum_{j=1}^{k} S(k,j)\, x^j$. Furthermore, the ρ-dependent parts of the results in the table are basically the generalized cosecant numbers except that each power of ρ is multiplied by a Bell polynomial of the first kind, whose order is the same as the power of ρ, viz. $B_k(az^\mu)$. In Chapter 2 we saw that the generalized cosecant numbers could be represented by $c_{\rho,k} = \sum_{j=1}^{k} C_{k,j}\rho^j$, where the coefficients $C_{k,j}$ could be expressed in terms of the Stirling numbers of the first kind as in (2.72), but eventually reduced to combinatorial expressions with a polynomial in k in the numerator. Therefore, according

Table 4.3 The coefficients $Z_{\rho,\mu,k}(a, z)$ in (4.44)

k	$Z_{\rho,\mu,k}(a, z)$
0	1
1	$\frac{1}{3!}\,\rho\mathcal{B}_1\!\left(az^\mu\right)$
2	$\frac{2}{6!}\left(2\rho\mathcal{B}_1\!\left(az^\mu\right) + 5\rho^2\mathcal{B}_2\!\left(az^\mu\right)\right)$
3	$\frac{8}{9!}\left(16\rho\mathcal{B}_1\!\left(az^\mu\right) + 42\rho^2\mathcal{B}_2\!\left(az^\mu\right) + 35\rho^3\mathcal{B}_3\!\left(az^\mu\right)\right)$
4	$\frac{2}{3\cdot10!}\left(144\rho\mathcal{B}_1\!\left(az^\mu\right) + 404\rho^2\mathcal{B}_2\!\left(az^\mu\right) + 420\rho^3\mathcal{B}_3\!\left(az^\mu\right) + 175\rho^4\mathcal{B}_4\!\left(az^\mu\right)\right)$
5	$\frac{4}{3\cdot12!}\left(768\rho\mathcal{B}_1\!\left(az^\mu\right) + 2288\rho^2\mathcal{B}_2\!\left(az^\mu\right) + 2684\rho^3\mathcal{B}_3\!\left(az^\mu\right) + 1540\rho^4\mathcal{B}_4\!\left(az^\mu\right)$ $+\, 385\rho^5\mathcal{B}_5\!\left(az^\mu\right)\right)$
6	$\frac{2}{9\cdot15!}\left(1061376\rho\mathcal{B}_1\!\left(az^\mu\right) + 3327584\rho^2\mathcal{B}_1\!\left(az^\mu\right) + 4252248\rho^3\mathcal{B}_3\!\left(az^\mu\right)$ $+\, 2862860\rho^4\mathcal{B}_4\!\left(az^\mu\right) + 1051050\rho^5\mathcal{B}_5\!\left(az^\mu\right) + 175175\rho^5\mathcal{B}_6\!\left(az^\mu\right)\right)$
7	$\frac{1}{27\cdot15!}\left(552960\rho\mathcal{B}_1\!\left(az^\mu\right) + 1810176\rho^2\mathcal{B}_2\!\left(az^\mu\right) + 2471456\rho^3\mathcal{B}_3\!\left(az^\mu\right)$ $+\, 1849848\rho^4\mathcal{B}_4\!\left(az^\mu\right) + 820820\rho^5\mathcal{B}_5\!\left(az^\mu\right) + 210210\rho^6\mathcal{B}_6\!\left(az^\mu\right) + 25025\rho^7\mathcal{B}_7\!\left(az^\mu\right)\right)$
8	$\frac{2}{45\cdot18!}\left(200005632\rho\mathcal{B}_1\!\left(az^\mu\right) + 679395072\rho^2\mathcal{B}_2\!\left(az^\mu\right) + 978649472\rho^3\mathcal{B}_3\!\left(az^\mu\right)$ $+\, 792548432\rho^4\mathcal{B}_4\!\left(az^\mu\right) + 397517120\rho^5\mathcal{B}_5\!\left(az^\mu\right) + 125925800\rho^6\mathcal{B}_6\!\left(az^\mu\right)$ $+\, 23823800\rho^7\mathcal{B}_7\!\left(az^\mu\right) + 2127125\rho^8\mathcal{B}_8\!\left(az^\mu\right)\right)$
9	$\frac{4}{81\cdot21!}\left(129369047040\rho\mathcal{B}_1\!\left(az^\mu\right) + 453757851648\rho^2\mathcal{B}_2\!\left(az^\mu\right) + 683526873856\rho^3\mathcal{B}_3\!\left(az^\mu\right)$ $+\, 589153364352\rho^4\mathcal{B}_4\!\left(az^\mu\right) + 323159810064\rho^5\mathcal{B}_5\!\left(az^\mu\right) + 117327450240\rho^6\mathcal{B}_6\!\left(az^\mu\right)$ $+\, 27973905960\rho^7\mathcal{B}_7\!\left(az^\mu\right) + 4073869800\rho^8\mathcal{B}_8\!\left(az^\mu\right) + 282907625\rho^9\mathcal{B}_9\!\left(az^\mu\right)\right)$
10	$\frac{2}{6075\cdot22!}\left(38930128699392\rho\mathcal{B}_1\!\left(az^\mu\right) + 140441050828800\rho^2\mathcal{B}_2\!\left(az^\mu\right)$ $+\, 219792161825280\rho^3\mathcal{B}_3\!\left(az^\mu\right) + 199416835425280\rho^4\mathcal{B}_4\!\left(az^\mu\right)$ $+\, 117302530691808\rho^5\mathcal{B}_5\!\left(az^\mu\right) + 47005085727600\rho^6\mathcal{B}_6\!\left(az^\mu\right)$ $+\, 12995644662000\rho^7\mathcal{B}_7\!\left(az^\mu\right) + 2422012593000\rho^8\mathcal{B}_8\!\left(az^\mu\right)$ $+\, 280078548750\rho^9\mathcal{B}_9\!\left(az^\mu\right) + 15559919375\rho^{10}\mathcal{B}_{10}\!\left(az^\mu\right)\right)$

to the results in Table 4.3, the coefficients $Z_{\rho,\mu,k}(a, z)$ can be expressed as

$$Z_{\rho,\mu,k}(a, z) = \sum_{j=1}^{k} C_{k,j}\rho^j \sum_{i=1}^{j} S(j, i)\left(az^\mu\right)^i. \qquad (4.45)$$

For the special case of $\mu = 0$, (4.44) reduces to

$$\exp\!\left(az^{-\rho}\sin^\rho z\right) \equiv e^a \sum_{k=0}^{\infty} z^{2k} \sum_{j=1}^{k} \mathcal{B}_j(a)\, C_{k,j}(k)\rho^j. \qquad (4.46)$$

Furthermore, for $a = 1$, this result becomes

$$\exp\!\left(z^{-\rho}\sin^\rho z\right) \equiv e \sum_{k=0}^{\infty} z^{2k} \sum_{j=1}^{k} \mathcal{B}_j\, C_{k,j}(k)\rho^j. \qquad (4.47)$$

In these results e is Euler's number/Napier's constant or the base of the natural logarithm. Note also that we can set $a = ia$ in (4.46), whereupon we obtain expansions for the oscillatory or trigonometric forms of the argument on the lhs by separating the rhs into real and imaginary parts.

Since the concept of differentiability was introduced when studying Faà di Bruno's formula, let us now turn our attention to the situation where the quotient of the pseudo-composite functions in Theorem 4.1 produces a smooth function. Thus, we arrive at the following corollary.

Corollary 1 to Theorem 4.1. If the functions $f(z)$ and $g(z)$ obey the same conditions as in Theorem 4.1 and the quotient of the pseudo-composite functions $g_a \circ f$ and $h_a \circ f$ represents a smooth function, $r(y)$, then for $p_0 \neq 0$,

$$r^{(k)}(0) = k! \sum_{\substack{\lambda_1, \lambda_2, \lambda_3, \ldots, \lambda_k = 0 \\ \sum_{i=1}^{k} i\lambda_i = k}}^{k, \lfloor k/2 \rfloor, \lfloor k/3 \rfloor, \ldots, 1} a^{N_k} F^{(N_k)}(ap_0) \prod_{i=1}^{k} \frac{p_i^{\lambda_i}}{\lambda_i!}, \qquad (4.48)$$

while for $p_0 = 0$,

$$r^{(k)}(0) = k! \sum_{\substack{\lambda_1, \lambda_2, \lambda_3, \ldots, \lambda_k = 0 \\ \sum_{i=1}^{k} i\lambda_i = k}}^{k, \lfloor k/2 \rfloor, \lfloor k/3 \rfloor, \ldots, 1} q_{N_k} a^{N_k} N_k! \prod_{i=1}^{k} \frac{p_i^{\lambda_i}}{\lambda_i!}. \qquad (4.49)$$

Furthermore, if $r(0) \neq 0$, then inversion of the quotient of pseudo-composite functions yields

$$\left(\frac{1}{r(y)} \right)^{(k)} \Bigg|_{y=0} = k! \frac{E_k(a)}{D_0(a)}, \qquad (4.50)$$

where the coefficients $E_k(a)$ are given by (4.7).

Proof. From the proof of Theorem 4.1, we know that the quotient of $g_a \circ f$ and $h_a \circ f$ is equivalent to a power series expansion in y. That is,

$$\frac{g_a \circ f}{h_a \circ f} \equiv \sum_{k=0}^{\infty} D_k(a) y^k, \qquad (4.51)$$

where the coefficients $D_k(a)$ are given by either (4.4) for $p_0 \neq 0$ or (4.5) for $p_0 = 0$. Because the quotient of the pseudo-composite functions yields a smooth or infinitely differentiable function $r(y)$ as stated in the corollary, it can also be expressed as a Taylor/Maclaurin series given by

$$\frac{g_a \circ f}{h_a \circ f} \equiv \sum_{k=0}^{\infty} r^{(k)}(0) \frac{y^k}{k!}, \qquad (4.52)$$

where the superscript (k) denotes the k-times differentiation of $r(y)$ with respect to y. Since the regularized value is unique as discussed in Appendix A and in Refs. [6,18,19,43,44], the rhs's of the two preceding equivalence statements are equal to one another. Moreover, y is arbitrary in the resulting equation, which means that we can equate like powers of y to one another. Consequently, we obtain (4.48) or (4.49) depending upon the value of p_0.

If the quotient of the pseudo-composite functions is inverted, then it will equal $1/r(y)$. For $r(0) \neq 0$, the function $1/r(y)$ can also be expressed as a Taylor/Maclaurin series since $r(y)$ is smooth. Therefore, we have

$$\frac{h_a \circ f}{g_a \circ f} \equiv \sum_{k=0}^{\infty} \left(\frac{d^k}{dy^k} \frac{1}{r(y)} \right)\Bigg|_{y=0} \frac{y^k}{k!}. \tag{4.53}$$

From Theorem 4.1 we know that the above quotient also represents the regularized value of a power series in y whose coefficients are equal to $E_k(a)/D_0(a)$ with the $E_k(a)$ given by (4.7). Moreover, since $r(0) \neq 0$, $D_0(a)$ does not vanish. As both power series possess the same regularized value, they are equal to one another for the same reason as in the first part of the proof. Moreover, because y is arbitrary, we can again equate like powers of y. Hence we obtain (4.50). This completes the proof of the corollary.

In an interesting twist Theorem 4.1 can also be generalized to the case where the quotient of the pseudo-composite functions is taken to an arbitrary power ρ by a further application of the partition method for a power series expansion as demonstrated by the following corollary.

Corollary 2 to Theorem 4.1. Given the same conditions on the pseudo-composite functions $(g_a \circ f)(y)$ and $(h_a \circ f)(y)$ as in Theorem 4.1, there exists a power series expansion for the quotient of the pseudo-composite functions raised to an arbitrary power, ρ, which is given by

$$\left(\frac{(g_a \circ f)(y)}{(h_a \circ f)(y)} \right)^{\rho} \equiv D_0^{\rho}(a) \left(1 + \sum_{k=1}^{\infty} D_k(a, \rho) y^k \right). \tag{4.54}$$

For $k \geq 1$, the generalization of the coefficients $D_k(a)$ in Theorem 4.1 or $D_k(a, \rho)$ as they are denoted in the above result are given by

$$D_k(a, \rho) = \sum_{\substack{\lambda_1, \lambda_2, \lambda_3, \ldots, \lambda_k = 0 \\ \sum_{i=1}^{k} i\lambda_i = k}}^{k, \lfloor k/2 \rfloor, \lfloor k/3 \rfloor, \ldots, 1} (-\rho)_{N_k} D_0^{-N_k}(a) \prod_{i=1}^{k} \frac{(-D_i(a))^{\lambda_i}}{\lambda_i!}, \tag{4.55}$$

and

$$D_k(a, -\rho) = \sum_{\substack{\lambda_1,\lambda_2,\lambda_3,...,\lambda_k=0 \\ \sum_{i=1}^{k} i\lambda_i=k}}^{k,\lfloor k/2\rfloor,\lfloor k/3\rfloor,...,1} N_k! \, D_0^{-N_k}(a, \rho) \prod_{i=1}^{k} \frac{(-D_i(a, \rho))^{\lambda_i}}{\lambda_i!}, \qquad (4.56)$$

where $(\rho)_{N_k}$ denotes the Pochhammer notation for $\Gamma(N_k + \rho)/\Gamma(\rho)$. In addition, for $\rho = \mu + \nu$, the coefficients satisfy the following recurrence relation:

$$D_k(a, \rho) = \sum_{j=0}^{k} D_j(a, \mu) \, D_{k-j}(a, \nu). \qquad (4.57)$$

Remark 1. By putting $\rho = -1$ in (4.55) we see immediately that the coefficients $D_k(a, -1)$ become the coefficients $E_k(1)$ in Theorem 4.1 given by (4.7), while for $\rho = 1$, we recover (4.1) since the $D_k(a, 1)$ yield $D_k(a)/D_0(a)$. In this case the only non-zero contribution comes from $N_k = 1$, which corresponds to the single part partition, {k} or $\lambda_k = 1$.

Remark 2. For ρ equal to a non-negative integer, j, (4.55) simplifies drastically because $(-\rho)_{N_k}$ vanishes for $N_k > j$, which means in turn that only partitions with j or less parts will contribute, whereas for all other values of ρ, the coefficients $D_k(a, \rho)$ will be affected by all partitions summing to k.

Proof. Since the pseudo-composite functions $(g_a \circ f)(y)$ and $(h_a \circ f)(y)$ are subject to the conditions given in Theorem 4.1, we know that there exists a power series expansion such that

$$\left(\frac{(g_a \circ f)(y)}{(h_a \circ f)(y)}\right)^{\rho} \equiv \left(\sum_{k=0}^{\infty} D_k(a) \, y^k\right)^{\rho}. \qquad (4.58)$$

We also know that there will be a region in the complex plane where the equivalence symbol can be replaced by an equals sign. Separating the zeroth order term in the above result yields

$$\left(\frac{(g_a \circ f)(y)}{(h_a \circ f)(y)}\right)^{\rho} \equiv D_0^{\rho}(a)\left(1 + \sum_{k=1}^{\infty}(D_k(a)/D_0(a)) \, y^k\right)^{\rho}. \qquad (4.59)$$

We can treat the series on the rhs as the variable in the regularized value of the binomial series. Then according to the lemma in Appendix A, we have

$$\left(\frac{(g_a \circ f)(y)}{(h_a \circ f)(y)}\right)^{\rho} \equiv D_0^{\rho}(a) \sum_{k=0}^{\infty} \frac{\Gamma(k-\rho)}{\Gamma(-\rho) k!}\left(-\sum_{j=1}^{\infty}(D_j(a)/D_0(a)) \, y^j\right)^k. \qquad (4.60)$$

The above result is isomorphic to (4.16), which means that we can apply the partition method for a power series expansion again. As stated in the proof to Theorem 4.1, the values that are assigned to each part i in the partitions are given by the coefficients of y^i in the inner series. In this case each part i is assigned a value of $-D_i(a)/D_0(a)$. To evaluate the contribution from each partition, we need to multiply the product of all the assigned values by the modified multinomial factor consisting of $N_k! \prod_{i=1}^{k} 1/\lambda_i!$ multiplied by $\Gamma(N_k - \rho)/N_k! \Gamma(-\rho)$. The latter factor arises from the fact that k in the coefficient of the outer series plays the role of N_k in the partition method for a power series expansion. Therefore, by introducing the Pochhammer notation of $(\rho)_{N_k}$ for $\Gamma(N_k + \rho)/\Gamma(\rho)$, we find that the contribution due to each partition is

$$C\left(\lambda_1, \lambda_2, \ldots, \lambda_k\right) = (-1)^{N_k} (-\rho)_{N_k} D_0^{-N_k}(a) \prod_{i=1}^{k} \frac{D_i(a)^{\lambda_i}}{\lambda_i!}. \qquad (4.61)$$

As expected, for $\rho = -1$, the above result reduces to (4.32). Furthermore, by summing over all partitions summing to k, we obtain the total coefficient in the resulting power series expansion, which is given by (4.55).

As a result of establishing (4.54), we have

$$\left(\frac{(g_a \circ f)(y)}{(h_a \circ f)(y)}\right)^{-\rho} \equiv D_0^{-\rho}(a)\left(1 + \sum_{k=1}^{\infty} D_k(a, -\rho)\, y^k\right). \qquad (4.62)$$

We also know that there is a region in the complex plane where the equivalence symbol can be replaced by an equals sign in (4.54). For these values of y we can invert the equivalence, thereby obtaining

$$\left(\frac{(h_a \circ f)(y)}{(g_a \circ f)(y)}\right)^{\rho} = \frac{D_0^{-\rho}(a)}{\left(1 + \sum_{k=1}^{\infty} (D_k(a, \rho)/D_0(a))\, y^k\right)}. \qquad (4.63)$$

Now Theorem 4.1 can be applied to the above result by treating the denominator in the rhs as the regularized value of the geometric series whose power variable is equal to the series. Therefore, we have $h(z) = 1$, $q_k = (-1)^k$, $a = 1$ and $p_k = D_k(a, \rho)/D_0(a, \rho)$ for $k \geq 0$, while $p_0 = 0$. The coefficients of the resulting power series expansion can again be obtained from (4.5). Moreover, the resulting power series expansion will be equal to the power series expansion on the rhs of (4.62) since both have the same regularized value. Hence by equating like powers in both expansions, we arrive at

$$D_k(a, -\rho) = \sum_{\substack{\lambda_1, \lambda_2, \lambda_3, \dots, \lambda_k=0 \\ \sum_{i=1}^{k} i\lambda_i = k}}^{k, \lfloor k/2 \rfloor, \lfloor k/3 \rfloor, \dots, 1} N_k! \, (-1)^{N_k} \prod_{i=1}^{k} \left(\frac{D_i(a, \rho)}{D_0(a, \rho)} \right)^{\lambda_i} \frac{1}{\lambda_i!}. \tag{4.64}$$

By taking the factor of $(-1)^{N_k}$ inside the product and $1/D_0(a, \rho)$ outside of it, we obtain (4.56).

Since $\rho = \mu + \nu$, we have

$$\left(\frac{(g_a \circ f)(\gamma)}{(h_a \circ f)(\gamma)} \right)^{\rho} = \left(\frac{(g_a \circ f)(\gamma)}{(h_a \circ f)(\gamma)} \right)^{\mu} \left(\frac{(g_a \circ f)(\gamma)}{(h_a \circ f)(\gamma)} \right)^{\nu}. \tag{4.65}$$

We also note that there will be a region of complex plane for γ in (4.54) where the equivalence symbol can be replaced by an equals sign. Therefore, for these values of γ, (4.65) yields

$$\sum_{k=0}^{\infty} D_k(a, \rho) \, \gamma^k = \sum_{k=0}^{\infty} D_k(a, \nu) \, \gamma^k \sum_{k=0}^{\infty} D_k(a, \mu) \, \gamma^k$$

$$= \sum_{k=0}^{\infty} \gamma^k \sum_{j=0}^{k} D_j(a, \nu) D_{k-j}(a, \mu). \tag{4.66}$$

Since there is an infinite number of values of γ for which the above equation holds, we can equate like powers of γ. Consequently, we arrive at (4.57). This completes the proof of the corollary.

Table 4.4 displays the coefficients $D_k(a, \rho)$ up to $k = 6$. As can be seen from the table, these coefficients are considerably more complicated than the $D_k(a)$ of Table 4.2. Hence it is not practical to display them beyond $k = 6$, although they are easily handled by Mathematica when p_i and q_i are substituted by p[k] and q[k]. For $\rho = 1$, all the Pochhammer terms, i.e. the terms $(-\rho + 1)_\ell$ with ℓ ranging from 0 to 5, vanish. Then the coefficients reduce to the $D_k(a)$ of Table 4.2 provided they are multiplied by D_0 or q_0. This factor is necessary because $D_0(a)^\rho$ has been taken out of the bracketed expression on the rhs of (4.59), whereas this is not the case with (4.9). On the other hand, D_0 has been retained here as it enables the various terms in the coefficients to be ordered.

Let us examine how these results can be implemented in Mathematica. First, p_i and q_i must be altered to p2[i] and q2[i]. This is necessary in order to eliminate the confusion by the other, but different, p_i and q_i of Table 4.1. Since we need them too, we shall denote them as p[i] and q[i]. Next we set p2[k_]:=-Dz[k,a]/D, where Dz[k,a] is set equal to the various coefficients

Table 4.4 Coefficients $D_k(a, \rho)$ given by (4.55)

k	$D_k(a, \rho)$
0	1
1	$\rho p_1 q_1 a / D_0$
2	$\frac{\rho a}{2D_0^2}\left(-(-\rho+1)_1 p_1^2 q_1^2 a + 2D_0 p_2 q_1 + 2D_0 p_1^2 q_2 a\right)$
3	$\frac{\rho a}{6D_0^3}\left((-\rho+1)_2 p_1^3 q_1^3 a^2 - 6aD_0(-\rho+1)_1(p_1 p_2 q_1^2 + p_1^3 q_1 q_2 a) + 6D_0^2(p_3 q_1\right.$ $\left. + 2p_1 p_2 q_2 a + p_1^3 q_3 a^2)\right)$
4	$\frac{\rho a}{24D_0^4}\left(-(-\rho+1)_3 p_1^4 q_1^4 a^3 + 12D_0(-\rho+1)_2(p_1^2 p_2 q_1^3 a^2 + p_1^4 q_1^2 q_2 a^3) - 12D_0^2(-\rho+1)_1\right.$ $\times (p_2^2 q_1^2 a + 2p_1 p_3 q_1^2 a + 6p_1^2 p_2 q_1 q_2 a^2 + p_1^4 q_2^2 a^3 + 2p_1^4 q_1 q_3 a^3) + 24D_0^3(p_4 q_1 + p_2^2 q_2 a$ $\left. + 2p_1 p_3 q_2 a + 3p_1^2 p_2 q_3 a^2 + p_1^4 q_4 a^3)\right)$
5	$\frac{\rho a}{5! D_0^5}\left((-\rho+1)_4 p_1^5 q_1^5 a^4 - 20D_0(-\rho+1)_3(p_1^3 p_2 q_1^4 a^3 + p_1^5 q_1^3 q_2 a^4) + 60D_0^2(-\rho+1)_2\right.$ $\times p_1 q_1(p_2 q_1 a + p_1^2 q_2 a^2)^2 + 60D_0^2(-\rho+1)_2 p_1^2 q_1^2(p_3 q_1 a^2 + 2p_1 p_2 q_2 a^3 + p_1^3 q_3 a^4)$ $- 120D_0^3(-\rho+1)_1(p_2 q_1 + p_1^2 q_2 a)(p_3 q_1 a + 2p_1 p_2 q_2 a^2 + p_1^3 q_3 a^3) - 120D_0^3(-\rho+1)_1$ $\times (p_1 p_4 q_1^2 a + p_1 p_2^2 q_1 q_2 a^2 + 2p_1^2 p_3 q_1 q_2 a^2 + 3p_1^3 p_2 q_1 q_3 a^3 + p_1^5 q_1 q_4 a^4) + 120D_0^4$ $\left. \times (p_5 q_1 + 2p_2 p_3 q_2 a + 2p_1 p_4 q_2 a + 3p_1 p_2^2 q_3 a^2 + 3p_1^2 p_3 q_3 a^2 + 4p_1^3 p_2 q_4 a^3 + p_1^5 q_5 a^4)\right)$
6	$\frac{\rho a}{6! D_0^6}\left(-(-\rho+1)_5 p_1^6 q_1^6 a^5 + 30D_0(-\rho+1)_4(p_1^4 p_2 q_1^5 a^4 + p_1^6 q_1^4 q_2 a^5) - 180D_0^2(-\rho+1)_3\right.$ $\times p_1^2 q_1^2 a^3(p_2 q_1 + p_1^2 q_2 a)^2 - 120D_0^2(-\rho+1)_3(p_1^3 p_3 q_1^4 a^3 + 2p_1^4 p_2 q_1^3 q_2 a^4 + p_1^6 q_1^3 q_3 a^5)$ $+ 120D_0^3(-\rho+1)_2(p_2^3 q_1^3 a^2 + 3a^3 p_1^2 p_2^2 q_1^2 q_2 a^3 + 3a^4 p_1^4 p_2 q_1 q_2^2 a^4 + p_1^6 q_2^3 a^5)$ $+ 360D_0^3(-\rho+1)_2(p_1^2 p_4 q_1^3 a^2 + p_1^2 p_2^2 q_1^2 q_2 a^3 + 2p_1^3 p_3 q_1^2 q_2 a^3 + 3p_1^4 p_2 q_1^2 q_3 a^4$ $+ p_1^6 q_1^2 q_4 a^5) + 720D_0^3(-\rho+1)_2(p_1 p_2 q_1^2 a + p_1^3 q_1 q_2 a^2)(p_3 q_1 a + 2p_1 p_2 q_2 a^2 + p_1^3 q_3 a^3)$ $- 360aD_0^4(-\rho+1)_1(p_3 q_1 + 2p_1 p_2 q_2 a + p_1^3 q_3 a^2)^2 - 720D_0^4(-\rho+1)_1(p_2 q_1 a$ $+ p_1^2 q_2 a^2)(p_4 q_1 + p_2^2 q_2 a + 2p_1 p_3 q_2 a + 3p_1^2 p_2 q_3 a^2 + p_1^4 q_4 a^3) - 720D_0^4(-\rho+1)_1$ $\times (p_1 p_5 q_1^2 a + 2p_1 p_2 p_3 q_1 q_2 a^2 + 2p_1^2 p_4 q_1 q_2 a^2 + 3p_1^2 p_2^2 q_1 q_1 q_3 a^3 + 3p_1^3 p_3 q_1 q_3 a^3$ $+ 4p_1^4 p_2 q_1 q_4 a^4 + p_1^6 q_1 q_5 a^5) + 720D_0^5(p_6 q_1 + p_3^2 q_2 a + 2p_2 p_4 q_2 a + 2p_1 p_5 q_2 a$ $+ p_2^3 q_3 a^2 + 3p_1^2 p_4 q_3 a^2 + 6p_1 p_2 p_3 q_3 a^2 + 6p_1^2 p_2^2 q_4 a^3 + 4p_1^3 p_3 q_4 a^3 + 5p_1^4 p_2 q_5 a^4$ $\left. + p_1^6 q_6 a^5)\right)$

listed in Table 4.1, D has replaced D_0 and q2[k_]:= Pochhammer[-ρ,k]/k!. If we wish to determine $D_4(a, \rho)$, then we require the first four coefficients in Table 4.1. Typically, these have to be introduced separately, e.g.,

Dz[2, a_] := Simplify[p[2] q[1] a + p[1]^(2) q[2] a^2],

which is, of course, the second result in Table 4.1. Therefore for $k = 4$, we type in

In[1]:= Dkarho[4],

which generates the following output:

Out[2]:= 1/(24 D^4) a ρ (a^3 (-3 + ρ) (-2 + ρ) (-1 + ρ) p[1]^4 q[1]^4 + 12 a^2 D (-2 + ρ) (-1 +ρ) p[1]^2 q[1]^2 (p[2] q[1] + a p[1]^2 q[2]) + 12 a D^2 (-1 + ρ) (p[2] q[1] + a p[1]^2 q[2])^2 + 24 a D^2 (-1 +ρ) p[1] q[1] (p[3] q[1] + a p[1] (2 p[2] q[2] + a p[1]^2 q[3])) + 24 D^3 (p[4] q[1] + a (p[2]^2 q[2] + 2 p[1] p[3] q[2] + 3 a p[1]^2 p[2] q[3] + a^2 p[1]^4 q[4])))

This result is very general, but let us see if we can recover the secant numbers. From Chapter 1 we know that the coefficients of the inner power series are given by p[k_]:= (-1)^(k + 1)/(2 k)!, while the coefficients of the outer power series are given by q[k_]:=1. By typing Dkarho[4] again with these statements, we find that

In[3] := Dkarho[4]

$$\text{Out}[4] := \frac{1}{40320D^4} \, a\rho\Big(-D^3 + 63aD^2(-1+2D+\rho) - 210a^2D(2 + 6D^2 + 6D(-1+\rho) - 3\rho + \rho^2) + 105a^3(-6+24D^3 + 36D^2(-1+\rho) + 11\rho - 6\rho^2 + \rho^3 + 12D(2-3\rho+\rho^2))\Big).$$

The output is still general since we have not specified values for ρ, D and a. Actually, this result is closer to the generalized secant numbers of Chapter 2 because ρ has not been specified. So, now we create a function F4[a_, ρ_, D_] and assign it to the above output. If we type in Simplify[F4[1,ρ,1], then Mathematica prints $d_{\rho,4}$ in Table 2.4. On the other hand, if we type in F4[1,1,1], then Mathematica prints out a value of 277/8064, which equals the $k = 4$ value of 1385/8! in the third column of Table 1.1.

Although the results in Theorem 4.1 and the corollaries are general, they can nevertheless be used in further applications. Therefore, to complete this chapter we shall consider an important application of the partition method for a power series expansion that requires the $D_k(a, \rho)$ in Corollary 2 to Theorem 4.1. In this application we replace a in the quotient of the pseudo-composite functions by b since there will be another occurrence of a in the analysis. Hence we arrive at the following theorem:

Theorem 4.2. The exponential of the quotient in Corollary 2 to Theorem 4.1 can be expressed via the partition method for a power series expansion as

$$\exp\left(a\left(\frac{(g_b \circ f)(y)}{(h_b \circ f)(y)}\right)^\rho\right) \equiv \exp\left(a(D_0(b))^\rho\right)\left(1 + \sum_{k=1}^{\infty} X_k(a, b, \rho)\, y^k\right), \quad (4.67)$$

where the coefficients $X_k(a, b, \rho)$ are given by

$$X_k(a, b, \rho) = \sum_{\substack{\lambda_1, \lambda_2, \lambda_3, \dots, \lambda_k = 0 \\ \sum_{i=1}^{k} i\lambda_i = k}}^{k, \lfloor k/2 \rfloor, \lfloor k/3 \rfloor, \dots, 1} a^{N_k} (D_0(b))^{\rho N_k} \prod_{i=1}^{k} \frac{(D_i(b, \rho))^{\lambda_i}}{\lambda_i!}. \tag{4.68}$$

Proof. First we introduce (4.54) with a replaced by b into the lhs of (4.67), which gives

$$\exp\left(a\left(\frac{(g_b \circ f)(\gamma)}{(h_b \circ f)(\gamma)} \right)^{\rho} \right) \equiv \exp\left(a(D_0(b))^{\rho} \right) \left(1 \right.$$
$$\left. + \sum_{k=1}^{\infty} \frac{\left(a(D_0(b))^{\rho} \right)^k}{k!} \left(\sum_{j=1}^{\infty} D_j(b, \rho) \, \gamma^j \right)^k \right). \tag{4.69}$$

By applying Theorem 4.1 to the second term on the rhs of (4.69), we observe that the coefficients of the inner series or p_k are equal to $a(D_0(b))^{\rho} D_k(b, \rho)$, where the $D_k(b, \rho)$ are displayed in Table 4.4. On the other hand, the coefficients of the outer series simply become those for the exponential power series, i.e., $q_k = 1/k!$. Alternatively, we can set the coefficients of the inner series equal to $D_k(b, \rho)$ with the coefficients of the outer series now becoming $a^k (D_0(b))^{\rho k}/k!$. Irrespective of either approach a in Theorem 4.1 is equal to unity. Therefore, applying either approach to (4.5) yields

$$X_k(a, b, \rho) = \sum_{\substack{\lambda_1, \lambda_2, \lambda_3, \dots, \lambda_k = 0 \\ \sum_{i=1}^{k} i\lambda_i = k}}^{k, \lfloor k/2 \rfloor, \lfloor k/3 \rfloor, \dots, 1} \prod_{i=1}^{k} \frac{(a D_0(b)^{\rho} D_i(b, \rho))^{\lambda_i}}{\lambda_i!}. \tag{4.70}$$

Since both a and $D_0(b)^{\rho}$ can be taken outside the product, we arrive at the result in Theorem 4.2, viz. (4.68). Table 4.5 presents the first five $X_k(a, b, \rho)$. More were actually generated in Mathematica, but they are simply too cumbersome to transcribe. Because of the number of variables/parameters, it is also difficult to devise a systematic ordering of the terms in the coefficients. Broadly speaking, they have been arranged according to highest powers of b to the lowest powers of b, although there may be powers of b inside the bracketed terms. The terms have also been ordered according to the lowest power a multiplying the power of b appearing first. Thus, the term given as $3ab^2(\rho - 1)_2 D_0(b)^{\rho} p_1^3 q_1^3$ appears after the term $b^2(\rho - 2)_2 p_1^3 q_1^3$, but before the term $a^2 b^2 \rho^2 D_0(b)^{2\rho} p_1^3 q_1^3$ in $X_3(a, b, \rho)$ in the table. Note also that since a can be purely imaginary, we can obtain oscillatory or basic trigonometric functions whose argument is the quotient

Table 4.5 Coefficients $X_k(a, b, \rho)$ given by (4.70)

k	$X_k(a, b, \rho)$
0	1
1	$ab\rho D_0(b)^{\rho-1}p_1q_1$
2	$(1/2)ab\rho D_0(b)^{\rho-2}\Big(b(\rho-1)p_1^2q_1^2 + ab\rho D_0(b)^\rho p_1^2q_1^2 + 2D_0(b)(p_2q_1 + bp_1^2q_2)\Big)$
3	$(1/6)ab\rho D_0(b)^{\rho-3}\Big(b^2(\rho-2)2p_1^3q_1^3 + 3ab^2(\rho-1)2D_0(b)^\rho p_1^3q_1^3$
	$+\, a^2b^2\rho^2 D_0(b)^{2\rho}p_1^3q_1^3 + 6b(\rho-1)D_0(b)p_1q_1(p_2q_1 + bp_1^2q_2)$
	$+\, 6ab\rho D_0(b)^{\rho+1}p_1q_1(p_2q_1 + bp_1^2q_2) + 6D_0(b)^2(p_3q_1 + bp_1(2p_2q_2 + bp_1^2q_3))\Big)$
4	$(1/24)ab\rho D_0(b)^{\rho-4}\Big(b^3(\rho-3)3p_1^4q_1^4 + a^3b^3\rho^3 D_0(b)^{3\rho}p_1^4q_1^4$
	$+\, 12b^2(\rho-2)_2 D_0(b)p_1^2q_1^2(p_2q_1 + bp_1^2q_2)$
	$+\, 6a^2b^2\rho^2 D_0(b)^{2\rho}p_1^2q_1^2(b(\rho-1)p_1^2q_1^2 + 2D_0(b)(p_2q_1 + bp_1^2q_2))$
	$+\, 12b(\rho-1)D_0(b)^2(p_2^2q_1^2 + 2p_1p_3q_1^2 + 6bp_1^2p_2q_1q_2 + b^2p_1^4(q_2^2 + 2q_1q_3))$
	$+\, 3ab\rho D_0(b)^\rho(b(\rho-1)p_1^2q_1^2 + 2D_0(b)(p_2q_1 + bp_1^2q_2))^2$
	$+\, 4ab\rho D_0(b)^\rho p_1q_1(b^2(\rho-2)2p_1^3q_1^3 + 6b(\rho-1)D_0(b)p_1q_1(p_2q_1 + bp_1^2q_2)$
	$+\, 6D_0(b)^2(p_3q_1 + bp_1(2p_2q_2 + bp_1^2q_3))) + 24D_0(b)^3(p_4q_1 + b(p_2^2q_2$
	$+\, 2p_1p_3q_2 + 3bp_1^2p_2q_3 + b^2p_1^4q_4))\Big)$

of the pseudo-composite functions raised to an arbitrary power by inserting *ia* into the results in Table 4.5 and equating the real and imaginary parts of both sides of (4.69). This not only completes the proof of the theorem, but also the general theory behind the partition method for a power series expansion.

CHAPTER 5

Programming the Partition Method for a Power Series Expansion

It has previously been stated that once one goes beyond the tenth order in the partition method for a power series expansion, it is no longer practical to carry out the evaluation of the coefficients by hand. This is because the partition method for a power series expansion relies on evaluating the contribution from each partition and the number of partitions $p(k)$ grows exponentially. In fact, $p(k)$ is $O(\exp(\pi\sqrt{2k/3})/4k\sqrt{3})$ [13,50,55], which represents the leading term in the Hardy-Ramanujan-Rademacher formula [12]. To circumvent this problem, it was suggested that a programming methodology was required, although any computer approach based on partitions will ultimately become very slow. Moreover, since all the partitions summing to the order of each power are involved, such an approach can be regarded as a brute-force approach to deriving power series expansions. Despite this drawback, determining power series expansions up to the 40-th order ($p(40) = 37\,338$) should be achievable in quick time with most number-crunching laptop computers around today. So, at least for intermediate values of k, developing a programming methodology is still of great benefit, particularly in the case of intractable functions where the partition method for a power series expansion is the only method that we have at our disposal.

By a programming methodology, it is meant that we need to analyse the problem by developing algorithms based on modern-day programming techniques, then design programs using appropriate languages and finally implement the programs on one or more platforms. We already have a starting point with the theory behind the method in Chapter 4, specifically Theorem 4.1. Furthermore, by incorporating the BRCP algorithm of Chapter 3 in this methodology, we will have satisfied the issue of developing algorithms based on modern-day programming techniques because of the algorithm's utilization of partition trees.

In regard to very high orders it should be noted that the partition method for a power series expansion does not actually use the partitions

The Partition Method for a Power Series Expansion.
DOI: http://dx.doi.org/10.1016/B978-0-12-804466-7.00005-X

themselves, only the composition or the parts with non-zero multiplicities. Such information can be stored in external arrays which can be called upon when one wishes to derive series expansions for different problems. Thus, there is no need to repeat the process of generating the partitions. In addition, the contributions due to many partitions will often be negligible even by today's computing standards. In those cases the calculation of the coefficients can be simplified yielding extremely accurate approximations. In other cases it is possible to sum the contributions in classes or groups, thereby avoiding the necessity of processing each partition separately. For example, we have seen that summing over partitions with only even integer parts can be reduced to summing over standard integer partitions by dividing each part by 2 and re-assigning their values. Finally, as a result of developing a programming methodology for the partition method for a power series expansion, we will be able to study different problems in the theory of partitions such as the evaluation of partitions with specific parts including those with discrete parts, doubly-restricted partitions and conjugate partitions. As we shall observe, such problems require minor changes to the BRCP algorithm as opposed to devising completely new programs. For example, in order to determine partitions with a fixed number of parts, Knuth presents a different algorithm based on the 18-th century dissertation of C.F. Hindenburg on p. 38 of Ref. [13]. In Chapter 7 this problem is solved simply by inserting a few lines into the BRCP algorithm in a similar manner to when we determined the rank of each partition at the end of Chapter 3.

Now that we have indicated the benefits of developing a programming methodology for the partition method for a power series expansion, we turn to the issue of the programming languages required for the task. The first point to be noted is that if we choose a standard high-level programming language like C/C++ or Fortran, then the values of the coefficients will inevitably become decimal numbers when they could be rational. Moreover, they will invariably be rounded off or worse still, may only yield zero when they are significantly smaller than the precision allowed by the computing system. In addition, the coefficients need not be numerical as exemplified by many of the applications in Chapters 1 and 2. Hence we require a mathematical software package such as Mathematica or Maple either to retain the rationality of the coefficients or to yield symbolic values, when required. However, programming the BRCP algorithm in these packages with its bivariate recursion is also a formidable task. Instead, we could combine the strengths of both C/C++ and Mathematica by writing the initial program in C/C++ whereby the coefficients are generated in a general symbolic form so that they can be transported to Mathematica. Then Mathematica's integer arithmetic routines could be invoked to calculate the coefficients without the round-off errors or its symbolic routines could be used to reduce all the terms generated by the high level code into simple mathematical forms such as polynomials.

Program No. 1 in Appendix B is the C/C++ program called **partmeth**, which prints out the coefficients D_k and E_k given in Theorem 4.1 in symbolic form. The code deals only with the important case of $p_0 = 0$ or (4.4) for the D_k, while the E_k are given by (4.7). The case of $p_0 \neq 0$ or (4.3) is left as an exercise for the reader. If we compare the program with the final program in Chapter 3, then we notice that the overall structure has remained the same. Specifically, there is a main function accompanied by the same two functions **termgen** and **brcp**. In fact, the function **brcp** which incorporates the BRCP algorithm, has not been altered at all, but **termgen** and **main** have been modified so that the symbolic forms for the coefficients are now generated. Besides calculating the execution time, **main** carries out the calculation of the coefficients via one for loop, which is limited by the variable *dim*. This represents the maximum value of k or the coefficient of the highest order term which the user must input. Within the for loop there are two calls to **brcp**, one of which applies to the calculation of the D_k and the other to the calculation of the E_k. Therefore, it is **termgen** that is doing the heavy work in the program. In fact, we shall observe in Chapter 7 that modifying **termgen** repeatedly will enable us to determine many classes of partitions, which, as mentioned previously, often require separate programs.

Within **termgen** we see that the D_k and E_k, which are represented by the variables $DS[k, n]$ and $ES[k, n]$ respectively, are evaluated depending upon the value of the variable *inv_case*. If this variable equals zero, then the D_k are evaluated, while if it equals unity, then the E_k are evaluated. In evaluating the latter there is an extra complication due to the phase factor of $(-1)^{N_k}$ in (4.7). Consequently, for this case **termgen** must also determine the number of distinct parts in each partition. When $dim = 4$, **partmeth** prints out the first four values of the E_k and D_k in symbolic form. For example, the $k = 4$ values printed out by the program are:

$DS[4,n\_] := p[4,n] \, q[1] \, a + p[1,n] \, p[3,n] \, q[2] \, a^\wedge(2) \, 2!$
$+ \, p[1,n]^\wedge(2) \, p[2,n] \, q[3] \, a^\wedge(3) \, 3!/2! + p[1,n]^\wedge(4) \, q[4] \, a^\wedge(4)$
$+ \, p[2,n]^\wedge(2) \, q[2] \, a^\wedge(2)$

ES[4,n_]:= -DS[0,0]^(-2) DS[4,n] + DS[0,0]^(-3) DS[1,n] DS[3,n] 2!
- DS[0,0]^(-4) DS[1,n]^(2) DS[2,n] 3!/2! + DS[0,0]^(-5) DS[1,n]^(4)
+ DS[0,0]^(-3) DS[2,n]^(2)

From these results we see that each coefficient is composed of five distinct terms corresponding to the fact that the number partitions summing to 4, viz. $p(4)$, is equal to five. These results allow for the cases where the coefficients of the inner series represented by $p[k, n]$ may be dependent upon another variable, viz. n, even though it may not be required.

Program **partmeth** is suitable for values up to and around $k = 20$. In fact, all the values of D_k and E_k for $k \leq 20$ are computed within one CPU second on the Sony VAIO laptop mentioned in Chapter 3. For $k \geq 20$, however, the expressions become unwieldy and thus, it is better to evaluate them separately so that each can be introduced directly into Mathematica. This amounts to removing the for loop in **main** and computing only for the value of k or dim. The second program listed in Appendix B, called **mathpm**, also computes the coefficient D_k in symbolic form. In order that the coefficients can be introduced directly in Mathematica, only three terms appear on each line of output with a plus sign as the last character except, of course, on the final line.

Let us now consider the evaluation of D_{30} via **mathpm**. Since $p(30) = 5604$, this will be the number of distinct terms when **mathpm** prints out for DS[30, n]. Even though the output file for DS[30, n] is very large, it can still be imported into Mathematica. Furthermore, if we set $p_k = (-1)^k/(2k + 1)!$, $q_k = 1(-1)^k$ and $a = 1$ by typing into Mathematica,

p[k_]:= (-1)^k /(2 k+1)!
q[k_]:= 1
a:=1 ,

which represent the inner and outer series for the cosecant numbers, then it takes 0.15 CPU sec to evaluate $c[30]$ or c_{30} in integer form on the same Sony VAIO laptop mentioned above. In this instance the numerator is given by a 60 digit integer, while the denominator is given by a 90 digit integer. In decimal notation the value of c_{30} is approximately 2.965×10^{-30}. If we use (1.16) to evaluate c_{30}, then we find that it takes close to zero CPU sec to evaluate the same result. On the other hand, if we set $p_k = (-1)^k/(2k)!$, which is the situation for the secant numbers d_k, then we find that it takes 0.14 CPU sec to evaluate d_{30} in integer form on the same laptop. In this case the numerator and denominator are respectively 67 and 78 digit numbers, while in decimal form d_{30} is approximately equal to 2.176×10^{-12}.

Unfortunately, if (1.30) is implemented in Mathematica, then we only obtain approximate values in decimal form for the secant numbers. Hence we need to implement a recurrence relation such as (1.32) in order to obtain them in integer form. When this is done, it is found that Mathematica takes 6548 CPU sec to compute d_{30} exactly.

If we set $p_k = (-1)^k/(k+1)$, which represents the situation for the reciprocal logarithm numbers, A_k, that were also introduced in Chapter 1, then we find that it takes only 0.1 CPU sec to determine A_{30} in integer form. In this instance the numerator and denominator are 35 and 38 digit numbers yielding an approximate decimal value of 1.474×10^{-3}, the slowest converging of the various sets of numbers considered so far. It was also stated that the reciprocal logarithm numbers can be evaluated by either relating them to the Stirling numbers of the first kind via (1.60) or by the recurrence relation given by (1.55). If (1.60) is introduced into Mathematica, then it takes 0.1 CPU sec to compute A_{30}, while with (1.55) it takes 5719 CPU sec. Therefore, we see that evaluating the coefficients of power series expansions via the partition method for a power series expansion can be vastly superior to using recurrence relations and is almost on a par with the cases where intrinsic forms have already been implemented within the mathematical software package. Furthermore, by simply altering the values for the coefficients of the inner and outer series in addition to changing a, we obtain the coefficients for different power series expansions.

As discussed previously, the coefficients of the inner and outer series do not need to be numbers as in the examples mentioned so far. If we now set $q_k = (\rho)_k/k!$, $p_k = (-1)^{k+1}/(2k+1)!$ and $a = 1$ by typing

```
p[k_]:= (-1)^k / (2 k+1)!
q[k_]:= Pochhammer[ρ,k]/k!
a:=1 ,
```

then $DS[30, n]$ becomes the generalized cosecant number $c_{\rho,30}$ as derived from (2.54). As a result, we find that it only takes 0.36 CPU sec to compute the resulting polynomial, whose degree is 30. By typing in the Simplify routine in Mathematica the polynomial can be arranged in increasing order of powers of ρ within another 0.36 CPU sec. Before this calculation was performed, the results for $DS[5, n]$ and $DS[8, n]$ had been found to agree with the generalized cosecant numbers, $c_{\rho,5}$ and $c_{\rho,8}$, displayed in Table 2.3.

The calculation of the coefficients via the forms generated by either of the first two programs in Appendix B can be continued beyond the thirtieth order, but eventually problems arise due to the combinatorial explosion occurring in the number of partitions. For example, there are $190\,569\,272$ partitions summing to 100, which means that this number of terms will be present in $DS[100, n]$. If **mathpm** is run for this case, then it takes around 600 CPU sec to compute $DS[100, n]$. Whilst this is not an overly long time of computation compared with the earlier results obtained via recurrence relations, it produces a file greater than 16 Gb. Files of this size are going to present a problem when imported into mathematical software packages such as Mathematica. For example, it appears that Mathematica is only able to import files with 2 Gb of data. One method of circumventing this problem would be to divide the file into smaller files so that each could be handled either by a different processor on a supercomputer or studied one by one. Then the results obtained from each file could then be combined to yield the final answer.

Another method of overcoming this problem is to introduce the values for p_k, q_k and a first and then evaluate the coefficient for a specific number of terms via **mathpm**. Once the limit point is reached, the values where this occurs would need to be stored. In terms of Fig. 1.2 this amounts to storing the values of both arguments in the function **brcp**. The partial value of the coefficient could be evaluated and stored, while all the terms printed out in running **mathpm** can be either deleted or overwritten in a re-run of **mathpm**. In the re-run of **mathpm** the code would not print out any terms until **brcp** reaches the values of the arguments of **brcp** stored from the first run. The code would either continue to print out the next limit of terms stopping at two new values of **brcp** or would terminate on reaching the central partition. Then the terms stored in the second run could be calculated and combined with the result obtained from the first run. If the central partition is not reached in the second run, then the process could continue until the central partition is eventually reached. Of course, the disadvantage in this approach is that we have lost the ability to evaluate a new coefficient by altering the p_k, q_k and a as we were able to do with $DS[30, n]$ above. This second method of avoiding very large data files produced by running **mathpm** is contingent on whether we can stop and re-start the program at specific points in the partition trees. This means that the BRCP algorithm needs to be adapted so that specific partitions can be determined, which is the subject of Chapter 7.

CHAPTER 6

Operator Approach

By presenting a programming methodology on the partition method for a power series expansion in the previous chapter, we have effectively by-passed having to calculate the sum over all partitions summing to k when determining the k-th order term in the resulting power series expansion. As a consequence, we can replace the sum by an operator, which will be referred to as the partition operator. This discrete operator is more intricate when compared, for example, with the differential operator or d/dz. Although taking the derivative of a function is viewed as an abstract operation, one has at least an understanding of the process because one can always calculate the limit of Newton's difference quotient provided, of course, it exists. As a result of this understanding, general shorthand rules such as $dz^k/dz = kz^{k-1}$ have evolved. As we shall see in this chapter, such rules/formulas can also be developed via the partition operator approach. Therefore, we shall define the new operator $L_P[\cdot]$ as

$$L_{P,k}[\cdot] := \sum_{\substack{\lambda_1,\lambda_2,\lambda_3,\dots,\lambda_k=0 \\ \sum_{i=1}^k i\lambda_i=k}}^{k,\lfloor k/2 \rfloor, \lfloor k/3 \rfloor,\dots,1} (\cdot). \tag{6.1}$$

The advantage of the above approach is that it will simplify results whenever the sum over partitions appears. It also means that in cases where one is interested in carrying out a summation of a class or subset of partitions, e.g., discrete or distinct partitions, one can define a new operator such as $L_{DP}[\cdot]$. The importance of the latter operator will become evident in later chapters.

Perhaps, the simplest example of the above definition in practice is the case where the partition operator acts on unity. From Chapter 3 we know that this yields the number of partitions summing to k or $p(k)$. Hence using the above definition we can represent this result as $L_{P,k}[1] = p(k)$. We shall return to this result in Chapter 8 when we study in detail the infinite product $P(z)$ given by

$$P(z) = \prod_{j=1}^{\infty} \frac{1}{(1-z^j)}. \tag{6.2}$$

The Partition Method for a Power Series Expansion.
DOI: http://dx.doi.org/10.1016/B978-0-12-804466-7.00006-1

In accordance with the rule in Chapter 1, the power series expansion of $P(z)$ is more commonly referred to as a generating function since its coefficients are equal to the partition-number function $p(k)$.

By putting $p_0 = 0$, $p_k = b^k$, $h(z) = 1$, $a = 1$, $q_k = 1$ and $y = z$, we find that the quotient of the pseudo-composite functions in Theorem 4.1 reduces to

$$g(f(z)) = 1 + \frac{bz}{1 - 2bz} \equiv 1 + \frac{1}{2}\sum_{k=0}^{\infty}(2bz)^{k+1}, \tag{6.3}$$

where according to (4.5)

$$L_{P,k}\left[N_k!\prod_{i=1}^{k}\frac{1}{\lambda_i!}\right] = 2^{k-1}. \tag{6.4}$$

If we set $p_k = (-b)^k$ and $q_k = (-1)^k$ instead, then following the same procedure leading to (6.4) we arrive at

$$L_{P,k}\left[(-1)^{N_k}N_k!\prod_{i=1}^{k}\frac{1}{\lambda_i!}\right] = 0, \tag{6.5}$$

where $k \geq 2$. This is an interesting result in which the partitions with an even number of parts cancel those with an odd number of parts when the multiplicities of the parts in the partitions are involved. Note also that this situation represents the case where all parts are assigned a value of -1.

The above result is not the only instance where the sum over all partitions summing to k vanishes. To observe this, we apply Theorem 4.1 to the function $f(z) = \exp(a\ln(1 + z))$ or $f(z) = (1 + z)^a$. Here, the coefficients of the inner series are given by $p_0 = 0$ and $p_k = (-1)^{k+1}/k$ for $k \geq 1$, while the coefficients of the outer series are given by $q_k = 1/k!$. Then we find that

$$D_k = (-1)^k L_{P,k}\left[(-1)^{N_k}a^{N_k}\prod_{i=1}^{k}\frac{1}{i^{\lambda_i}\lambda_i!}\right]. \tag{6.6}$$

Moreover, from Lemma A.1 in Appendix A, we know that

$$\sum_{k=0}^{\infty}\frac{\Gamma(k-a)}{\Gamma(-a)k!}(-z)^k \equiv (1 + z)^a. \tag{6.7}$$

Since the series is absolutely convergent for $|z| < 1$, we can replace the equivalence symbol by an equals sign. In addition, because the D_k given by (6.6) are the coefficients of the power series expansion in z, they are equal to the coefficients in the above result. Therefore, we end up with

$$L_{P,k}\left[(-a)^{N_k}\prod_{i=1}^{k}\frac{1}{i^{\lambda_i}\lambda_i!}\right] = \frac{\Gamma(k-a)}{\Gamma(-a)k!}. \tag{6.8}$$

It should also be mentioned that the results appearing in Theorem 4.1 are actually more general than (6.8). Consequently, for $p_0 = 0$, they can be expressed in terms of the operator approach as

$$D_k = L_{P,k}\left[q_{N_k}\, a^{N_k}\, N_k!\, D_0^{-N_k} \prod_{i=1}^{k} \frac{p_i^{\lambda_i}}{\lambda_i!}\right],\tag{6.9}$$

and

$$E_k = L_{P,k}\left[(-1)^{N_k} N_k!\, D_0^{-N_k} \prod_{i=1}^{k} \frac{D_i^{\lambda_i}}{i^{\lambda_i}\,\lambda_i!}\right].\tag{6.10}$$

When $a = 1$, we have $f(z) = 1 + z$, which, in turn, means that $D_k = 0$ for $k > 1$. Consequently, from (6.8) we obtain

$$L_{P,k}\left[(-1)^{N_k} \prod_{i=1}^{k} \frac{1}{i^{\lambda_i}\,\lambda_i!}\right] = 0,\tag{6.11}$$

for $k > 1$. On the other hand, when $a = -1$, we have $f(z) = 1/(1 + z)$, which represents the regularized value of the geometric series. Since the coefficients of this series equal $(-1)^k$, (6.8) becomes

$$L_{P,k}\left[\prod_{i=1}^{k} \frac{1}{i^{\lambda_i}\,\lambda_i!}\right] = 1.\tag{6.12}$$

More importantly, the above results can be generalized by letting $a = l$, where l is an arbitrary (positive) integer. Then $f(z) = (1 + z)^l$, whose coefficients, according to the binomial theorem [10], are equal to $\binom{l}{k}$ when $k \le l$ and vanish for the remaining values of l. As a result, (6.8) yields

$$L_{P,k}\left[(-l)^{N_k} \prod_{i=1}^{k} \frac{1}{i^{\lambda_i}\,\lambda_i!}\right] = \begin{cases} 0, & k > l, \\ (-1)^k \binom{l}{k}, & k \le l. \end{cases}\tag{6.13}$$

If $-a$ is replaced by α in (6.8), then it reduces to

$$k!\, L_{P,k}\left[\alpha^{N_k} \prod_{i=1}^{k} \frac{1}{i^{\lambda_i}\,\lambda_i!}\right] = (\alpha)_k.\tag{6.14}$$

From (1.56) we arrive at

$$L_{P,k}\left[\alpha^{N_k} \prod_{i=1}^{k} \frac{1}{i^{\lambda_i}\,\lambda_i!}\right] = \frac{(-1)^k}{k!} \sum_{j=0}^{k} (-1)^j s_k^{(j)} \alpha^j.\tag{6.15}$$

We can equate like powers of α by fixing j and N_k so that they equal each other. This means the partition operator must be altered so that only those parts with j parts in them are considered. Thus, we require a new discrete operator that only deals with a fixed number of parts in the partitions. This is defined as

$$L_{P,k}^{(j)}[\cdot] := \sum_{\substack{\lambda_1,\lambda_2,\lambda_3,\dots,\lambda_k=0 \\ \sum_{i=1}^{k} i\lambda_i=k, \ \sum_{i=1}^{k} \lambda_i=j}}^{k,\lfloor k/2\rfloor,\lfloor k/3\rfloor,\dots,1} . \tag{6.16}$$

It should be noted that the upper limits have been retained even though they may never be attained because of the second constraint. For example, when $j < k$, λ_1 can only equal $j-1$ at the most, while λ_{k+1-j} will be equal to unity in this partition represented as $\{1_{j-1}, k+1-j\}$.

The above operator can be realized by introducing into the **partgen** program of Chapter 3, a global integer variable called *numparts*, which represents the desired number of parts in the partitions, and a local variable called *sumparts*, which sums the parts in each partition scanned in the function **termgen**. When both variables are equal to one another, **termgen** prints out the partition. Otherwise, the partition is discarded. This program will be presented and discussed in more detail in the following chapter. In addition, the new operator is related to the partition operator by

$$L_{P,k}[\cdot] := \sum_{j=1}^{k} L_{P,k}^{(j)}[\cdot] . \tag{6.17}$$

As mentioned in Chapter 3, the number of partitions summing to k with exactly j parts is denoted by $\left|\begin{matrix} k \\ j \end{matrix}\right|$, which means in turn that

$$L_{P,k}^{(j)}[1] = \left|\begin{matrix} k \\ j \end{matrix}\right| , \tag{6.18}$$

while the recurrence relation given by (3.1) can be expressed as

$$L_{P,k}^{(j)}[1] = L_{P,k-1}^{(j-1)}[1] + L_{P,k-j}^{(j)}[1] . \tag{6.19}$$

In Chapter 1 an expression in the form of nested sums was given for the Stirling numbers of the first kind, namely (1.58), which had been derived in Ref. [4] originally. By using this result we were able to derive formulas for $s_k^{(k-\ell)}$ for ℓ ranging from zero to four as given by (1.59). Unfortunately, for higher values of ℓ, (1.58) becomes increasingly cumbersome, whereas this is not the case with (6.15) excluding very large values. This is specially the case if we have at our disposal the above-mentioned program that determines all the partitions with a fixed number of parts in them. To observe this more clearly, let us consider $s_k^{(k-5)}$ or $\ell = 5$. From (6.15) and (6.16) we obtain

$$L_{P,k}^{(j)}\left[\prod_{i=1}^{k}\frac{1}{i^{\lambda_i}\,\lambda_i!}\right] = \frac{(-1)^{j+k}}{k!}\,s_k^{(j)}. \tag{6.20}$$

Note the similarity of this result with (2.9) for the Stirling numbers of the second kind. The major difference occurs in the assigned values to the parts, i in the above result compared with $i!$ for $S(k, j)$. As a consequence, it is easier to calculate the Stirling numbers of the first kind. A simple check of this result is to consider the nine parts summing to 10 with four parts in them. These are: $\{1_3, 7\}$, $\{1_2, 2, 6\}$, $\{1_2, 3, 5\}$, $\{1_2, 4_2\}$, $\{1, 2_2, 5\}$, $\{1, 2, 3, 4\}$, $\{1, 3_3\}$, $\{2_3, 4\}$ and $\{2_2, 3_2\}$. Then the lhs becomes

$$L_{P,k}^{(j)}\left[\prod_{i=1}^{k}\frac{1}{i^{\lambda_i}\,\lambda_i!}\right] = \frac{1}{7 \cdot 3!} + \frac{1}{2 \cdot 6 \cdot 2!} + \frac{1}{3 \cdot 5 \cdot 2!} + \frac{1}{4^2 \cdot 2! \cdot 2!} + \frac{1}{2^2 \cdot 5 \cdot 2!}$$
$$+ \frac{1}{1 \cdot 2 \cdot 3 \cdot 4} + \frac{1}{3^3 \cdot 3!} + \frac{1}{2^3 \cdot 4 \cdot 3!} + \frac{1}{2^2 \cdot 3^2 \cdot 2! \cdot 2!} = \frac{4523}{22680}. \tag{6.21}$$

This is indeed equal to the value on the rhs of (6.20), which can be obtained by using the StirlingS1 routine in Mathematica [32]. For example, one types in

In[1]:= S1[k_, j_] := (-1)^(j + k) StirlingS1[k, j]/k!
In[2]:= S1[10, 4] ,

and the following output is generated

Out[3]:= $\frac{4523}{22680}$.

Thus we have derived a different kind of combinatorial identity for calculating the Stirling numbers of the first kind or $s_k^{(j)}$.

In actual fact (6.20) is homologous to cycles in permutations where permutations are regarded as derangements of a standard order such as the numerical or counting order of positive integers. E.g., 265431 represents a permutation of 123456. Moreover, this permutation is composed of cycles [61] since the derangement of 123456 to 265431 means that 1 is replaced by 2, which in turn is replaced by 6, but 6 is replaced by 1. Therefore adopting the convention that the smallest element appears first in a cycle, the permutation of 265431 possesses the cycle (126). In addition, there is another cycle where 3 is replaced by 5 and 5 by 3, while 4 remains fixed and is a unit cycle. Therefore, 265431 in terms of cycles is represented as (126)(35)(4). A permutation with λ_1 unit cycles, λ_2 two-element cycles etc., is said to be of cycle class $(\lambda_1, \lambda_2, \cdots)$. For example, the permutations 132, 213 and 321 are of cycle class $(1, 1, 0)$, while 231 and 312 are of cycle class $(0, 0, 1)$. The identity permutation 123 is of class $(3, 0, 0)$. If $C(\lambda_1, \lambda_2, \ldots, \lambda_k)$ represents the number of permutations of the k elements of a cycle class such that $k = \sum_{i=1}^{k} i\lambda_i$, then it is given by

$$C(\lambda_1, \lambda_2, \ldots, \lambda_k) = \prod_{i=1}^{k} \frac{k!}{i^{\lambda_i} \lambda_i!} . \tag{6.22}$$

The $k!$ in the numerator comes from the fact that there are $k!$ ways of permuting k elements. On the other hand, the denominator is the product of the total number of duplications of the above cycle class, viz. $\prod_{i=1}^{k} i^{\lambda_i}$, by the number of ways the λ_j j-cycles can be permuted, viz. $\lambda_j!$. Therefore, we see that length of the cycles is analogous to the value of the parts in a partition, while the number of cycles of a specific length in the cycle class is analogous to the multiplicity of the part with that length. Furthermore, according to Corollary 12.1 in Ref. [37], the number of permutations of a finite set of k elements decomposed into j cycles is equal to $|s_k^{(j)}|$. In other words, $C(\lambda_1, \ldots, \lambda_k) = (-1)^{k-j} s_k^{(j)}$, which is identical to (6.20).

In Chapter 1 general formulas were presented for the Stirling numbers of the first kind when j was reasonably close to k. For example, when $j = k - 1$, $s_k^{(k-1)} = -\binom{k}{2}$ according to (1.59). This means that

$$L_{P,k}^{(k-1)} \left[\prod_{i=1}^{k} \frac{1}{i^{\lambda_i} \lambda_i!} \right] = \frac{1}{k!} \binom{k}{2} = \frac{1}{2(k-2)!} . \tag{6.23}$$

For $j = k - 5$, (6.20) reduces to

$$s_k^{(k-5)} = -k! \, L_{P,k}^{(k-5)} \left[\prod_{i=1}^{k} \frac{1}{i^{\lambda_i} \lambda_i!} \right] . \tag{6.24}$$

Here we need to determine all the partitions summing to k with $k - 5$ parts. First, there is the partition with $k - 6$ ones and one six or $\{1_{k-6}, 6\}$. Next come the partitions with $k - 7$ ones and those partitions with two parts (excluding unity) that sum to 7 of which there are two, $\{2, 5\}$ and $\{3, 4\}$. Hence the two partitions with $k - 7$ ones in them are $\{1_{k-7}, 2, 5\}$ and $\{1_{k-7}, 3, 4\}$. With $k - 8$ ones we need to consider all those partitions that sum to 8 with three parts, but with no part equal to unity. In this case there are only two partitions summing to 8 under these conditions, $\{2_2, 4\}$ and $\{2, 3_2\}$. Then there is the partition with $k - 9$ ones with the remaining four parts summing to 9, but with neither part equal to unity. This results

Table 6.1 The polynomials $r_\ell(k)$ in the Stirling numbers of the first kind

ℓ	$r_\ell(k)$
1	1
2	$\frac{1}{4}(3k-1)$
3	$\frac{1}{2}k(k-1)$
4	$\frac{1}{48}(15k^3 - 30k^2 + 5k + 2)$
5	$\frac{1}{16}k(k-1)(3k^2 - 7k - 2)$
6	$\frac{1}{576}(63k^5 - 315k^4 + 315k^3 + 91k^2 - 42k - 16)$
7	$\frac{1}{144}k(k-1)(9k^4 - 54k^3 + 51k^2 + 58k + 16)$
8	$\frac{1}{3840}(135k^7 - 1260k^6 + 3150k^5 - 840k^4 - 2345k^3 - 540k^2 - 66262636k$ $+ 596367504)$
9	$\frac{1}{768}k(k-1)(15k^7 - 180k^6 + 630k^5 - 448k^4 - 665k^3 + 100k^2 + 404k - 144)$
10	$\frac{1}{9216}(99k^9 - 1485k^8 + 6930k^7 - 8778k^6 - 8085k^5 + 8195k^4 + 11792k^3$ $+ 2068k^2 - 2288k - 768)$

in only one partition $\{1_{k-9}, 2_3, 3\}$. Finally, there is the partition with $k - 10$ ones and five parts summing to 10 with neither part equal to unity. Again, there is only one partition $\{1_{k-10}, 2_5\}$. Altogether we have seven partitions with $k - 5$ parts, which corresponds to $p(5)$. Introducing the partitions into (6.24) yields

$$s_k^{(k-5)} = -\binom{k}{6}\left(5! + 132(k-6) + \frac{85}{2}(k-6)(k-7) + 5(k-6)(k-7)(k-8)\right.$$
$$\left. + \frac{3}{16}(k-6)(k-7)(k-8)(k-9)\right) = -\frac{1}{16}\binom{k}{6}\left(3k^4 - 10k^3 + 5k^2 + 2k\right).$$

$$(6.25)$$

As a check, if we set the arguments in the StirlingS1 routine in Mathematica equal to 50 and 45, then we obtain a value of $-17\,392\,967\,051\,250$, which agrees with the $k = 50$ value of (6.25).

From (1.59) and (6.25) we see that the Stirling numbers of the first kind can be represented as $s_k^{(k-\ell)} = (-1)^\ell \binom{k}{\ell+1} r_\ell(k)$, where the $r_\ell(k)$ are polynomials of degree $\ell - 1$. This is a similar situation to the Stirling numbers of the second kind discussed in Chapter 2, which were described in terms of the polynomials $R_\ell(k)$. Table 6.1 displays the polynomials $r_\ell(k)$ up to $\ell = 10$. Unlike the $R_\ell(k)$, the coefficients of the $r_\ell(k)$ do not toggle in sign as the order of k decrements, although the degree of corresponding polynomials is the same. For odd values of ℓ the polynomials possess a common external factor of $k(k-1)$, whereas for the $R_\ell(k)$, we found that the external factor was dependent upon ℓ, viz. $(k-\ell)(k-\ell+1)$. Moreover, the coefficients of the $r_\ell(k)$ are generally not as large in magnitude as their corresponding values in the $R_\ell(k)$.

It should also be mentioned that the results given in (1.59) and (6.25) can be obtained from the results for $S_1(n,k)$ given on p. 192 of Ref. [38]. This quantity is related to the Stirling numbers of the first kind by (14.5) on p. 193 of the same reference, where it is given by

$$S_1(n,k) = \sum_{1 \le j_1 < j_2 < \cdots < j_n \le n} j_1 j_2 \cdots j_k = (-1)^k s_{n+1}^{(n+1-k)} . \tag{6.26}$$

When $j = k - \ell$ in (6.20), we can determine the large k-behaviour of the Stirling numbers of the first kind by adopting the same approach as in Chapter 2 for the Stirling numbers of the second kind. There we found that the leading order term in k was determined by the partition with the least number of ones when the total number of parts was equal to $k - \ell$. This was the partition $\{1_{k-2\ell}, 2_\ell\}$. The contribution from this partition in (6.20) is identical to (2.21) except that is multiplied by $(-1)^\ell$ externally, which is due to the fact that 2! is equal to 2. The next leading order term is determined by the partition $\{1_{k-2\ell+1}, 2_{\ell-2}, 3\}$. The contribution from this partition is given by

$$C_{\{1_{k-2\ell+1}, 2_{\ell-2}, 3\}} = \frac{(-1)^\ell}{3 \cdot 2^{\ell-2}} \frac{k!}{(k-2\ell+1)!(\ell-2)!} = \frac{(-1)^\ell}{3 \cdot 2^{\ell-2}(\ell-2)!} \left(k^{2\ell-1} \right.$$
$$\left. - (2\ell-1)(\ell-1) k^{2\ell-2} + \cdots \right). \tag{6.27}$$

Again, this result is similar to its analogue for the Stirling numbers of the second kind, viz. (2.22). The only difference is that the 3! in the denominator has been replaced by 3 and the external phase factor of $(-1)^\ell$ appears. Therefore, to obtain the coefficient of $k^{2\ell-1}$ in the Stirling numbers of the first kind, we must combine the $k^{2\ell-1}$ term in the above result with the $k^{2\ell-1}$ term in (2.21) after it is multiplied by $(-1)^\ell$.

The $k^{2\ell-2}$ term in the Stirling numbers of the first kind is determined by combining the $k^{2\ell-2}$ terms in the previous contributions with those from the partitions $\{1_{k-2l+2}, 2_{\ell-3}, 4\}$ and $\{1_{k-2l+2}, 2_{\ell-4}, 3_2\}$. The contribution from the first of these partitions is

$$C_{\{1_{k-2\ell+2}, 2_{\ell-3}, 4\}} = \frac{(-1)^\ell}{4 \cdot 2^{\ell-3}} \frac{k!}{(k-2\ell+2)!(\ell-3)!} , \tag{6.28}$$

while that from the second partition is

$$C_{\{1_{k-2\ell+2}, 2_{\ell-4}, 3_2\}} = \frac{(-1)^\ell}{2 \cdot 3^2 \cdot 2^{\ell-4}} \frac{k!}{(k-2\ell+2)!(\ell-4)!} . \tag{6.29}$$

Therefore, the three highest order terms in k for the Stirling numbers of the first kind with the upper subscript equal to $k - \ell$ are found to be

$$s_k^{(k-\ell)} = \frac{(-1)^\ell}{2^\ell \ell!} k^{2\ell} - \frac{(-1)^\ell (2l+1)}{3 \cdot 2^\ell (\ell-1)!} k^{2\ell-1}$$

$$+ \frac{(-1)^\ell (4\ell^2 + 4\ell + 3)}{18 \cdot 2^\ell (\ell-2)!} k^{2\ell-2} + O\left(k^{2\ell-3}\right). \qquad (6.30)$$

As in the case of the Stirling numbers of the second kind, the coefficients of the powers of k are only dependent upon ℓ.

From the recurrence relation for the Stirling numbers of the first kind, viz. (1.57), we obtain

$$(k+1) L_{P,k+1}^{(j)} \left[\prod_{i=1}^{k+1} \frac{1}{i^{\lambda_i} \lambda_i!} \right] = L_{P,k}^{j-1} \left[\prod_{i=1}^{k} \frac{1}{i^{\lambda_i} \lambda_i!} \right] + k L_{P,k}^{j} \left[\prod_{i=1}^{k} \frac{1}{i^{\lambda_i} \lambda_i!} \right]. \qquad (6.31)$$

In addition, specific results for the Stirling numbers of the first kind when j is relatively small are derived in Refs. [3] and [53]. For example, these references show that $s_k^{(2)} = (-1)^k \Gamma(k) H_1(k)$, where $H_1(k) = \sum_{j=1}^{k-1} 1/j$. Therefore, by using (6.20), we find that

$$L_{P,k}^{(2)} \left[\prod_{i=1}^{k} \frac{1}{i^{\lambda_i} \lambda_i!} \right] = \frac{1}{k} H_1(k). \qquad (6.32)$$

Another fundamental result can be obtained by applying Theorem 4.1 to $\exp(-z)$. By writing the function as $1/\exp(z)$ and introducing the power series expansion for $\exp(z)$ in the denominator, we see that the coefficients of the inner series, viz. p_k, are equal to $1/k!$ for $k \geq 1$, while $p_0 = 0$. Moreover, the outer series is now given by the geometric series so that $q_k = (-1)^k$. The coefficients of the power series for $\exp(-z)$ are $(-1)^k/k!$, which are also equal to the D_k obtained via (4.5). Then we have

$$L_{P,k} \left[(-1)^{N_k} N_k! \prod_{i=1}^{k} \frac{1}{(i!)^{\lambda_i} \lambda_i!} \right] = \frac{(-1)^k}{k!}. \qquad (6.33)$$

In the above result we can remove the $i = k$ term in the product since this term equals $-1/k!$. Hence (6.33) reduces to

$$L_{P,k} \left[(-1)^{N_k} N_k! \prod_{i=1}^{k-1} \frac{1}{(i!)^{\lambda_i} \lambda_i!} \right] = \frac{(-1)^k + 1}{k!}. \qquad (6.34)$$

Note in the above result that even though the partition $\{k\}$ has been excluded, the partition operator is still constrained by k. In addition, we see that when k is odd, the rhs vanishes.

We continue with the operator approach by considering the cosecant numbers c_k presented in Chapter 1. These were derived by applying the partition method for a power series expansion to $\csc z$ as in Theorem 1.1. Alternatively, Theorem 4.1 can be applied to (1.5), in which case we set $h(z) = z$, $\gamma = z^2$, and $a = 1$, while $p_k = (-1)^{k+1}/(2k+1)!$, $p_0 = 0$, and $q_k = 1$. The last result arises because the outer series corresponds to the geometric series. Hence we arrive at (1.3). Moreover, expressing (1.4) in terms of the partition operator yields

$$(-1)^k c_k = L_{P,k}\left[N_k! \prod_{i=1}^{k} \left(\frac{-1}{(2i+1)!} \right)^{\lambda_i} \frac{1}{\lambda_i!} \right]. \tag{6.35}$$

This result can also be expressed in terms of even integer values of the Riemann zeta function via (1.16). Then we obtain

$$L_{P,k}\left[(-1)^{N_k} N_k! \prod_{i=1}^{k} \left(\frac{1}{(2i+1)!} \right)^{\lambda_i} \frac{1}{\lambda_i!} \right] = 2(-1)^k \left(1 - 2^{1-2k} \right) \frac{\zeta(2k)}{\pi^{2k}}. \tag{6.36}$$

Therefore, altering $i!$ to $(2i+1)!$ in the argument of the partition operator changes the result dramatically, resulting in the emergence of even integer values of the Riemann zeta function, i.e. $\zeta(2k)$, divided by π^{2k}.

We now invert the power series expansion given by (1.3) and apply Theorem 4.1 again. In this case $p_k = -c_k$ with $p_0 = 0$, while the q_k remain equal to the coefficients of the geometric series. As expected, the resulting power series expansion becomes the standard Taylor/Maclaurin power series for $\sin(z)/z$. That is, we obtain $\sum_{k=0}^{\infty}(-1)^k z^{2k}/(2k+1)!$, which is convergent for all values of z, despite the fact that (1.3) has a radius of absolute convergence of π. This corroborates the earlier remark stating that the resulting power series expansion obtained via Theorem 4.1 can turn out to be convergent everywhere even though the inner or the outer series may be divergent in a sector of the complex plane. Therefore, from (4.5) we arrive at

$$L_{P,k}\left[N_k! \prod_{i=1}^{k} (-c_i)^{\lambda_i} \frac{1}{\lambda_i!} \right] = \frac{(-1)^k}{(2k+1)!}. \tag{6.37}$$

Alternatively, we could have used (4.7). In this instance the $D_i(a)$ are given by the lhs of (6.35), i.e., $D_i(a) = -c_i$. Introducing this result in (4.7) yields (6.37) when the power series for $\sin(z)/z$ replaces the rhs of (4.6).

As stated in Chapter 1, several recurrence relations for the cosecant numbers are derived in Ref. [5], two of which have been given as (1.21) and (1.22). If we introduce (6.35) into (1.21), then we find that

$$\sum_{j=0}^{k-1} \frac{1}{(2k-2j+1)!} \, L_{P,j} \left[(-1)^{N_j} N_j! \prod_{i=1}^{j} \left(\frac{1}{(2i+1)!} \right)^{\lambda_i} \frac{1}{\lambda_i!} \right]$$

$$= -L_{P,k} \left[(-1)^{N_k} N_k! \prod_{i=1}^{k} \left(\frac{1}{(2i+1)!} \right)^{\lambda_i} \frac{1}{\lambda_i!} \right], \tag{6.38}$$

while if it is introduced into (1.22), we obtain

$$\sum_{j=0}^{k-1} \frac{1}{(2k-2j)!} \, L_{P,j} \left[(-1)^{N_j} N_j! \prod_{i=1}^{j} \left(\frac{1}{(2i+1)!} \right)^{\lambda_i} \frac{1}{\lambda_i!} \right]$$

$$= -\left(\frac{2-2^{1-2k}}{1-2^{1-2k}} \right) L_{P,k} \left[(-1)^{N_k} N_k! \prod_{i=1}^{k} \left(\frac{1}{(2i+1)!} \right)^{\lambda_i} \frac{1}{\lambda_i!} \right]. \tag{6.39}$$

Thus we see that for specific arguments the partition operator can be represented by a sum of the partition operators where the partitions sum to j with j ranging from zero to $k-1$. Similar results can also be obtained by using the other recurrence relations in Ref. [5]. Furthermore, there are instances where the partition operator on being applied to one set of partitions can equal the partition operator being applied to another of set of partitions. For example, we have already seen from (1.83) that partitions summing to $2k$ can be related to partitions summing to k. If we apply the operator approach to this result, then we arrive at

$$L_{P,2k} \left[(-1)^{N_{2k}} N_{2k}! \prod_{i=1}^{2k} \left(\frac{1}{(i+1)!} \right)^{\lambda_i} \frac{1}{\lambda_i!} \right]$$

$$= \frac{1}{(2-2^{2k})} L_{P,k} \left[(-1)^{N_k} N_k! \prod_{i=1}^{k} \left(\frac{1}{(2i+1)!} \right)^{\lambda_i} \frac{1}{\lambda_i!} \right]. \tag{6.40}$$

Consequently, we do not need to consider all the partitions up to $2k$ when calculating the sum on the lhs, which is a significant reduction in computational effort.

If we set $y = z^2$, $p_0 = 0$, $a = 1$ and $q_k = 1$ in Theorem 4.1 as in the case of the cosecant numbers, but now, set $p_k = (-1)^{k+1}/(2k)!$, then we obtain the secant numbers as given by (1.24). Expressing them in terms of the partition operator gives

$$(-1)^k d_k = L_{P,k} \left[N_k! \prod_{i=1}^{k} \left(-\frac{1}{(2i)!} \right)^{\lambda_i} \frac{1}{\lambda_i!} \right]. \tag{6.41}$$

To obtain the result for the inverted function, we apply Theorem 4.1 with the D_i set equal to the secant numbers, d_i in (4.7), while the lhs, which is now equal to $\cos z$, is replaced by its Taylor/Maclaurin series. Hence the coefficients E_k are simply equal to $(-1)^k/(2k)!$, thereby yielding

$$\frac{(-1)^k}{(2k)!} = L_{P,k}\left[N_k!\prod_{i=1}^{k}(-d_i)^{\lambda_i}\frac{1}{\lambda_i!}\right]. \tag{6.42}$$

Again, we observe the interplay between the arguments of (6.41) and (6.42) as occurred with (6.35) and (6.37) for the cosecant numbers, which appears to be, if not intriguing, at least unexpected behaviour occurring with such sums over partitions.

Like the cosecant numbers, the secant numbers were found to obey recurrence relations, although not as many as their cosecant counterparts [5]. In fact, the analogue of (1.21) has already been given as (1.32). By introducing (6.41) into this result, we find that

$$\sum_{j=0}^{k-1}\frac{1}{(2k-2j)!}L_{P,j}\left[(-1)^{N_j}N_j!\prod_{i=1}^{j}\left(\frac{1}{(2i)!}\right)^{\lambda_i}\frac{1}{\lambda_i!}\right]$$
$$= -L_{P,k}\left[(-1)^{N_k}N_k!\prod_{i=1}^{k}\left(\frac{1}{(2i)!}\right)^{\lambda_i}\frac{1}{\lambda_i!}\right] \tag{6.43}$$

which is virtually identical to (6.38) except that there are now no "+1's" in the denominators of the products. The similarity between (6.38) and (6.43) is perhaps an indication that something profound is occurring with particular arguments inside the partition operator.

More sophisticated results involving both the secant and cosecant numbers can also be derived. From No. 1.518(2) of Ref. [14] we have

$$\ln\sec(\pi z) = \sum_{k=1}^{\infty}\frac{(2^{2k}-1)}{k(2k)!}2^{2k-1}|B_{2k}|(\pi z)^{2k}, \tag{6.44}$$

which is absolutely convergent for $|z| < 1/2$. With the aid of (1.25) we can write the lhs of (6.44) as

$$\ln\sec(\pi z) = \ln\left(1 + \sum_{k=1}^{\infty}d_k(\pi z)^{2k}\right). \tag{6.45}$$

We now apply Theorem 4.1 to the rhs. This means that the logarithm is expanded in terms of its Taylor/Maclaurin series, in which case $q_k = (-1)^{k+1}/k$, while $y = z^2$ and $p_k = d_k\pi^{2k}$. Then we equate the resulting power series expansion to like powers of z^2 or y in (6.45). By expressing the Bernoulli numbers in terms of the cosecant numbers via (1.15), we arrive at

$$L_{P,k}\left[(-1)^{N_k}(N_k-1)!\prod_{i=1}^{k}\frac{d_i^{\lambda_i}}{\lambda_i!}\right]=\frac{1}{2k}\frac{(1-2^{2k})}{(1-2^{1-2k})}c_k. \tag{6.46}$$

Once again, we have another strange argument inside the partition operator yielding an interesting finite quantity for all values of k. In principle, this means that tables of such results involving the partition operator can be developed much like tables of integrals or combinatorial identities.

In order for the reader to become familiar with results like (6.46), let us check it for $k=3$. To evaluate the lhs, we require d_1, d_2 and d_3, which from Table 1.1 are equal to $1/2$, $5/24$ and $61/720$, respectively, while for the rhs, we require c_3, which equals $31/3\cdot 7!$ according to the same table. There are three partitions summing to 3, viz. $\{1_3\}$, $\{1,2\}$ and $\{3\}$, whose values of N_3 are 3, 2 and 1, respectively. Hence the lhs of (6.46) becomes

$$L_{P,3}\left[(-1)^{N_3}(N_3-1)!\prod_{i=1}^{3}\frac{d_i^{\lambda_i}}{\lambda_i!}\right]=-(d_1^3/3-d_1d_2+d_3)=-\frac{1}{45}. \tag{6.47}$$

For $k=3$ the rhs yields

$$\frac{1}{2k}\frac{(1-2^{2k})}{(1-2^{1-2k})}c_k=\frac{1}{6}\frac{(-63)}{(1-1/32)}\frac{31}{3\cdot 7!}=-\frac{1}{45}, \tag{6.48}$$

which, as expected, agrees with (6.47).

In Chapter 1 the partition method for a power series expansion was applied to the reciprocal of the logarithmic function, viz. $1/\ln(1+z)$, which produced a power series expansion with special coefficients referred to as the reciprocal logarithm numbers or A_k. These fractions, which oscillate in sign, can be obtained via Theorem 4.1 by setting $f(z)=\ln(1+z)$ and $g(z)=1/z$. Hence the coefficients of the inner series, viz. p_k, are equal to $(-1)^{k+1}/(k+1)$ for $k>0$ and $p_0=0$. The resulting denominator can be regarded as the regularized value of the geometric series, which means that the q_k and a are again equal to $(-1)^k$ and unity, respectively. Hence we obtain (1.42), which in terms of the operator approach becomes

$$(-1)^k A_k=L_{P,k}\left[N_k!\prod_{i=1}^{k}\left(-\frac{1}{i+1}\right)^{\lambda_i}\frac{1}{\lambda_i!}\right]. \tag{6.49}$$

The inverse of this result is obtained by putting $D_0=1$ and $D_k=A_k$ in (4.7), while the E_k in Theorem 4.1 equal the coefficients in the Taylor/Maclaurin series for $\ln(1+z)$, namely $(-1)^k/(k+1)$. Thus, we find that

$$\frac{(-1)^k}{k+1} = L_{P,k}\left[N_k!\prod_{i=1}^{k}\frac{(-A_i)^{\lambda_i}}{\lambda_i!}\right]. \tag{6.50}$$

With the aid of Corollary 2 to Theorem 4.1 we can generalize the preceding examples to situations where the generating functions are raised to an arbitrary power ρ. For example, the quotient in (6.3) raised to an arbitrary power ρ becomes

$$g(f(z))^\rho = \left(\frac{1-bz}{1-2bz}\right)^\rho \equiv \sum_{k=0}^{\infty}(bz)^k\sum_{j=0}^{k}\frac{\Gamma(j-\rho)}{\Gamma(-\rho)j!}\frac{2^{k-j}\Gamma(k-j+\rho)}{\Gamma(\rho)(k-j)!}, \tag{6.51}$$

where we have used the regularized value for the binomial series given in Lemma A.1 of Appendix A. Alternatively, we can express the lhs of the above result as

$$g(f(z))^\rho \equiv \left(1+\frac{1}{2}\sum_{k=1}^{\infty}(2bz)^k\right)^\rho. \tag{6.52}$$

From (4.54), we see that $D_0 = 1$ and $D_k = 2^{k-1}b^k$ for $k \geq 1$, while $\gamma = z$ and $a = 1$. Hence according to (4.55), the coefficients of the resulting power series expansion become

$$D_k(\rho) = (2b)^k L_{P,k}\left[(-1/2)^{N_k}(-\rho)_{N_k}\prod_{i=1}^{k}\frac{1}{\lambda_i!}\right]. \tag{6.53}$$

Equating like powers of z on the rhs's of the preceding equivalence statements, viz. (6.51) and (6.52), gives the following identity:

$$L_{P,k}\left[(-1/2)^{N_k}(-\rho)_{N_k}\prod_{i=1}^{k}\frac{1}{\lambda_i!}\right] = \sum_{j=0}^{k}\frac{\Gamma(j-\rho)}{\Gamma(-\rho)j!}\frac{\Gamma(k-j+\rho)}{2^j\Gamma(\rho)(k-j)!}. \tag{6.54}$$

The rhs can be simplified by introducing the reflection formula for the gamma function [16], which is

$$\Gamma(s)\Gamma(1-s) = \frac{\pi}{\sin(\pi s)}. \tag{6.55}$$

Then we find that

$$L_{P,k}\left[\left(-\frac{1}{2}\right)^{N_k}(-\rho)_{N_k}\prod_{i=1}^{k}\frac{1}{\lambda_i!}\right] = \rho\sum_{j=0}^{k}\frac{\Gamma(k-j+\rho)}{\Gamma(1-j+\rho)j!(k-j)!}\left(-\frac{1}{2}\right)^j. \tag{6.56}$$

As a check, let us consider $k = 5$ in the above result. Hence the lhs yields

$$L_{P,5}\left[\left(-\frac{1}{2}\right)^{N_5}(-\rho)_{N_5}\prod_{i=1}^{5}\frac{1}{\lambda_i!}\right] = \left(-\frac{1}{2}\right)^5\frac{(-\rho)_5}{5!} + \left(-\frac{1}{2}\right)^4\frac{(-\rho)_4}{3!}$$

$$+ 2\left(-\frac{1}{2}\right)^3\frac{(-\rho)_3}{2!} + 2\left(-\frac{1}{2}\right)^2(-\rho)_2 + \left(-\frac{1}{2}\right)(-\rho)_1. \tag{6.57}$$

This is actually an unusual result since the contributions from the partitions $\{1, 2_2\}$ and $\{1_2, 3\}$ are identical as are the contributions from the partitions $\{2, 3\}$ and $\{1, 4\}$. Previously, the contributions from the partitions were always distinct. Consequently, a factor of 2 appears twice in (6.57) and the number of distinct terms becomes five instead of $p(5) = 7$. If the above result is introduced into Mathematica [32], then by using the Expand routine one obtains

$$L_{P,5}\left[\left(-\frac{1}{2}\right)^{N_5}(-\rho)_{N_5}\prod_{i=1}^{5}\frac{1}{\lambda_i!}\right] = \frac{31}{160}\rho + \frac{29}{128}\rho^2$$

$$+ \frac{55}{768}\rho^3 + \frac{1}{128}\rho^4 + \frac{1}{3840}\rho^5. \tag{6.58}$$

On the other hand, the rhs of (6.56) can be implemented as a one-line statement in Mathematica for all values of ρ and k as follows:

PSum[ρ_, k_] := Expand[FullSimplify[ρ Sum[Gamma[k-j+ρ]]
(-1/2)^j/(j! (k - j)! Gamma[1-j + ρ]), {j, 0, k}]]]

If k is set equal to 5 and PSum[ρ,5] is entered into Mathematica, then one obtains the same result given by (6.58).

We have already observed that taking (1.3) to the arbitrary power of ρ resulted in another power series expansion whose coefficients were referred to as the generalized cosecant numbers and were studied extensively in Chapter 2. They too can be derived by setting D_k equal to the cosecant numbers c_k in Theorem 4.1. If we denote $D_k(\rho)$ by $c_{\rho,k}$ in Corollary 2 to Theorem 4.1, the generalized cosecant numbers can be expressed as

$$c_{\rho,k} = L_{P,k}\left[(-1)^{N_k}(-\rho)_{N_k}\prod_{i=1}^{k}\frac{c_i^{\lambda_i}}{\lambda_i!}\right] = L_{P,k}\left[(\rho+1-N_k)_{N_k}\prod_{i=1}^{k}\frac{c_i^{\lambda_i}}{\lambda_i!}\right], \tag{6.59}$$

where the identity $(-\rho)_N = (-1)^N(\rho+1-N)_N$ has been used to obtain the second version of the above result. Moreover, the lhs of the power series expansion or generating function can be written in alternate form as

$$s^\rho \csc^\rho s = \left(1 + \sum_{k=1}^{\infty} \frac{(-1)^k x^{2k+1}}{(2k+1)!}\right)^{-\rho}. \tag{6.60}$$

Now the $D_k(a)$ on the rhs of (4.55) are equal to $(-1)^k/(2k+1)!$, while ρ has changed sign. Hence the generalized cosecant numbers can also be expressed as

$$c_{\rho,k} = (-1)^k L_{P,k}\left[(-1)^{N_k}(\rho)_{N_k}\prod_{i=1}^{k}\left(\frac{1}{(2i+1)!}\right)^{\lambda_i}\frac{1}{\lambda_i!}\right]. \tag{6.61}$$

Thus, we observe that when the argument of the partition operator in (6.59) is altered so that ρ becomes $-\rho$ and the assigned values of the parts become $(-1)^i/(2i+1)!$ instead of c_i, we obtain virtually identical results. The above form for the generalized cosecant numbers has been used to derive the closed form solutions of a finite sum of different inverse powers of cosines as explained in Ref. [6].

In a similar manner we can apply the operator approach to the generalized secant numbers. By taking the ρ-th power of (1.25) and setting $D_i(a)$ in (4.55) to d_i, we find that the generalized secant numbers can be expressed in operator form as

$$d_{\rho,k} = L_{P,k}\left[(-1)^{N_k}(-\rho)_{N_k}\prod_{i=1}^{k}\frac{d_i^{\lambda_i}}{\lambda_i!}\right]. \tag{6.62}$$

Alternatively, we can set $D_k(a)$ and ρ equal to $(-1)^i/(2i)!$ and $-\rho$, respectively or use (2.114) directly. Consequently, we arrive at

$$d_{\rho,k} = (-1)^k L_{P,k}\left[(-1)^{N_k}(\rho)_{N_k}\prod_{i=1}^{k}\left(\frac{1}{(2i)!}\right)^{\lambda_i}\frac{1}{\lambda_i!}\right]. \tag{6.63}$$

Similarly, the operator approach can be applied to the generalized reciprocal logarithm numbers. In this instance the $D_i(a)$ in (4.55) are set equal to A_i. Therefore, we obtain

$$A_k(\rho) = L_{P,k}\left[(-1)^{N_k}(-\rho)_{N_k}\prod_{i=1}^{k}\frac{A_i^{\lambda_i}}{\lambda_i!}\right]. \tag{6.64}$$

Alternatively, they can be determined by taking the $-\rho$-th power of the Taylor/Maclaurin series for $\ln(1 + z)$ and setting the $D_i(a)$ in (4.55) to $(-1)^i/(i+1)$, which yields (2.141) in Theorem 2.3. In terms of the partition operator, the generalized reciprocal logarithm numbers become

$$A_k(\rho) = (-1)^k L_{P,k}\left[(-1)^{N_k}(\rho)_{N_k}\prod_{i=1}^{k}\left(\frac{1}{i+1}\right)^{\lambda_i}\frac{1}{\lambda_i!}\right]. \qquad (6.65)$$

At the end of Chapter 3 an alternative interpretation was presented for generating partitions via partition trees, which results in a different formulation of the partition method for a power series expansion. Before introducing this approach, we need to create a new operator by amending the partition operator $L_{P,k}$ so that subtrees can be determined. For example, the entire subtree emanating from $\{1,5\}$ in Fig. 1.2 of Chapter 1 represents the partition tree for all partitions summing to 5. To obtain the partitions summing to 6 from this subtree, we need to increment λ_1 in all the partitions summing to 5 by one. This will, of course, affect the contributions to the coefficients. In addition, the partitions emanating from $\{2,4\}$ represent all the partitions summing to 4 excluding those with unity in them. To obtain the partitions summing to 6, we need to include an extra two, thereby incrementing λ_2 by unity. The partitions emanating from $\{3_2\}$ in Fig. 1.1 are the partitions summing to 3 with no ones and twos. In this case, λ_3 is incremented to yield partitions summing to 6. Because partitions are being excluded in the subtrees, we require the following definition:

$$L_{RP,k,i}\left[\cdot\right] := \sum_{\substack{\lambda_i=1,\lambda_{i+1},\ldots,\lambda_{k-i}=0 \\ \sum_{j=i}^{k-i} j\lambda_j=k}}^{1+\lfloor(k-i)/i\rfloor,\lfloor(k-i)/(i+1)\rfloor,\ldots,1}. \qquad (6.66)$$

We shall refer to the above definition as the restricted partition (RP) operator since it excludes all partitions summing to k with at least one part less than i or it only considers those partitions where the parts are greater than or equal to i. Note also that the operator only makes sense if $(k-i) \geq i$.

The major differences between the above operator and the partition operator given by (6.1) are: (1) the multiple sum begins at λ_i rather than at λ_1 due to the exclusion of parts less than i in the subtree emanating from $\{i, k-i\}$, (2) λ_i begins with unity rather than zero as for the other parts, which accounts for the fact that the part i appears at least once in each partition of the subtree and (3) the maximum part in the partitions of the subtrees becomes $k-i$, thereby altering the upper limit of each sum. On the other hand, whilst the number of sums has reduced, the value to which the partitions are summed is still equal to k.

With the above definition we can now implement the algorithm described at the end of Chapter 3. As a consequence, (4.5) can be expressed in operator form alternatively as

$$L_{P,k}\left[q_{N_k}\, a^{N_k}\, N_k! \prod_{i=1}^{k} \frac{p_i^{\lambda_i}}{\lambda_i!}\right] = q_1\, a\, p_k$$

$$+ \sum_{j=1}^{\lfloor k/2 \rfloor} L_{RP,k,j}\left[q_{N_{j,k-j}}\, a^{N_{j,k-j}}\, N_{j,k-j}! \prod_{i=j}^{k-j} \frac{p_i^{\lambda_i}}{\lambda_i!}\right], \qquad (6.67)$$

where $N_{i,l} = \sum_{j=i}^{l} \lambda_j$. For the specific case of $i=1$ and $l=k$, we put $N_{1,k} = N_k$ as it has been used throughout this book. The first term on the rhs corresponds to the contribution due to the partition $\{k\}$, which is excluded in the sum containing the restricted partition operators. When $p_0 \neq 0$ in Theorem 4.1, we must use (4.4) instead, which amounts to replacing $q_1 a p_k$ and $q_{N_{j,k-j}}\, a^{N_{j,k-j}}\, N_{j,k-j}!$ in the above result by $a\, F^{(k)}(ap_0)$ and $a^{N_{j,k-j}}\, F^{N_{j,k-j}}(ap_0)$ respectively.

The above algorithm means that the calculation of all the contributions due to the partitions summing to k can be divided into smaller blocks of calculations. This is particularly important for the calculation of the coefficients at very high orders in power series expansions where combinatorial explosion becomes an issue. For example, if one wishes to determine the coefficient of the two-hundredth order term, then one must calculate $p(200)$ or 3972999029388 contributions. Clearly, to embark on such a calculation, one needs to divide the calculation into smaller blocks. On systems with limited memory, one could calculate the contribution from one block first and store the final result in an array. Because the calculations are now redundant, they can be deleted and one can proceed to the second block. The result from the second block could then be added to the stored result from the first block. Then the calculations due to the second block could be deleted and the third block considered. This process would continue until the final block was reached. It may, however, become time-consuming in which case the calculations could be sped up by getting different processors on a supercomputer to evaluate each block separately.

CHAPTER 7

Classes of Partitions

In the previous chapter the partition operator $L_{P,k}[\cdot]$ was introduced in order to replace the sum over all partitions summing to k appearing in the coefficients obtained by the partition method for a power series expansion. Later, it became necessary to introduce another discrete operator $L_{P,k}^{(j)}$, which only calculated the contributions from those partitions for a fixed number of parts j. This operator was required, for example, when we wished to calculate the Stirling numbers of the first kind via the partition method as given by (6.20). Thus, this was an example where a class of the partitions summing to k was required instead of all the partitions. Consequently, we see that there are problems where the partition operator, which has been programmed in Chapter 5, will need to be modified so that only classes or particular types of partitions are needed in calculations. Another example appearing in the next chapter is the inverted form of $P(z)$ whose generating function coefficients are the partition–number function. There we shall encounter the discrete partition operator $L_{DP,k}[\cdot]$, which only deals with those partitions possessing discrete or distinct parts. At this stage, however, we cannot introduce such an operator because we do not have a method of determining this type or class of partition. Nevertheless, it should be possible because the partition operator deals with all the partitions summing to k, while the operators mentioned here refer to a subset of them. Moreover, since the BRCP algorithm considers all the partitions summing to k, it follows that all we need to do is modify the algorithm so that it only accesses the class of partitions of interest to us. Therefore, the aim of this chapter is to investigate how the BRCP algorithm can be modified to solve problems involving specific classes of partitions. In particular, we shall consider the following problems: (1) partitions with a fixed number of parts, (2) doubly-restricted partitions, (3) discrete partitions, (4) perfect partitions, (5) conjugate partitions and (6) partitions with specific parts in them. All these problems are well-known within the theory of partitions [12], but in the past, solving them has invariably required the creation of different algorithms or programs for each problem as can be seen by Refs. [13], [50] and [54]. Here we shall see that these problems and others can be solved with relatively minor modifications to the BRCP algorithm, once again highlighting its versatility. As we shall be modifying

The Partition Method for a Power Series Expansion.
DOI: http://dx.doi.org/10.1016/B978-0-12-804466-7.00007-3

the program **partgen** of Chapter 3, we shall still generate the partitions in the compact multiplicity representation, although it should be mentioned that the various algorithms presented in this section only require minor modification to the function **termgen** in order to generate them in the standard or lexicographic representation.

Of all the problems mentioned above, perhaps the simplest one to consider is the determination of those partitions with a specific part(s). For example, suppose our aim is to determine all those partitions summing to 15 with a five in them. From the discussion at the beginning of Chapter 3, we know that the total number of partitions for this problem is equal to the number of partitions summing to 10 or $p(10)$. Furthermore, the partitions generated by a code solving this problem should be the same as those generated by running the various codes in Chapter 3 except that each partition generated by the new code will now print out an extra five. We shall see that this is indeed the case, although the order in which the partitions are printed out will be different.

In order to modify **partgen** in Chapter 3 so that it generates only those partitions with a specific part, first we need to alter the function **main**. This is necessary in order to enable users to input the specific value of the part that they wish to appear in the partitions generated by the new program. Consequently, the function **main** becomes

```c
int main ( int argc , char *argv [] )
{
int i;
if (argc !=3) printf ("usage: specpart <partition sum>
     <sp_val> \n");
else {
        tot=atoi (argv [1]);
        sp_val=atoi (argv [2]);
        part =(int *) malloc (tot*sizeof (int ));
        if (part == NULL) printf ("unable to allocate
        array \n");
        else {
               for (i =0;i<tot ;i++) part [i]=0;
               brcp (tot ,1);
               free (part );
            }

    }
printf ("\n");
return (0);
}
```

Here we see that the modified program called **specpart** has a global variable called *sp_val*, which represents the specific part that will appear at least once in each partition generated by the program.

As discussed in Chapter 3, the function **brcp(tot,1)** scans over all the partitions summing to *tot*, while the function **termgen**, which is called in **brcp**, is responsible for generating or printing out partitions. Therefore, in order to determine those partitions with a specific part or parts in them, we need to modify **termgen** because we still need to scan over all partitions when calling **brcp(tot,1)**. In fact, aside from making minor modifications to **main**, we shall observe that to solve all the problems mentioned above, we only need to modify **termgen**, which basically acts as a filter.

In Chapter 3 **termgen** was responsible for printing out all the partitions summing to *tot* in the multiplicity representation. This was achieved by processing the array *part*, which stored the multiplicities of the parts in each partition. Thus, the variable *freq* was used to represent the multiplicity of the part $i + 1$ in the partition with i ranging from 0 to *tot* -1. Now that we wish to determine those partitions with a specific part(s) in them, we need to restrict the partitions that are printed out by **termgen**. This is accomplished simply by introducing a local variable called *freq_spval*, which evaluates the multiplicity of *sp_val* in each partition. If this value is non-zero, then we know that there is at least one occurrence of *sp_val* in the partition and the partition can then be printed out in the same manner as in Chapter 3. If *freq_spval* is zero, then the partition is ignored. Therefore, **termgen** for program **specpart** becomes

```
void  termgen ()
{
int  freq ,i ,freq_spval ;

freq_spval=part[sp_val −1];
if (freq_spval){
        printf("%ld:  ",term ++);
        for  (i =0;i<tot ;i++){
                freq=part[i ];
                if (freq)  printf("%i(%i)  ",freq ,i +1);
                        }
        printf("\n");
                }
}
```

The output produced by running **specpart** with *tot* and *sp_val* set equal to 11 and 6, respectively, is:

1: 5(1) 1(6)
2: 3(1) 1(2) 1(6)
3: 2(1) 1(3) 1(6)
4: 1(1) 2(2) 1(6)
5: 1(1) 1(4) 1(6)
6: 1(2) 1(3) 1(6)
7: 1(5) 1(6) .

Hence we observe that the number of partitions is once again 7 or $p(5)$. If, however, we remove one six from each partition in the above output, then we do obtain the same partitions as those generated by **partgen** except that the order in which they appear is different. Moreover, as a result of the above code, we can now define the specific part operator $L_{SP,k,j}[\cdot]$ as

$$L_{SP,k,j}\left[\cdot\right] := \sum_{\substack{\lambda_1,\dots,\lambda_{j-1}=0,\lambda_j=1,\lambda_{j+1},\dots,\lambda_k=0 \\ \sum_{i=1}^{k} i\lambda_i=k}}^{k,\dots,\lfloor k/(j-1)\rfloor,\lfloor k/j\rfloor,\lfloor k/(j+1)\rfloor\dots,1} , \qquad (7.1)$$

where the specific part j appears at least once in each partition. Hence the lower limit of its multiplicity, i.e., λ_j, has been adjusted to 1. That is, the program still scans all partitions, but partitions such as $\{1_k\}$ will not be printed out due to the fact that $\lambda_j = 0$. From the preceding discussion it follows that

$$L_{SP,k,j}\left[1\right] = L_{P,k-j}\left[1\right] = p(k-j) . \qquad (7.2)$$

As a result of the preceding analysis, it is now a simple matter to consider partitions with more than one specific part. All we need to do is introduce more values in **main** and then create local variables like *freq_spval* to represent the multiplicities of each of these values. As before, each multiplicity is set equal to *part[sp_val-1]* in **termgen**. Next, we modify the if statement to include all the multiplicities. For example, if we wish to determine all the partitions with two specific parts, viz. *sp_val* and *sp_val2*, then the if statement becomes

```
if (freq_spval && freq_spval2){   etc.
```

On the other hand, we may want to select partitions with at least one occurrence of either *sp_val* or *sp_val2*. Then all that is required is to replace the logical AND operator by the logical OR operator in the above if statement.

By making a few minor modifications, mostly to **termgen**, we have been able to solve several problems involving specific partitions. Now we consider determining the partitions with a fixed number of parts in them, whose operator has already been introduced in the previous chapter as (6.16). Previously, it was stated that the solution to this problem often results in a completely different algorithm. For example, on p. 38 of Ref. [13] Knuth presents a different algorithm compared with the earlier algorithm that he uses to generate partitions in reverse lexicographic form. In the case of the BRCP algorithm, however, the number of parts in a partition is determined by the number of branches along its path prior to termination in a partition tree. E.g., the partitions with only two parts in them in Fig. 1.2 are obtained by searching for the terminating tuples appearing in the third column, which are $(0, 5)$, $(0, 4)$ and $(0, 3)$. In other words, the number of 2-part partitions is determined by counting the number of the terminating tuples two branches away from the seed number. When we include the first value of the tuples in the second column of the partition tree together with the second values in the terminating tuples of the third column, we obtain the partitions $\{1, 5\}$, $\{2, 4\}$ and $\{3_2\}$. Since partitions with a fixed number of parts are easily determined from a partition tree, it means that only minor modifications to **partgen** need to be carried out again rather than having to create an entirely different algorithm.

As in the previous example, we modify **main** so that the user can input the number of parts of interest to them. This is represented by the global variable *numparts*, which is also the name given to the complete code. Then we modify **termgen** as presented below. There we see that a new local variable called *sumparts* has been introduced into **termgen**. This variable determines the number of parts in each partition. When this number equals *numparts*, the partition is printed out. Otherwise, it is discarded.

```
void  termgen ()
{
int  freq ,i ,sumparts =0;
/* sumparts is the number of parts or elements in a
    partition */

for ( i =0;i <tot ; i ++){
        sumparts= sumparts+part [i ];
                }
if ( sumparts  ==  numparts ){
    printf ("%ld :  ",term ++);
    for ( i =0;i <tot ; i ++){
                freq=part [i ];
                if ( freq )  printf ("%i (%i )  ",freq ,i +1);
                }
```

```
printf("\n");
                            }
}
```

By running program **numparts** to determine those partitions with 5 parts summing to 10, i.e. where *tot* and *numparts* are set equal to 10 and 5, respectively, we obtain the following output:

1: 4(1) 1(6)
2: 3(1) 1(2) 1(5)
3: 3(1) 1(3) 1(4)
4: 2(1) 2(2) 1(4)
5: 2(1) 1(2) 2(3)
6: 1(1) 3(2) 1(3)
7: 5(2) .

Therefore, we observe that the number of partitions in this instance is 7, which can be expressed as either $\left| \begin{matrix} 10 \\ 5 \end{matrix} \right|$ according to the definition immediately above (3.1) or $L^5_{P,10}[1]$ via (6.20).

It should also be mentioned that if the condition

```
if(sumparts==numparts){  etc.,
```

in **numparts** is replaced by

```
if(sumparts<=numparts){  etc.,
```

then the resulting code generates all those partitions summing to *tot* with at most *numparts* parts. For example, the number of partitions summing to 10 with at most 5 parts in them is equal to 30. From Chapter 3 this is equivalent to $\left| \begin{matrix} 15 \\ 5 \end{matrix} \right|$. This, of course, only applies to the number of partitions, not the actual partitions, in both cases. As expected, the partitions generated by program **numparts** for *tot* and *numparts* equal to 15 and 5 respectively, are different from those generated by the code with the modified if statement.

Doubly-restricted partitions are those partitions where all the parts are greater than a particular value and less than another value. Since this represents a combination of two separate conditions, we need to modify **partgen** so that it generates only those partitions where the parts are either greater than or lower than a specified value. Therefore, let us consider first the case where all the parts in the partitions are less than or equal to a value, which will be represented by the global variable *largest_elt*. As in the previous examples, this value will need to be introduced into **main** of **partgen**. We now consider the modification of **termgen**.

To modify **termgen** so that only partitions with parts greater than *smallest_prt* are generated, we need to introduce an extra for loop. This is required so that if a partition is encountered where a part is less than or equal to *smallest_prt*, it is discarded via a goto statement as demonstrated by the modified version of **termgen** below. Although goto statements are generally frowned upon by programmers, the goto statement here is being used to abandon unnecessary processing in a nested structure of two for loops. In fact, the code behaves much like **partgen** when all the parts are greater than *smallest_prt*. However, when a part is less than or equal to *smallest_prt*, the goto statement discards the partition by branching to the statement labelled end.

```
void termgen ()
{
int f,i;
for (i=0;i<smallest_prt;i++){
      f=part[i];
      if(f)
          goto end;
                                    }
printf("%ld: ",term++);
for (i=0;i<dim;i++){
      f=part[i];
      if(f) printf("%i(%i) ",f,i+1);
                      }
printf("\n");
end:;
}
```

When the above code is run for partitions summing to 14 in where all the parts are greater than 3, we obtain the following output:

1: 1(14)
2: 1(4) 1(10)
3: 2(4) 1(6)
4: 1(4) 2(5)
5: 1(5) 1(9)
6: 1(6) 1(8)
7: 2(7) .

Hence we see that there are only seven partitions summing to 14, where all the parts are greater than 3.

As a result of the above code, it is now a simple matter to consider the case where all the parts are greater than or equal to another value specified by the user. In this instance we simply alter the condition in the first for loop of the previous version of **termgen**. That is, the first for loop in the preceding version of **termgen** simply becomes

```
for (i=0;i<smallest_prt -1;i++){    etc.
```

Consequently, when the new code is run for *tot* and *smallest_prt* set equal to 10 and 3 respectively, it prints out

1: 1(10)
2: 1(3) 1(7)
3: 2(3) 1(4)
4: 1(4) 1(6)
5: 2(5) .

Therefore, partitions with threes now appear in the output.

For doubly-restricted partitions, where the parts are greater than or equal to one value and less than or equal to another (larger) value, all we need to do is incorporate two for loops that branch to the statement labelled end. For this problem **termgen** becomes

```
void termgen()
{
int f,i;
for (i=0;i<smallestpart -1;i++){
        f=part[i];
        if (f)
            goto end;
                            }
for(i=largestpart;i<dim;i++){
        f=part[i];
        if(f)
            goto end;
                            }
        printf("%ld: ",term++);
        for(i=0;i<dim;i++ ){
            f=part[i];
            if(f) printf("%i(%i) ", f,i+1);
                            }
printf("\n");
end:;
}
```

When the above code is run for partitions summing to 13 with the parts greater than or equal to 3 and less than or equal to 9, the following output is produced:

1: 2(3) 1(7)
2: 3(3) 1(4)
3: 1(3) 1(4) 1(6)
4: 1(3) 2(5)
5: 1(4) 1(9)
6: 2(4) 1(5)
7: 1(5) 1(8)
8: 1(6) 1(7) .

Thus, we see that there are 8 partitions with all parts lying in the interval [3, 9].

In Ch. 3 of Ref. [12] Andrews defines restricted partitions differently from the definition given by (6.66). According to his definition, restricted partitions are those where the parts are less than a value, say *prt_max*, while the number of parts is less than or equal to another value, which we take to be *numparts* again as in the previous examples. Although there is now a condition dealing with the number of parts, determining this class of partitions is actually similar to the doubly-restricted case studied above. First, we introduce *prt_max* into **main** in addition to *numparts*. Then we need to insert an extra for loop into **termgen** so that it can make use of the different condition. The new loop appears first since if it is true, we immediately by-pass any action to process the current partition. Therefore, this modified version of **termgen** becomes

```
void  termgen ()
{
int  freq , i , sumparts =0;
for ( i =0; i <tot ; i ++){
        if ( i > prt_max −1 && part [ i ] >0)  goto  end ;
        }
/*(1)  sumparts  is  the  number  of  parts  in  a  partition
   (2)  all  parts  are  now  less  than  or  equal  to  prt_max
*/

for ( i =0; i <tot ; i ++){
        sumparts= sumparts+part [ i ];
        }
if ( sumparts  <=  numparts ){
    printf ("%ld :  " , term ++);
    for ( i =0; i <tot ; i ++){
                freq=part [ i ];
                if ( freq )  printf ("%i (%i )  " , freq , i +1);
```

```
                              }
        printf("\n");
                          }
end:  ;
}
```

When this code is run with *tot*, *numparts* and *prt_max* set equal to 10, 3 and 5 respectively, the following output is generated:

1: 1(1) 1(4) 1(5)
2: 1(2) 1(3) 1(5)
3: 1(2) 2(4)
4: 2(3) 1(4)
5: 2(5) .

Hence we see that there are 5 partitions summing to 10 with at most 3 parts and all parts less than or equal to 5.

The number of partitions summing to k with at most M parts and each part less than or equal to N is represented by $p_G(N, M, k)$ in Chapter 3 of Ref. [12]. The subscript G has been introduced here so that the reader will not be confused with notation appearing in the next chapter. From the above example we have $p_G(5, 3, 10) = 5$. If $k > MN$, then $p_G(N, M, k)$ vanishes, while $p_G(N, M, NM) = 1$. These numbers also appear as the coefficients in the generating function for Gaussian polynomials, which is given by

$$G(N, M; q) = \prod_{i=1}^{N} \frac{(1 - q^{M+i})}{(1 - q^i)} = 1 + \sum_{k=1}^{NM} p_G(N, M, k) \, q^k . \qquad (7.3)$$

Hence Gaussian polynomials are polynomials in q of degree NM. Moreover, to avoid confusion with the restricted partitions studied earlier, we shall refer to the partitions displayed in the above output as Gaussian partitions. As a consequence, we define the Gaussian partition operator $L_{GP,k,N,M}[\cdot]$ as

$$L_{GP,k,N,M}\left[\cdot\right] := \sum_{\substack{\lambda_1, \lambda_2 ..., \lambda_N = 0 \\ \sum_{i=1}^{N} i\lambda_i = k, \ \sum_{i=1}^{N} \lambda_i \leq M}}^{M, \text{Min}\{\lfloor k/2 \rfloor, M\}, ..., \text{Min}\{\lfloor k/i \rfloor, M\}} . \qquad (7.4)$$

Consequently, we have $L_{GP,k,N,M}[1] = p_G(N, M, k)$. Andrews also presents the recurrence relation for $p_G(N, M, k)$, which is

$$p_G(N, M, k) - p_G(N, M - 1, k) = p_G(N - 1, M, k - M) . \qquad (7.5)$$

We can verify (7.5) by running the preceding code for $p_G(5, 3, 10)$. Then the second quantity on the lhs of the recurrence relation becomes $p_G(5, 2, 10)$, which is equal to one when the above code is run. This is simply the partition $\{5_2\}$. In actual fact, the entire lhs of (7.5) is a genuine quantity on its own. It represents the number of partitions summing to k with only M parts and with each part less than or equal to N. On the other hand, the rhs of (7.5) is given by $p_G(4, 3, 7)$, which should equal 4. If we run the above code with *tot*, *numparts* and *prt_max* set equal to 7, 3 and 4 respectively, then the following output is printed out

1: 1(1) 1(2) 1(4)
2: 1(1) 2(3)
3: 2(2) 1(3)
4: 1(3) 1(4) .

Thus, we see that there are indeed four partitions summing to 7 with at most 3 parts and each part less than or equal to 4, as predicted by the lhs of (7.5). These are: $\{1, 2, 4\}$, $\{1, 3_2\}$, $\{2_2, 3\}$ and $\{3, 4\}$. On the other hand, the partitions summing to 10 with only three parts and each part less than or equal to 4 are: $\{1, 4, 5\}$, $\{3_2, 4\}$, $\{2, 3, 5\}$ and $\{2, 4_2\}$.

We now turn our attention to a more complicated example— the problem of determining discrete or distinct partitions. By discrete partitions, we mean those partitions in which the parts appear at most once, if at all. As in all the preceding examples, they too represent a subset or class of the set of integer partitions. As we shall see in the next chapter, these partitions figure prominently in the theory of partitions. Because of their importance, we define the discrete partition operator, $L_{DP,k}[\cdot]$, as

$$L_{DP,k}\left[\cdot\right] := \sum_{\substack{\lambda_1,\lambda_2\ldots,\lambda_k=0 \\ \sum_{i=1}^{k} i\lambda_i=k}}^{1,1,\ldots,1} . \qquad (7.6)$$

Hence the only difference between this operator and the partition operator presented in the previous chapter is that the upper limits of the summations are now restricted to unity, whereas previously they were set to $\lfloor k/i \rfloor$ for each part *i*.

The third program in Appendix B called **dispart** is a program that generates discrete partitions via the BRCP algorithm. In this program **termgen** has an extra for loop than in **partgen**. This for loop checks each multiplicity in a partition and if it is greater than unity, then the partition is discarded by branching to the statement labelled END. If all the

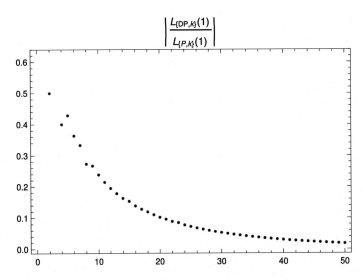

Figure 7.1 The number of discrete partitions to the total number of partitions versus the summation value, k.

multiplicities are less than or equal to one, then the partition is processed in the second for loop.

As stated previously, the number of partitions summing to 100 or $p(100)$ equals 190 569 272. Because of this exponential increase, a programming methodology was created in Chapter 4 so that the 100-th coefficient in the partition method for a power series expansion could be evaluated. However, if we run **dispart** to determine the discrete partitions summing to 100, then we find that it only takes 5 CPU seconds to generate the 444 793 partitions, which is 0.233 per cent of $p(100)$. Moreover, we can run the code for a large number of values and then plot the ratio. Fig. 7.1 displays the ratio of the number of discrete partitions or $L_{DP,k}[1]$ to the number of integer partitions or $L_{P,k}[1]$ versus k for $k \leq 50$. From the figure we see that the ratio decreases monotonically for $k \geq 10$, reaching a value of $0.0179\cdots$ when $k = 50$.

According to Ref. [62], a perfect partition is a partition of a number k whose parts uniquely generate any number from 1 to k. For example, the partition $\{1, 2, 3\}$ generates all the numbers between 1 and 6, but it is not perfect because 3 can be obtained from the single part partition of $\{3\}$ and also from the partition $\{1, 2\}$. Hence the parts in the partition do not uniquely generate 3. On the other hand, the partition $\{1, 2, 4\}$ is perfect. Furthermore, there is always at least one perfect partition for any value of

k because the partition $\{1_k\}$ is perfect. Other examples of perfect partitions are: (1) $\{1, 2_\ell\}$ for any positive integer ℓ, (2) $\{2^0, 2^1, \ldots, 2^{k-1}\}$, which sum to $2^k - 1$ and (3) $\{1_k, k+1\}$, which sum to $2k + 1$. It is also obvious that each perfect partition must have at least one part equal to unity. Therefore, perfect partitions will only appear within the subtree emanating from the tuple $(1, k-1)$ in a partition tree, which, as stated previously, represents the partition tree for partitions summing to $k - 1$.

Riordan shows on p. 123 or Section 5 of Chapter 6 in Ref. [61] that in addition to possessing at least one part with unity, a perfect partition obeys the following condition:

$$\prod_{i=1}^{k}(\lambda_i + 1) = k + 1 . \tag{7.7}$$

For the perfect partition of $\{2^0, 2^1, 2^2, 2^3\}$ or $\{1, 2, 4, 8\}$, the product on the lhs of (7.7) yields 2^4 since $\lambda_1 = \lambda_2 = \lambda_4 = \lambda_8 = 1$. This is indeed equal to $k + 1$ as the partition sums to 15. However, this condition is not sufficient. If a code is constructed using the above condition as the sole criterion for obtaining perfect conditions, one finds that the code generates eleven partitions instead of eight perfect partitions. The three extra partitions are: $\{1_2, 2, 7\}$, $\{1_2, 4, 5\}$ and $\{1, 2_3, 4\}$. Thus, another condition is required.

The extra condition is that the running total of the preceding parts is always less than the next part in the partition. For example, the partition $\{1, 2_3, 4\}$ has a running total of unity after the first part has been parsed and a running total of seven after second part has been parsed. However the next part is four in the partition, which is less than the running total. Hence we exclude it. The same applies to $\{1_2, 4, 5\}$. The remaining partition of $\{1_2, 2, 7\}$ has a running total of two after the first part has been parsed, which equals the next part in the partition. Hence it must be excluded.

As a consequence of the above conditions, we can now modify **dispart** to generate perfect partitions. Basically, all we need to do is alter the for loop at the beginning of **termgen** so that it becomes

```
void termgen ()
{
    int freq ,i ,prod , run_tot =0;

    prod =1;
    for ( i =0;i<tot ; i++){
        freq=part [ i ];
        if ( i !=0 && (( i+1)<=run_tot) && freq !=0)
            goto end;
```

```
\* Note that in the code i+1 represents the value of
   the part              */
             if (i==0 && freq==0) goto end;
             prod=prod*(freq+1);
             if(freq >0) run_tot=(i+1)*freq + run_tot;
                 }
         if(prod-1 != tot) goto end;
         printf("%ld: ",term++);
         for(i=0;i<tot;i++){
                 freq=part[i];
                 if(freq) printf("%i(%i) ",freq ,i+1);
                 }
         printf("\n");
end:;
}.
```

As can be seen, two local variables called *prod* and *run_tot* have been created. The first represents the product of the multiplicities of each partition
when unity is added to them, while the second represents the running total
as each part is parsed in a partition. The condition where the running total
must be less than the next part in a partition appears after the multiplicity
or *freq* has been determined. Then the condition that each perfect partition
must have at least one part equal to unity appears next followed by the calculation of the product *prod*. The condition given by (7.7) is tested in the if
statement before entry to the second for loop. If it is false, then the program
ignores the partition by going to the statement labelled end. Otherwise, the
partition is printed out after being processed in the second for loop. When
the program is run for partitions summing to 11, the following output is
produced:

1: 5(1) 1(6)
2: 11(1)
3: 3(1) 2(4)
4: 2(1) 1(3) 1(6)
5: 2(1) 3(3)
6: 1(1) 2(2) 1(6)
7: 1(1) 5(2)
8: 1(1) 2(3) 1(4) .

Thus, we find that there are eight perfect partitions summing to 11 including $\{1_{11}\}$, $\{1, 2_5\}$ and $\{1_5, 6\}$ as mentioned previously.

The final problem in this chapter is the determination of the conjugate
partition as a partition is generated. According to Ref. [13], the conjugate

partition is obtained by transposing the rows and columns of the corresponding Ferrers diagram for the original partition. In a Ferrers diagram a partition is represented by an array of dots in which the first part, say a_1, is allocated a row of a_1 dots, while the next part (a_2) is represented by another row of a_2 dots immediately below the first row of dots. This process of allocating rows of dots for each part in a partition is continued up to the final part in the partition. The conjugate partition or α^T of the partition $\{a_1, a_2, \ldots, a_k\}$ is obtained by transposing the rows and columns of its Ferrers diagram. For example, the Ferrers diagram for the partition $\{1_2, 2, 3_2, 4\}$ has two rows with one dot, a row with two dots followed by two rows with three dots and finally a row with four dots. The transpose is obtained by counting the dots in the vertical columns of the Ferrers diagram. Hence α^T for the example is found to be $\{6, 4, 3, 1\}$ or $\{1, 3, 4, 6\}$. Note that the conjugate partition does not necessarily possess the same number of parts as the original partition.

A code that determines the conjugate partition for each partition based on **partgen** is presented as Program No. 4 in Appendix B. As in the other classes of partitions studied previously, the code called **transp** has been created by modifying **termgen** yet again. Moreover, the major difficulty with this code as compared with the others in this chapter is that one is now required to create and allocate space for a two-dimensional array *ferrers* of size *tot* × *tot* in addition to creating a one-dimensional array of pointers called *rptr* to it. In both cases the arrays are of integer type. In reality they should be of type char since a true Ferrers diagram consists of dots. That is, instead of allocating dots **transp** allocates ones when constructing a Ferrers diagram. Nevertheless, by adding the ones vertically rather than dots, one obtains the conjugate partition. For the purists it is a simple matter to modify the **termgen** in **transp** to count dots rather than ones. As an aside, it should be mentioned that the Combinatorica package in Mathematica [32] is also able to create the standard Ferrers diagram and determine the transpose partition by means of its FerrersDiagram and TransposePartition routines.

When **transp** is run with *tot* set equal to 4, the following output is generated:

Partition 1 is: 1(4) and its conjugate is: 4(1)
Partition 2 is: 1(1) 1(3) and its conjugate is: 1(2) 2(1)
Partition 3 is: 2(1) 1(2) and its conjugate is: 1(3) 1(1)
Partition 4 is: 4(1) and its conjugate is: 1(4)
Partition 5 is: 2(2) and its conjugate is: 2(2)

As can be seen from the output given above, the conjugate partitions are printed out in the reverse order from the ascending order of the BRCP algorithm. That is, the partitions are in ascending order, while their conjugates are in descending order. For example, the conjugate partition to Partition 2 or $\{1, 3\}$ is printed out as 1(2) 2(1) or $\{2, 1_2\}$, whereas this partition via the BRCP algorithm, viz. Partition 3, is 2(1) 1(2) or $\{1_2, 2\}$. Throughout this work we have not been concerned with the order of the parts in partitions. When the order of the parts becomes an issue, however, integer partitions are referred to as or become known as compositions [12]. For example, the three compositions of the partition $\{1_2, 2\}$ are: (112), (121) and (211). We shall not be concerned with compositions at any stage in this book. Moreover, when generating partitions, we have tended to print them out in ascending order.

There is, however, one problem where generating partitions in descending order is more advantageous than in ascending order. This is the Durfee square, which is defined as the largest square inside a Ferrers diagram as described on p. 28 of Ref. [12] or Ref. [57]. When the Ferrers diagram is constructed for partitions in descending order, this square appears in the upper left hand corner, whereas with partitions in ascending order, it appears in the bottom left hand corner. As a consequence, it is easier to devise an algorithm to determine the size of the square or the side d. One simply sets d equal to zero initially and then determines whether the entries in the second row and column of the Ferrers diagram have a dot (in our case a 1) or not. If a dot or a 1 is missing in an entry, then the code stops processing and the existing value of d becomes the side. If all the entries possess a dot or 1, then the side is incremented and the code processes the third row and column in the same manner. This procedure continues until the first entry is found without a dot or 1 in it.

Conjugate partitions also possess interesting properties. There is always at least one partition whose conjugate is itself, although it can yield a different composition. For example, when handling partitions summing to a square of an integer k, the same-part partition $\{k_k\}$ is a self-conjugate as can be seen with $\{2_2\}$ when **transp** is run with $tot = 4$ as above. On the other hand, if the partitions should sum to $k(k + 1)/2$, then the partition $\{1, 2, \ldots, k\}$ is a self-conjugate, although the composition is different again. For partitions summing to $2k$ for $k > 2$, the partition given by $\{1_{k-2}, 2, k\}$ is also a self-conjugate, while if they sum to $2k + 1$, where $k \geq 7$, then the partition given by $\{1_{k-6}, 2, 3, k - 2\}$ is a self-conjugate. In fact, the number of self-conjugate partitions is the same as the number of partitions with distinct odd parts [12,55].

There are other classes of partitions that can be determined by modifying **termgen**. One such class is partitions with only odd parts in them. As a consequence, we define the odd-part partition (OP) operator as

$$
L_{OP,2k+1}[\cdot] := \sum_{\substack{\lambda_1,\lambda_3,\ldots,\lambda_{2k+1}=0 \\ \sum_{i=0}^{k}(2i+1)\lambda_{2i+1}=2k+1}}^{2k+1,\lfloor(2k+1)/3\rfloor,\ldots,1} \ , \tag{7.8}
$$

which is only valid for partitions summing to $2k+1$, while for partitions summing to $2k$, the operator is defined as

$$
L_{OP,2k}[\cdot] := \sum_{\substack{\lambda_1,\lambda_3,\ldots,\lambda_{2k-1}=0 \\ \sum_{i=1}^{k}(2i-1)\lambda_{2i-1}=2k}}^{2k,\lfloor 2k/3\rfloor,\ldots,1} \ . \tag{7.9}
$$

To obtain partitions with only odd parts, all we need to do is insert the following for loop at the beginning of **termgen** just after the type declarations:

```
for (i=0;i<tot;i++){
            freq=part[i];
            if (i % 2 && freq > 0) goto end;
            }
printf("%ld: ",term++);           .
```

If the upper limits in the above definitions for $L_{OP,2k+1}[\cdot]$ and $L_{OP,2k}[\cdot]$ are set equal to unity, then we obtain the number of partitions with discrete odd parts in them. Moreover, if we wish to generate partitions with discrete odd parts in them, all we need to do is introduce the above for loop at the beginning of **termgen** in program **dispart** in Appendix B. On the other hand, if we wish to investigate whether the number of partitions with discrete odd parts is the same as the number of self-conjugate partitions as mentioned above, then the program **transp** in Appendix B would need to be modified so that the original partition is stored in a temporary array before it undergoes conjugation. Then a test is required to see if both partitions are identical to each other. If they are, then the partition is printed out. Otherwise, it is discarded. This problem is left as an exercise for the reader.

In a similar manner we can define an even-part partition (EP) operator that only applies to those partitions summing to $2k$ with even parts. This is defined as

$$L_{EP,2k}[\cdot] := \sum_{\substack{\lambda_2,\lambda_4,\dots,\lambda_{2k}=0 \\ \sum_{i=1}^{k} i\lambda_{2i}=k}}^{k,\lfloor k/2 \rfloor,\dots,1} . \tag{7.10}$$

The number of partitions generated by this operator is equal to the number of partitions summing to k. Hence we arrive at

$$L_{EP,2k}[1] = L_{P,k}[1] . \tag{7.11}$$

In Chapter 1 and Ref. [5] the cosecant numbers or c_k were first derived in terms of the partitions summing to $2k$ with only even parts before the form in terms of the partition operator given by (1.4) was derived. From these results following identity is obtained

$$L_{P,k}\left[(-1)^{N_k} N_k! \prod_{i=1}^{k} \left(\frac{1}{(2i+1)!} \right)^{\lambda_i} \frac{1}{\lambda_i!} \right]$$

$$= L_{EP,2k}\left[(-1)^{N_k^*} N_k^*! \prod_{i=1}^{k} \left(\frac{1}{(2i+1)!} \right)^{\lambda_{2i}} \frac{1}{\lambda_{2i}!} \right], \tag{7.12}$$

where $N_k^* = \sum_{i=1}^{k} \lambda_{2i}$.

Another class of partitions are flushed partitions, which are discussed on pp. 30–31 of Ref. [12]. A flushed or Sylvestered partition, as Andrews prefers to call them, is a partition, where the smallest part possesses an even–numbered multiplicity and is also an even number. These conditions should be able to be coded following the various examples appearing above. This, too, is left as an exercise for the reader.

In this chapter we have developed various programs for obtaining special classes/types of partitions, which represent subsets of the total number of partitions summing to a particular value. Consequently, they can all be regarded as restricted cases of the partition operator $L_{P,k}[\cdot]$. In Chapter 5 we developed a programming methodology for the partition method for a power series expansion over the entire set of partitions summing to an integer k, which effectively implemented the partition operator. As a result of the material in this chapter it should now be possible to determine the contributions that these classes of partitions make to the partition method for a power series expansion. Moreover, in the next chapter we shall see that the partition-number function $p(k)$ can be evaluated by applying the partition method for a power series expansion over another restricted set of the partitions summing to k, although this restricted set will be more difficult to generate than most of the examples presented in this chapter. Finally, it should be noted that by possessing the capability to adapt the partition method for a power series expansion over classes of partitions, one is now able to determine which partitions make the largest contribution to the coefficients in Theorem 4.1. This may be useful in developing accurate approximations to the coefficients at very high orders to avoid combinatorial explosion.

CHAPTER 8

The Partition-Number Generating Function and Its Inverted Form

An important topic in the theory of partitions is the analysis of the generating functions whose coefficients are dependent on the properties of partitions. One of the greatest achievements in this regard is the derivation of the asymptotic formula for the partition-number function, $p(k)$, by analysing the infinite product $P(z)$ given by (6.2). The first step that led to this formula was the derivation of a remarkable formula for $P(z)$ by Dedekind. As described in Ref. [13], this result can be derived by the application of standard analytic techniques, namely Poisson's summation formula, to the logarithm of $P(z)$. By studying the behaviour of Dedekind's formula for $\ln P(e^{-t})$ with $\Re t > 0$, Hardy and Ramanujan [63] were able with amazing insight [12] to deduce the asymptotic behaviour of the partition-number function $p(k)$ for large k. Later, the asymptotic behaviour of $p(k)$ was evaluated completely by Rademacher [64], which culminated in the now famous Hardy-Ramanujan-Rademacher formula discussed at the beginning of Chapter 5.

Although we shall not reach such lofty heights, we shall in the remaining chapters of this book turn our attention to how the partition method for a power series expansion can be applied in the analysis of the various generating functions that appear in the theory of partitions and to extensions or generalizations thereof. We begin this chapter by applying the partition method for a power series expansion to $P(z)$, but before embarking upon this task, we need to determine precisely when the power series expansion or generating function is convergent. That is, we shall determine the values of z for which (6.2) is convergent or becomes an equation. According to Knuth [13], it was Euler who first noticed that the coefficient of z^n in the infinite product of $(1 + z + z^2 + z^3 + \cdots + z^j + \cdots)(1 + z^2 + z^4 + \cdots + z^{2k} + \cdots)(1 + z^3 + z^6 + \cdots + z^{3k} + \cdots)\cdots$ is the number of non-negative solutions to $k + 2k + 3k + \cdots = n$ or the partition-number function $p(n)$ and that $1 + z^m + z^{2m} + \cdots$ equals $1/(1 - z^m)$. As a result, he arrived at (6.2) except that the equivalence symbol was replaced by an equals sign, which is not entirely correct as can be seen from the following theorem.

The Partition Method for a Power Series Expansion.
DOI: http://dx.doi.org/10.1016/B978-0-12-804466-7.00008-5

Theorem 8.1. The equivalence statement relating the infinite product $P(z)$ given by (6.2) to the generating function consisting of a power series expansion with coefficients equal to $p(k)$, i.e.

$$P(z) \equiv \sum_{k=0}^{\infty} p(k) \, z^k , \qquad (8.1)$$

is absolutely convergent for $|z| < 1$, in which case the equivalence symbol can be replaced by an equals sign. For $|z| > 1$, however, it is divergent. Then $P(z)$ represents the regularized value of the series on the rhs. For $|z| = 1$, the generating function is singular.

Proof. The reason for the appearance of the equivalence symbol in (8.1) is due to Euler's second observation concerning the geometric series. Replacing the series by its limit value of $1/(1 - z^m)$ as Euler did is strictly not valid for all values of z as described in Appendix A and Refs. [3], [4], [18], [43] and [44]. There it is found that the standard geometric series, i.e. $\sum_{k=0}^{\infty} z^k$, is divergent for $\Re\, z > 1$, absolutely convergent for $|z| < 1$ and conditionally convergent for $|z| > 1$ and $\Re\, z < 1$. Regardless of the type of convergence, the limit value of the series yields a finite value of $1/(1 - z)$. For $\Re\, z > 1$, however, summing the series also yields an infinity. If this infinity is removed or regularized, then the remaining finite part is found to equal $1/(1 - z)$, the same as when the series is absolutely convergent. Hence for $z > 1$, the regularized value of the geometric series is equal to $1/(1 - z)$. Moreover, along the line $\Re\, z = 1$, the limit of the series is undefined or indeterminate, while at the point $z = 1$, where the line is tangent to the unit disk of absolute convergence, it is singular. This behaviour is expected since $\Re\, z = 1$ represents the border between convergence on the left side and divergence on the right side. Nonetheless, since the limit value is the same on both sides after regularization, we set the regularized value to $1/(1 - z)$ along $\Re\, z = 1$ as discussed in Appendix A.

In regard to (6.2) we are dealing with an infinite product of geometric series, each one involving a different power of z in its limit value. Despite this, we can use the above knowledge of the geometric series to determine where the series on the rhs of (6.2) is convergent and where it is divergent. For the product on the lhs to equal the series on the rhs of (6.2), all geometric series must be convergent, which means in turn that for all positive integer values of l, $\Re\, z^l < 1$. For $l = 1$ we end up with the standard geometric series, but for $l = 2$, the series will now only be convergent for $\Re\, z^2 < 1$ or $-1 < \Re\, z < 1$. Thus, the range of values for z has changed, which means that the convergence of the series of the series on the rhs of (6.2) will be affected by each value of l or each series in the product.

Let us examine the third series in the product, whose limit is $1/(1-z^3)$. In order to analyse this version of the geometric series, we write the limit value as

$$\frac{1}{1-z^3} = \frac{1}{(1-z)(1-ze^{2i\pi/3})(1-z^{-2i\pi/3})} \,. \tag{8.2}$$

Decomposing the rhs into partial fractions, we see that this version of geometric series is actually the sum of three geometric series, each with a different limit. The first yields the standard geometric series discussed above. The second series has a limit of $1/(1-z\exp(2i\pi/3))$. In this case we replace z by $z\exp(2i\pi/3)$ and continue with the same analysis. Then the second series is convergent for $\Re(z\exp(2i\pi/3)) < 1$ or $y < (2-x)/\sqrt{3}$ when $z = x + iy$, while it is divergent for $y > (2-x)/\sqrt{3}$. That is, the line $\Re z = 1$ separating the regions or planes of convergence and divergence has been rotated by $2\pi/3$ in a clockwise direction. The "left side" of the line representing where the series is convergent is now given by $y < (2-x)/\sqrt{3}$. On the other hand, it is divergent for $y > (2-x)/\sqrt{3}$ in which case the limit becomes the regularized value of the second series. The third series, whose limit is $1/(1-z\exp(-2i\pi/3))$, represents the opposite of the previous series. That is, the singularity at $z = 1$ in the standard geometric series has now been rotated by $2\pi/3$ in an anti-clockwise direction. Hence the third series is convergent for $\Re(z\exp(-2i\pi/3)) < 1$ or $y > -(x+2)/\sqrt{3}$ when $z = x + iy$. This represents the "left" side, while the "right" side or where it is divergent is given by $y < -(x+2)/\sqrt{3}$. For these values of z the limit represents the regularized value of the series.

It is the intersection of the "left" sides for the three series that represent the region of the complex plane for which the third series in the product or $\sum_{k=0}^{\infty} z^{3k}$, is convergent. Outside this region the series is divergent. The intersection of the tangent lines yields an equilateral triangle, where the mid-points of the edges coincide with the three singular points of the component geometric series on the circle $|z| = 1$. Moreover, the unit disk of absolute convergence is circumscribed by this triangle. Those parts of the triangle not in the unit disk of absolute convergence represent the regions of the complex plane for which the third series in the product is conditionally convergent. In total they are significantly less than the corresponding region of the complex plane for the second series in the product, which we found is given by the region outside the unit disk of absolute convergence in the plane $-1 < \Re z < 1$.

If we consider the fourth series in the product, i.e. $\sum_{k=0}^{\infty} z^{4k}$, then decomposing its limit into partial fractions yields four distinct geometric series with the singularities situated again on the unit circle, but at ± 1 and $\exp(\pm i\pi/2)$. If we draw tangent lines through each of these singularities, then we find the common region or the intersection of their "left" sides is now a square circumscribing the unit disk of absolute convergence. The singularities in the component geometric series appear at the mid-points of the square's sides. Outside the square, the fourth series in the product is divergent. Then the regularized value of the series is obtained by combining the limits of the component series. The regions inside the square, but outside the unit disk, represent the values of z for which the fourth series in the product is conditionally convergent. As expected, these regions in total are less than either of the regions of conditional convergence for the second and third series in the product.

If we continue this analysis to the l-th series in the product, i.e. for $\sum_{k=0}^{\infty} z^{lk}$, then we find that the intersection of tangent lines yields an l-sided polygon that circumscribes the unit of disk of absolute convergence. For the values of z outside the polygon the series will be divergent, while for those values of z within the polygon, but outside of the unit disk, the series will be conditionally convergent. Furthermore, as higher values of l are considered, the number of tangent lines not only increases, but also the total region of conditional convergence contracts. In the limit as $l \to \infty$ we will be left with the unit disk as the sole region where the series is (absolutely) convergent, while outside the disk the series is divergent. Since the product in (6.2) includes all values of l, the $l = \infty$ limit by virtue of the fact that it possesses the smallest region of convergence in the complex plane determines the values of z, where the series on the rhs of (6.2) is convergent. This means that the series or generating function, which we shall call from here on the partition-number series, is equal to the product on the lhs only for z situated inside the unit disk. Outside the unit disk, the lhs represents the regularized value of the series and thus, an equivalence symbol must be used instead of an equals sign. Finally, the circle $|z| = 1$ represents a ring of singularity separating the divergent values of the partition-number series from the absolutely convergent values in stark contrast to the standard geometric series, where there is only one singular point that separates absolutely convergent region from the divergent region, namely $z = 1$. Elsewhere, the line $\Re z = 1$ separates conditionally convergent values from divergent values. As mentioned above, the limit for the geometric series is indeterminate along the line, but is assigned a regularized value of $1/(1-z)$. However, for $z = 1$ the regularized value yields infinity. Hence it can be seen that there is a difference between separating absolutely convergent values from divergent values and separating conditionally convergent values from divergent values. This completes the proof of the theorem.

For $|z| < 1$, we can invert (6.2), thereby obtaining

$$\frac{1}{P(z)} = \prod_{k=1}^{\infty}\left(1 - z^k\right) = \frac{1}{1 + \sum_{k=1}^{\infty} p(k)z^k} . \tag{8.3}$$

Since the above product produces a power series that is valid for all values of z, we can write it as

$$(z; z)_{\infty} = \prod_{k=1}^{\infty}\left(1 - z^k\right) = 1 + \sum_{k=1}^{\infty} q(k)\, z^k . \tag{8.4}$$

The leftmost expression is a special case of the q-Pochhammer symbol [65], which is defined as

$$(a; z)_n := \prod_{k=0}^{n-1}\left(1 - az^k\right) . \tag{8.5}$$

According to Knuth [13], it was Euler, who discovered that much cancellation occurs when multiplying the various terms in the infinite product of (8.5). Specifically, he found that

$$\prod_{m=1}^{\infty}(1 - z^m) = 1 + \sum_{k=1}^{\infty}(-1)^k \left(z^{(3k^2 - k)/2} + z^{(3k^2 + k)/2} \right) . \tag{8.6}$$

Therefore, comparing the above result with rhs of (8.4) we see that the $q(k)$ are frequently equal to zero. When they are non-zero, i.e. for k equal to 1, 5, 12, 22, 35, etc., they are either equal to 1 or -1. The values of k for which the $q(k)$ do not vanish are known today as the pentagonal numbers [66,67]. They are themselves a particular case of a broader class of numbers known as the figurate or figural numbers [68,69]. By applying Theorem 4.1 to (6.2), we see that the coefficients of the outer series, viz. q_k, are equal to $(-1)^k$, while those for the inner series or the p_k are equal to the partition-number function $p(k)$ for $k \geq 1$ and zero for $k = 0$. Therefore, from (4.5) we have

$$q(k) = L_{P,k}\left[(-1)^{N_k} N_k! \prod_{i=1}^{k} \frac{p(i)^{\lambda_i}}{\lambda_i!} \right] . \tag{8.7}$$

Since the $q(k)$ are non-zero when $k = (3j^2 \pm j)/2$, (8.7) can also be expressed as

$$L_{P,k}\left[(-1)^{N_k} N_k! \prod_{i=1}^{k} \frac{p(i)^{\lambda_i}}{\lambda_i!}\right] = \begin{cases} (-1)^j, & k = (3j^2 \pm j)/2, \\ 0, & \text{otherwise}. \end{cases} \tag{8.8}$$

Although they do not give the actual number of discrete partitions, we shall refer to the $q(k)$ as the discrete partition numbers. Shortly, we shall see how these coefficients are related to the number of discrete partitions. They do, however, have an interesting connection with the partition-number function, which follows when both power series expansions on the rhs's of (6.2) and (8.4) are multiplied by one another. Then we find that

$$\sum_{k=0}^{\infty} z^k \sum_{j=0}^{k} p(j)\, q(k-j) = 1, \tag{8.9}$$

where $p(0) = q(0) = 1$. Again, since z is fairly arbitrary, like powers of z can be equated on both sides of the above equation. For $k \geq 1$, we obtain the following recurrence relation:

$$\sum_{j=0}^{k} p(j)\, q(k-j) = 0. \tag{8.10}$$

This is simply Euler's recurrence relation for the partition-number function, which is given as (20) on p. 42 of Ref. [13]. Occasionally, it is referred to as MacMahon's recurrence relation as stated in Ref. [70]. Because most of the discrete partition numbers or $q(k)$ vanish, it means that only a few of the previous values of the partition-number function are required to evaluate the latest value of the partition-number function given by $p(k)$ in (8.10).

Although (8.7) is an interesting result, it is not very practical for determining the discrete partition numbers, when it is realized that $p(k)$ grows exponentially. We can, however, use the partition method for a power series expansion to derive another result for the discrete partition numbers. First, by assuming that $|z| < 1$, we write the inverted form of $P(z)$ as

$$\frac{1}{P(z)} = \exp\left(\sum_{m=1}^{\infty} \ln\left(1 - z^m\right)\right) = \exp\left(-\sum_{m=1}^{\infty} \sum_{j=1}^{\infty} (z^m)^j / j\right). \tag{8.11}$$

The Taylor/Maclaurin series for the logarithm has been introduced in this result, since it too is absolutely convergent for $|z| < 1$ according to Refs. [3, 6,17,43,44]. Furthermore, the double sum can be expressed as a single sum by noting that the coefficients of z can be written as a sum over the divisors or factors of the power [71]. Thus we arrive at

$$P(z)^{-1} = 1 + \sum_{k=1}^{\infty} \frac{(-1)^k}{k!} \left(z + \frac{3z^2}{2} + \frac{4z^3}{3} + \frac{7z^4}{4} + \cdots + \gamma_j z^j + \cdots \right)^k, \quad (8.12)$$

where $\gamma_j = \sum_{d|j} d/j$ and d represents a divisor of j. That is, the sum is only over the divisors of j. Some values of the γ_j are: $\gamma_1 = 1$, $\gamma_2 = 3/2$, $\gamma_3 = 4/3$, $\gamma_4 = 7/4$, and $\gamma_5 = 6/5$. More explicitly, we find that $\gamma_6 = 1/6 + 1/3 + 1/2 + 1 - 2$, while for the case of $j = \rho^m$, where ρ is a prime number, the sum yields

$$\gamma_{\rho^m} = \frac{(1 - 1/\rho^{m+1})}{(1 - 1/\rho)}. \quad (8.13)$$

This result can be derived simply by using the limit for the geometric series. In addition, from (8.12) we obtain

$$P(z) = 1 + \sum_{k=1}^{\infty} \frac{1}{k!} \left(z + \frac{3z^2}{2} + \frac{4z^3}{3} + \frac{7z^4}{4} + \cdots + \gamma_j z^j + \cdots \right)^k. \quad (8.14)$$

Now we are in a position to apply Theorem 4.1 to (8.12), whereupon we see that the coefficients of the inner series p_k equal γ_k for $k \geq 1$, while $p_0 = 0$. On the other hand, the coefficients of the outer series q_k are equal to $(-1)^k/k!$. Then from (4.5) the discrete partition numbers are given by

$$q(k) = L_{P,k}\left[(-1)^{N_k} \prod_{i=1}^{k} \frac{\gamma_i^{\lambda_i}}{\lambda_i!} \right], \quad (8.15)$$

which is an entirely different version of (8.7). Moreover, whenever $k = (3j^2 \pm j)/2$, for j an integer, this result can be expressed as

$$L_{P,(3j^2 \pm j)/2}\left[(-1)^{N_{(3j^2 \pm j)/2}} \prod_{i=1}^{(3j^2 \pm j)/2} \frac{\gamma_i^{\lambda_i}}{\lambda_i!} \right] = (-1)^j. \quad (8.16)$$

For all other values of k, the sum over all partitions in (8.15) vanishes. Therefore, to evaluate $q(6)$, we require all the γ_j ranging from $j = 1$ to 6, which have already been given above. Summing over the eleven partitions summing to 6 in (8.15) yields

$$q(6) = -2 + 6/5 + 21/8 - 7/8 + 16/18 - 2 + 4/18$$
$$- 27/48 + 9/16 - 3/48 + 1/6! = 0, \quad (8.17)$$

which is indeed the value of this discrete partition number. By using these results, the reader can readily verify that $q(0) = 1$, $q(1) = -1$, $q(2) = -1$,

$q(5) = 1$ and $q(3) = q(4) = 0$. Moreover, since D_0 in Theorem 4.1 is non-zero in this case due to the fact that $q(0) = 1$, we can use (4.7) to determine the coefficients of the inverted power series expansion or the generating function for $P(k)$. Hence we find that

$$p(k) = L_{P,k}\left[(-1)^{N_k} N_k! \prod_{i=1}^{k} \frac{q(i)^{\lambda_i}}{\lambda_i!}\right]. \tag{8.18}$$

This result, which represents the inverse of (8.7), incorporates much redundancy since the $q(i)$ are only non-zero for specific values of i. Consequently, both the sum over the partitions and the product are only non-zero for those values of i, which are of the form of $(3j^2 - j)/2$ or $(3j^2 + j)/2$, where j is an integer ranging from 1 to $j_m = \lfloor(1 + \sqrt{1 + 24k})/6\rfloor$. In addition, when the $q(i)$ are non-zero, they are only equal to unity in magnitude. Therefore, it is the factor of $N_k!$ that is responsible for the exponential increase in the partition-number function as k increases, although this factor will often be countered by the $1/\lambda_i!$ terms in the denominator of the product. For example, if we wish to determine $p(6)$, there will not be any sums over λ_3, λ_4 and λ_6 in the above equation since we have seen that $q(3)$, $q(4)$ and $q(6)$ vanish, while in the product, $\lambda_3!$, $\lambda_4!$ and $\lambda_6!$ will equal unity (0!). Furthermore, since $q(1) = q(2) = -1$ and $q(5) = 1$, (8.18) yields

$$p(6) = 1 + (-1)^5(-1)^4(-1) \, 5!/4! + (-1)^4(-1)^2(-1)^2 4!/(2! \cdot 2!)$$
$$+ (-1)^3(-1)^3 \, 3!/3! + (-1)^2(-1)2! = 11. \tag{8.19}$$

Thus, we see that there are five contributions to $p(6)$ from the partitions whose parts are composed of pentagonal numbers only. These are the partitions, $\{1_6\}$, $\{1_4, 2\}$, $\{1_2, 2_2\}$, $\{2_3\}$ and $\{1, 5\}$. All except for the last partition produce a positive contribution to $p(6)$. It is, therefore, obvious that only partitions with pentagonal numbers will make contributions to (8.18). Hence the new class of partitions will be referred to as the pentagonal number partitions.

On the other hand, if we apply Theorem 4.1 to (8.14), then the only difference to the previous evaluation of the discrete partition numbers is that the coefficients of the outer series q_k are now equal to $1/k!$. That is, the coefficients of the outer series are still equal to γ_k. This means that

$$p(k) = L_{P,k}[1] = L_{P,k}\left[\prod_{i=1}^{k} \frac{\gamma_i^{\lambda_i}}{\lambda_i!}\right]. \tag{8.20}$$

Consequently, we have an entirely different means of evaluating the partition-number function with the sum of the reciprocals of the divisors/

factors of each part being the assigned values in the partition method for a power series expansion. In this instance, $p(6)$ becomes

$$p(6) = 1/6! + (3/2)/4! + (3/2)^2/(2! \cdot 2!) + (3/2)^3/3! + (4/3)/3!$$
$$+ (3/2)(4/3) + (4/3)^2/2! + (7/4)/2! + (3/2)(7/4) + 6/5 + 2 = 11 \,. \tag{8.21}$$

Moreover, by using (8.15) and (8.20), we can write (8.10) as

$$\sum_{j=0}^{k} L_{P,j} \left[\prod_{i=1}^{j} \frac{\gamma_i^{\lambda_i}}{\lambda_i!} \right] L_{P,k-j} \left[(-1)^{N_{k-j}} \prod_{i=1}^{k-j} \frac{\gamma_i^{\lambda_i}}{\lambda_i!} \right] = 0 \,. \tag{8.22}$$

In the above result there will be much cancellation. For example, the $j = 0$ and $j = k$ terms yield $L_{P,k} \left[\prod_{i=1}^{k} \frac{\gamma_i^{\lambda_i}}{\lambda_i!} \right] + L_{P,k} \left[(-1)^{N_k} \prod_{i=1}^{k} \frac{\gamma_i^{\lambda_i}}{\lambda_i!} \right]$. All the partitions summing with an odd number of parts will cancel because of the phase factor, $(-1)^{N_k}$, while those with an even number of parts will double.

In deriving these new results for the partition-number function $p(k)$, we have encountered a completely different and more complex class of partitions than those studied in the previous chapter. That is, we only require partitions whose parts are pentagonal numbers or are of the form of $(3j \pm 1)j/2$, where j is any integer lying between zero and j_m. In view of the importance of (8.16), let us consider introducing modifications to the program **partgen** presented in Chapter 3 to generate such partitions. As in the examples of the previous chapter, most of the modifications will be made to the function **termgen**.

The first modification is that we need to make is to introduce the math library with the other header files. This is necessary so that the floor function can be called to evaluate the maximum value of j, viz. j_m, when the value of k or rather *tot* is typed in by the user. Since j_m is called in both **main** and **termgen**, it must be declared as a global variable. Once these modifications are carried out, we can concentrate on the changes that are required for **termgen**, which is displayed below:

```
void termgen()
{
int freq,i,j,jval;

for (i=0;i<tot;i++){
            jval=0;
            freq=part[i];
            if (freq>0){
                for (j=1;j<=j_m;j++){
```

```
                    if (i == ((3*j-1)*j-2)/2) jval=j;
                    if (i == ((3*j+1)*j-2)/2) jval=j;
                                }
                    if ((jval==0) && (freq >0)) goto end;
                            }

                    }
printf("%ld: ",term++);
for(i=0;i<tot;i++){
            freq=part[i];
            if(freq) printf("%i(%i) ",freq,i+1);
                }
printf("\n");
end:         ;
}
```

Broadly speaking, the above version of **termgen** is similar to the other codes presented in the previous chapter. That is, before a partition is printed out, testing is done within the function to see if each partition belongs or conforms to the particular class of partitions under consideration. In this case our aim is to print out only those partitions whose parts can be expressed in the pentagonal number forms of $(3j^2-j)/2$ and $(3j^2+j)/2$, where j is any integer. This is accomplished by introducing another for loop in **termgen**, which evaluates the value of j called $jval$, for the appropriate parts. If a part is not of the required form, then $jval$ remains zero. Otherwise, it is non-zero. If $jval$ is zero for a part, then the partition is examined to see if there are any occurrences of the part by checking the variable $freq$. If it is greater than zero, then the entire partition is discarded by the goto statement. This test is carried out on all parts in the partition. The same procedure is then applied to all partitions summing to tot. Only those partitions whose parts are of the required form are printed out by the code. As an example, when $k=6$, the output for this program called **partfn** is:

1: 1(1) 1(5)
2: 4(1) 1(2)
3: 6(1)
4: 2(1) 2(2)
5: 3(2)

These are the five pentagonal number partitions that contribute to the calculation of $p(6)$ via (8.20). Interestingly, when the code is run for partitions summing to 100, it only prints out 42205 partitions, which represent about 0.02 percent of the total number of partitions given by $p(100)$. In addition,

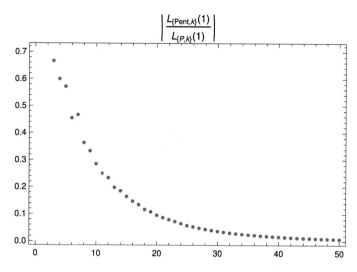

Figure 8.1 The ratio of the number of pentagonal number partitions to the total number of partitions versus k.

the number of pentagonal number partitions summing to k or $L_{Pent,k}[1]$ starts off greater than the number of discrete partitions for the same value of k, but drops away when $k > 17$. Fig. 8.1 displays the ratio of $L_{Pent,k}[1]$ to the number of partitions summing to k for $k \leq 50$. Once again, it is monotonically decreasing for k beyond a certain value, which in this case is about 8.

It was also stated at the end of the previous chapter that the partition method for a power series expansion can be adapted to handle situations where only a subset or class of the total number of partitions is required as in the above example. We shall demonstrate this by modifying the code in Chapter 5 so that the partition–number function $p(k)$ can be evaluated via (8.18). The resulting code called **pfn** is presented in its entirety as Program 5 in Appendix B. Basically, it expresses the partition–number function in a symbolic form where the final values can be evaluated by introducing the output into Mathematica. As expected, **termgen** has the same for loop displayed above. The interesting feature about the code is that this for loop appears twice at the beginning of **termgen**. This is necessary because the first step prints out $p(k) :=$ before considering the single part partition of $\{k\}$ in the first branch of the if statement. Clearly, the single part partition may not be of the required form, which explains why it appears first in the for loop. All the other partitions are processed by the second branch of the if statement, which means that the same for loop appears in this part of the program in order to determine which partitions are of the required form.

The code is also capable of expressing the final symbolic form for the partition–number function in two different forms. In the first form a printf statement prints the final form for $p(k)$ in terms of the $q(i)$ as they appear in (8.18). This statement has been commented out in favour of the version of the code appearing in the appendix. In the second version of the code the value of each $q(i)$ is evaluated. That is, the code prints out -1 to the power of *jval* for each part in a pentagonal number partition. In order to accomplish this, the for loop mentioned in the previous paragraph has had to be introduced into the latter part of **termgen** again. Finally, the quantity $q[N_k]a^{N_k}$ appearing in the output of the code discussed in Chapter 5 has been replaced $(-1)^\wedge(N_k)$ in **pfn**. Therefore, if the version of **pfn** presented in Appendix B is run for *tot* $= 6$, then the following output is produced:

p[6]:= ((-1)^(1)) ((-1)^(2)) (-1)^(2) 2! + ((-1)^(1))^(4) ((-1)^(1)) (-1)^(5) 5!/4! + ((-1)^(1))^(6) (-1)^(6) + ((-1)^(1))^(2) ((-1)^(1))^(2) (-1)^(4) 4!/(2! 2!) + ((-1)^(1))^(3) (-1)^(3)

Time taken to compute the coefficient is 0.000000 seconds.

This output can be imported into Mathematica whereupon it gives the correct value of 11 for $p(6)$. When the code is run for *tot* equal to 100, it takes 34 seconds to compute the symbolic form for $p(100)$ or $L_{p,100}[1]$ and only 0.27 seconds to produce the value of 190 569 292 in Mathematica on the same Sony VAIO laptop mentioned in previous chapters. However, whilst this is not too bad, the fastest method of obtaining $p(k)$ or $L_{p,k}[1]$ from the various programs considered in this book is to comment out **termgen** and introduce the statement

    term++;

immediately below in **partgen** of Chapter 3. Then *term* needs to be initialized to zero and a printf statement introduced into **main**. When the resulting program is run for *tot* equal to 100, it only takes 3 seconds to compute $p(100)$. Nevertheless, both methods for computing the partition–number function will slow down dramatically as *tot* continues to increase due to combinatorial explosion. This is where either (8.10) or even the Hardy-Ramanujan-Rademacher formula should be used. In fact, we may write

$$\lim_{k\to\infty} L_{P,k}\left[1\right] \to \frac{1}{4\sqrt{3}\,k}\, \exp(\sqrt{2k/3}\,\pi)\,, \tag{8.23}$$

or

$$\lim_{k\to\infty} L_{P,k}\left[(-1)^{N_k} N_k! \prod_{i=1}^{k} \frac{q(i)^{\lambda_i}}{\lambda_i!}\right] \to \frac{1}{4\sqrt{3}\,k} \exp(\sqrt{2k/3}\,\pi)\,. \qquad (8.24)$$

The arrow symbols in the above results are not really necessary and can be replaced by equals signs. For finite large values of k, the rhs becomes an approximation for the quantities on the lhs of both results.

8.1 GENERALIZATION OF THE INVERTED FORM OF $P(z)$

In the previous section we demonstrated how the partition method for a power series expansion can be used to derive different forms of the generating functions for the infinite product $P(z)$ given by (6.2) and its inverted or reciprocal form. In the case of $P(z)$ two different forms for the coefficients of the generating function or the partition–number function were obtained. Both employed the partition operator $L_{P,k}[\cdot]$, but operated on different arguments. The first result given by (8.18) involved the discrete partition numbers $q(k)$ in its argument, which represent the coefficients of the generating function for the inverted form of $P(z)$, while the argument of the second form given by (8.20) involved the sum over the inverted divisors/factors of each part i in a partition. In the case of the inverted form of $P(z)$, two different forms were also obtained, but then the partition operator was found to operate on either the partition–number function as in (8.7) or the same sum involving the inverted divisors but accompanied by an extra phase factor of $(-1)^{N_k}$ as in (8.15). The extra phase factor is responsible for ensuring that the discrete partition numbers equal either ± 1 when k is a pentagonal number and zero, otherwise. Because the phase factor does not appear in (8.20) compared with (8.15), the partition–number function experiences exponential growth as can be seen from the Hardy-Ramanujan-Rademacher formula given above. Nevertheless, it was possible to derive all four results because the coefficients of the powers of z in $P(z)$ are simple, namely equal to -1. Here we aim to generalize $P(z)$ to the situation where the coefficients of z^k are now equal to C_k. We begin by examining the extension of the inverted form of $P(z)$. Hence we arrive at the following theorem.

Theorem 8.2. The infinite product defined by

$$H(z) = \prod_{i=1}^{\infty} \left(1 + C_i z^i\right)\,, \qquad (8.25)$$

can be written as a power series expansion or generating function of the form:

$$H(z) = 1 + \sum_{k=1}^{\infty} h_k z^k , \qquad (8.26)$$

where the coefficients h_k can be expressed in terms of the discrete partition operator defined by (7.6) and are given by

$$h_k = L_{DP,k} \left[\prod_{i=1}^{k} C_i^{\lambda_i} \right] . \qquad (8.27)$$

Proof. If we expand the lowest order terms in z in the product given in (8.25), then as in the proof of Theorem 4.1, we obtain a power series expansion. The zeroth order term in the resulting power series is unity, while the first and second order terms are respectively $C_1 z$ and $C_2 z^2$. Once we proceed beyond the second order term, the coefficients become more complex to evaluate due to an ever-increasing number of terms having to be counted in them. For example, the third, fourth and fifth order coefficients are equal to $C_3 + C_1 C_2$, $C_4 + C_1 C_3$, and $C_5 + C_1 C_4 + C_2 C_3$, respectively. In fact, on close inspection we find that the coefficient of z^k in the power series depends on the number of discrete partitions summing to k. For example, the third and fourth order coefficients are composed of only two terms because there are only two discrete partitions for $k = 3$, viz. $\{1, 2\}$ and $\{3\}$, and $k = 4$, viz. $\{1, 3\}$ and $\{4\}$, whereas the fifth order term is composed of three terms since there are three discrete partitions summing to 5, viz. $\{5\}$, $\{1, 4\}$ and $\{2, 3\}$. Therefore, instead of summing over all the partitions summing to k as we did in Theorem 4.1, we need only sum over the class consisting of the discrete partitions summing to k, which we have already seen from Fig. 7.1 is significantly less than $p(k)$. Hence the sum over all partitions in (4.5) simplifies drastically as all the multiplicities lie between zero and unity, not between zero and $\lfloor k/i \rfloor$ for each part i. Furthermore, since there is no outer series in the expansion of the product, we can drop q^{N_k} and a^{N_k} in the same equation. In addition, there is no multinomial factor. Consequently, we find that according to (4.5), the coefficients h_k in the power series for $H(z)$ are given by

$$h_k = \sum_{\substack{\lambda_1,\lambda_2,\lambda_3,\dots,\lambda_k=0 \\ \sum_{i=1}^{k} i\lambda_i=k}}^{1,1,1,\dots,1} \prod_{i=1}^{k} C_i^{\lambda_i} . \qquad (8.28)$$

The sum over partitions in the above result is simply the discrete partition operator as defined by (7.6). When it is introduced into (8.28), we obtain (8.27). This completes the proof of the theorem.

It should also be mentioned that we can invert the rhs of (8.26) and apply Theorem 4.1 again. Then we obtain a power series expansion for $1/H(z)$. Consequently, we find that

$$\frac{1}{H(z)} \equiv 1 + \sum_{k=0}^{\infty} H_k z^k , \tag{8.29}$$

where

$$H_k = L_{P,k}\left[(-1)^{N_k} N_k! \prod_{i=1}^{k} \frac{h_i^{\lambda_i}}{\lambda_i!} \right]. \tag{8.30}$$

Moreover, from (4.7), we obtain

$$h_k = L_{DP,k}\left[\prod_{i=1}^{k} C_i^{\lambda_i} \right] = L_{P,k}\left[(-1)^{N_k} N_k! \prod_{i=1}^{k} \frac{H_i^{\lambda_i}}{\lambda_i!} \right]. \tag{8.31}$$

Since Theorem 8.2 is quite general, it means conversely that any power series expansion can be expressed as an infinite product of the form given by the rhs of (8.26). For example, as discussed on p. 111 of Ref. [13], the geometric series can be represented by the following product:

$$\sum_{k=0}^{\infty} z^k = \prod_{i=0}^{\infty} \left(1 + z^{2^i} \right). \tag{8.32}$$

In this example $C_{2^i} = 1$, while the other C_i simply vanish. Thus, the product of $\prod_{i=1}^{k} C_i^{\lambda_i}$ in the formula for h_k is either zero or unity. Furthermore, since the coefficients of the geometric series are equal to unity, we have $L_{DP,k}[\prod_{i=1}^{k} C_i^{\lambda_i}] = 1$. Because all the terms within the discrete partition operator are either equal to zero and unity, this means that there can only be one discrete partition where all the parts are of the form 2^j with each value of j lying between zero and $\lfloor \log k / \log 2 \rfloor$. For $k = 5$ and $k = 7$, these discrete partitions are respectively $\{2^0, 2^2\}$ or $\{1, 4\}$ and $\{2^0, 2^1, 2^2\}$ or $\{1, 2, 4\}$, while for $k = 11$ and $k = 13$, they are $\{2^0, 2^2, 2^3\}$ or $\{1, 2, 8\}$ and $\{2^0, 2^2, 2^3\}$ or $\{1, 4, 8\}$.

Another example of a well-known power series that can be expressed as an infinite product is the power series for the exponential function, which can be expressed as

$$e^y = (1+y)(1+y^2/2)(1-y^3/3)(1+3y^4/8)(1-y^5/5)(1+13y^6/72)$$
$$\times (1-y^7/7)(1+27y^8/128)(1-8y^9/81)(1+91y^{10}/800)\cdots. \tag{8.33}$$

From (8.33) we see that $C_1 = 1$, while the other C_i can be determined by the following recurrence relation involving the divisors or factors of i:

$$\sum_{d|i} \frac{(-1)^{d+1}}{d} C_{i/d}^d = 0.$$ (8.34)

This means that whenever i is a prime number greater than 2, say p, $C_p = -1/p$ as can be seen by the coefficients of y^3, y^5 and y^7 in (8.33). For $i = 4$ the divisors are 1, 2 and 4. Hence (8.34) yields

$$C_4 = \frac{1}{2} C_2^2 + \frac{1}{4} C_1^4 = \frac{1}{2}\left(\frac{1}{2}\right)^2 + \frac{1}{4} = \frac{3}{8}.$$ (8.35)

The derivation of (8.34) is left as an exercise for the reader as are the following results:

$$\sin y = y(1 - y^2/3!)(1 + y^4/5!)(1 + 6y^6/7!)(1 + 73y^8/9!)\cdots,$$ (8.36)

and

$$\cos y = (1 - y^2/2!)(1 + y^4/4!)(1 + 7y^6/360)(1 + 393y^8/8!)$$
$$\times (1 + 1843y^{10}/453600)\cdots.$$ (8.37)

By taking the logarithm of both (8.25) and (8.26), we can derive other interesting results such as

$$\ln\left(1 + \sum_{k=1}^{\infty} h_k z^k\right) = \sum_{i=1}^{\infty} \ln\left(1 + C_i z^i\right).$$ (8.38)

Then introducing the Taylor/Maclaurin series for logarithm into the rhs yields

$$\ln\left(1 + \sum_{k=1}^{\infty} h_k z^k\right) = \sum_{i=0}^{\infty}\sum_{j=1}^{\infty} \frac{(-1)^{j+1}}{j} C_i^j z^{ij},$$ (8.39)

where we have assumed that $|z| < 1$. Now we apply Theorem 4.1 to the lhs of this result. The inner series coefficients are given by $p_k = h_k$, while the outer series coefficients are given by $q_k = (-1)^{k+1}/k$. By equating like powers of z in the resulting equation, we find that

$$L_{P,k}\left[(-1)^{N_k}(N_k - 1)! \prod_{i=1}^{k} \frac{h_i^{\lambda_i}}{\lambda_i!}\right] = \sum_{i|k} \frac{(-1)^i}{i} C_{k/i}^i.$$ (8.40)

As a check, let us put $k = 3$. Then the lhs becomes

$$L_{P,3}\left[(-1)^{N_3}(N_3 - 1)! \prod_{i=1}^{3} \frac{h_i^{\lambda_i}}{\lambda_i!}\right] = -\frac{1}{3} h_1^3 + h_1 h_2 - h_3.$$ (8.41)

From the proof of Theorem 8.1, we have $h_1 = C_1$, $h_2 = C_2$ and $h_3 = C_1 C_2 + C_3$. Introducing these values into (8.41) yields

$$L_{P,3}\left[(-1)^{N_3}(N_3 - 1)!\prod_{i=1}^{3}\frac{h_i^{\lambda_i}}{\lambda_i!}\right] = -\frac{1}{3}C_1^3 - C_3.\qquad(8.42)$$

On the other hand, the two divisors of three for the sum on the rhs of (8.40) are 1 and 3. Hence there are only two contributions to the sum over the divisors, both of which are negative. For $i = 1$, $C_{3/i}^i/i = C_3$, while for $i = 3$, $C_{3/i}^i/i = C_1^3/3$. Therefore, the rhs produces the same result as $L_{P,3}\left[(-1)^{N_3}(N_3 - 1)!\prod_{i=1}^{3}h_i^{\lambda_i}/\lambda_i!\right]$. For the particular case of the geometric series, viz. (8.32), $h_k = 1$, while only those values of $C_{k/i}$, where $k/i = 2^{\ell}$ and ℓ is a non-negative integer, contribute and are equal to unity. By introducing these results into (8.40), we obtain

$$L_{P,2^j}\left[(-1)^{N_{2^j}}(N_{2^j} - 1)!\prod_{i=1}^{2^j}\frac{1}{\lambda_i!}\right] = -2^{-j}.\qquad(8.43)$$

If we put the coefficients C_i equal to unity in (8.25)–(8.27), then we find that

$$\prod_{k=1}^{\infty}\left(1 + z^k\right) = 1 + \sum_{k=1}^{\infty}L_{DP,k}\left[1\right]z^k.\qquad(8.44)$$

In other words, the coefficients of the generating function for $1/P(z)$ are equal to the sequence consisting of the number of discrete partitions summing to the power of z. That is, according to (8.26), we have

$$h_k = L_{DP,k}\left[1\right].\qquad(8.45)$$

Alternatively, if we put the $C_i = -1$, then we obtain

$$q(k) = L_{DP,k}\left[(-1)^{N_k}\right].\qquad(8.46)$$

Prior to (8.7) it was mentioned that the $q(k)$ or discrete partition numbers are equal to $(-1)^j$ whenever k is a pentagonal number [66,67] or is equal to $j(3j \pm 1)/2$. Hence (8.46) can be expressed as

$$L_{DP,k}\left[(-1)^{N_k}\right] = \begin{cases}(-1)^j, & \text{for } k = j(3j \pm 1)/2, \\ 0, & \text{otherwise}.\end{cases}\qquad(8.47)$$

The above result tells us that the number of discrete partitions with an odd number of parts is equal to the number of discrete partitions with an even number of parts when k is not a pentagonal number. It also tells us that the discrete partition numbers $q(i)$ come in pairs of either 1 or -1, the former corresponding to a pentagonal number derived from an even number, i.e. by setting j equal to an even number and the latter to a pentagonal number derived by an odd number. A value of 1 in the above result or for j an even number, means that the number of discrete partitions with an even number of parts is one more than the number of discrete partitions with an odd number of parts. A value of -1 or j odd, represents the opposite situation. This result is proved in Ch. 1 of Ref. [12] by means of Ferrers diagrams.

If we now put $C_i = \omega$ in (8.25), then the generating function for the ensuing infinite product becomes

$$Q(z, \omega) = \prod_{k=1}^{\infty} \left(1 + \omega z^k\right) = \frac{(\omega; z)_\infty}{(1 + \omega)} = 1 + \sum_{k=1}^{\infty} q(k, \omega) z^k, \qquad (8.48)$$

where $(\omega; z)_\infty$ is again the q-Pochhammer symbol as defined by (8.5) and

$$q(k, \omega) = L_{DP,k}\left[\omega^{N_k}\right]. \qquad (8.49)$$

Thus, we see that $q(k, -1)$ reduces to $q(k)$ or the discrete partition numbers according to (8.46). Note also that the power of ω represents the number of parts or length of the partitions, while summing over all partitions for the same power of ω, say ω^n, in $q(k, \omega)$ gives the total number of discrete partitions with n parts. A similar situation will arise when we study the inverse of (8.48). We shall refer to the $q(k, \omega)$ as the discrete partition polynomials. From (8.44) we have seen that the $\omega = 1$ case yields the coefficients that represent the total number of discrete partitions summing to k, viz. $L_{DP,k}[1]$. Consequently, we see that the difference between the discrete partition numbers and the number of discrete partitions summing to k is that in the case of the latter the discrete partition operator operates on unity, while in the case of the former, it operates on the phase factor of the length of the discrete partitions, namely $(-1)^{N_k}$. That is, the difference between them is due to the phase difference in the number of parts in the discrete partitions.

The discrete partition polynomials up to $k = 15$ are displayed in the second column of Table 8.1. As expected, for $\omega = -1$ these results reduce to the discrete partition numbers $q(k)$, while for $\omega = 1$ they yield the number of discrete partitions. From the table it can be seen that the discrete partition polynomials are polynomials of degree n, where $n(n + 1)/2 \leq k < (n + 1)(n + 2)/2$ since the partition with the most discrete parts when $k = n(n + 1)/2$ is $\{1, 2, 3, \ldots, n\}$. The lowest order term

Table 8.1 Discrete partition polynomials $q(k, \omega)$ and partition function polynomials $p(k, \omega)$

k	$q(k, \omega)$	$p(k, \omega)$
0	1	1
1	ω	ω
2	ω	$\omega^2 + \omega$
3	$\omega^2 + \omega$	$\omega^3 + \omega^2 + \omega$
4	$\omega^2 + \omega$	$\omega^4 + \omega^3 + 2\omega^2 + \omega$
5	$2\omega^2 + \omega$	$\omega^5 + \omega^4 + 2\omega^3 + 2\omega^2 + \omega$
6	$\omega^3 + 2\omega^2 + \omega$	$\omega^6 + \omega^5 + 2\omega^4 + 3\omega^3 + 3\omega^2 + \omega$
7	$\omega^3 + 3\omega^2 + \omega$	$\omega^7 + \omega^6 + 2\omega^5 + 3\omega^4 + 4\omega^3 + 3\omega^2 + \omega$
8	$2\omega^3 + 3\omega^2 + \omega$	$\omega^8 + \omega^7 + 2\omega^6 + 3\omega^5 + 5\omega^4 + 5\omega^3$ $+ 4\omega^2 + \omega$
9	$3\omega^3 + 4\omega^2 + \omega$	$\omega^9 + \omega^8 + 2\omega^7 + 3\omega^6 + 5\omega^5 + 6\omega^4 + 7\omega^3$ $+ 4\omega^2 + \omega$
10	$\omega^4 + 4\omega^3 + 4\omega^2 + \omega$	$\omega^{10} + \omega^9 + 2\omega^8 + 3\omega^7 + 5\omega^6 + 7\omega^5 + 9\omega^4$ $+ 8\omega^3 + 5\omega^2 + \omega$
11	$\omega^4 + 5\omega^3 + 5\omega^2 + \omega$	$\omega^{11} + \omega^{10} + 2\omega^9 + 3\omega^8 + 5\omega^7 + 7\omega^6 + 10\omega^5$ $+ 11\omega^4 + 10\omega^3 + 5\omega^2 + \omega$
12	$2\omega^4 + 7\omega^3 + 5\omega^2 + \omega$	$\omega^{12} + \omega^{11} + 2\omega^{10} + 3\omega^9 + 5\omega^8 + 7\omega^7$ $+ 11\omega^6 + 13\omega^5 + 15\omega^4 + 12\omega^3 + 6\omega^2 + \omega$
13	$3\omega^4 + 8\omega^3 + 6\omega^2 + \omega$	$\omega^{13} + \omega^{12} + 2\omega^{11} + 3\omega^{10} + 5\omega^9 + 7\omega^8$ $+ 11\omega^7 + 14\omega^6 + 18\omega^5 + 18\omega^4 + 14\omega^3$ $+ 6\omega^2 + \omega$
14	$5\omega^4 + 10\omega^3 + 6\omega^2 + \omega$	$\omega^{14} + \omega^{13} + 2\omega^{12} + 3\omega^{11} + 5\omega^{10} + 7\omega^9$ $+ 11\omega^8 + 15\omega^7 + 20\omega^6 + 23\omega^5 + 23\omega^4$ $+ 16\omega^3 + 7\omega^2 + \omega$
15	$\omega^5 + 6\omega^4 + 12\omega^3 + 7\omega^2 + \omega$	$\omega^{15} + \omega^{14} + 2\omega^{13} + 3\omega^{12} + 5\omega^{11} + 7\omega^{10}$ $+ 11\omega^9 + 15\omega^8 + 21\omega^7 + 26\omega^6 + 30\omega^5$ $+ 27\omega^4 + 19\omega^3 + 7\omega^2 + \omega$

in ω corresponds to the only single-part partition, viz. $\{k\}$. If we run the program **dispart**, which is presented as Program 3 in Appendix B, then we find that there are 4 discrete partitions summing to 6, which are $\{6\}$, $\{1, 5\}$, $\{2, 4\}$ and the self-conjugate $\{1, 2, 3\}$. By referring to Fig. 1.2, we see that the first partition is just one branch from the seed number, the second and third partitions are two branches away and the third is three branches away. Hence we observe that $q(6, \omega) = \omega + 2\omega^2 + \omega^3$, where the magnitude of the coefficients of ω^i represent the number of distinct partitions with i parts. That is, if the number of distinct partitions summing to k with i parts is denoted by $q_i(k)$, then the discrete partition polynomials can be expressed as

$$q(k, \omega) = \sum_{i=1}^{n} q_i(k)\omega^i , \qquad (8.50)$$

where $n(n+1)/2 \le k < (n+1)(n+2)/2$. Since we have already noted that $q(k, 1) = L_{DP,k}[1]$, we obtain the trivial equation of $q(k, 1) = \sum_{i=1}^{n} q_i(k)$. Furthermore, the lower bound on k gives us a limit as to the maximum number of parts that can appear in a discrete partition, which is given by

$$i_{max} = [(\sqrt{8k+1} - 1)/2] . \qquad (8.51)$$

By fixing the number of parts to i in the discrete partition operator so that it becomes

$$L_{DP,k,i}\left[\cdot \right] := \sum_{\substack{\lambda_1, \lambda_2 ..., \lambda_k = 0 \\ \sum_{j=1}^{k} j\lambda_j = k, \ \sum_{j=1}^{k} \lambda_j = i}}^{1,1,...,1} , \qquad (8.52)$$

we arrive at $L_{DP,k,i}[1] = q_i(k)$. This means that we need to insert another value into the program, which represents the number of parts the user wants. If this is set equal to a global variable called *numparts*, then the only changes to be made to **dispart** are: (1) introduce a local variable *sumpart*, which adds all the values of *freq* in the first for loop and (2) insert the following if statement before anything is printed out:

```
if(sumparts != numparts) goto end;
```

When these modifications are implemented and the resulting code run for the discrete partitions summing to 100 with the number of parts, i.e. *numparts*, set equal to 5, one finds that there are 25,337 5-part discrete partitions beginning with $\{1, 2, 3, 4, 90\}$ and ending with $\{18, 19, 20, 21, 22\}$. That is, $q_5(100)$ or $L_{DP,100,5}[1]$ is equal to 25 337. According to (8.51), the maximum number of parts in the discrete partitions summing to 100 is 13. When the code is run for *numparts* set equal to 13, 30 ($= q_{13}(100)$) partitions are printed out beginning with $\{1, 2, 3, 4, ..., 12, 22\}$ and ending with $\{1, 2, 3, 4, 6, 7, ..., 13, 14\}$. Running the code for higher values of *numparts* with $k = 100$ does not result in any partitions being printed out. Hence $q_i(100) = 0$ for $i > 13$, which is consistent with (8.51).

From (8.48) we have

$$\prod_{k=1}^{\infty} \left(1 - \omega^2 z^{2k}\right) = \frac{(-\omega; z^2)_\infty}{(1 - \omega^2)} = 1 + \sum_{k=1}^{\infty} q(k, -\omega^2) z^{2k} . \qquad (8.53)$$

The product on the lhs of this result can also be written as

$$\prod_{k=1}^{\infty}\left(1-\omega^2 z^{2k}\right)=\prod_{k=1}^{\infty}\left(1-\omega z^k\right)\left(1+\omega z^k\right).\tag{8.54}$$

Introducing the rhs of (8.48) into the rhs of the above result yields

$$\prod_{k=1}^{\infty}\left(1-\omega^2 z^{2k}\right)=\sum_{k=1}^{\infty}z^k\sum_{j=0}^{k}q(j,\omega)\,q(k-j,-\omega).\tag{8.55}$$

The equals sign is only valid in (8.53) and (8.55) for $|\omega|<1$ and $|z|<1$. For these values of ω and z, we can equate like powers of z on the rhs's of these equations, which gives for k equal to the odd number, $2m+1$,

$$\sum_{j=0}^{2m+1}q(j,\omega)\,q(2m+1-j,-\omega)=0.\tag{8.56}$$

On the other hand, for $k=2m$, we obtain

$$\sum_{j=0}^{2m}q(j,\omega)\,q(2m-j,-\omega)=q(m,-\omega^2).\tag{8.57}$$

When $\omega=1$, (8.57) reduces to

$$\sum_{j=0}^{2m}L_{DP,j}[1]\,L_{DP,2m-j}\left[(-1)^{N_{2m-j}}\right]=L_{DP,m}\left[(-1)^{N_m}\right].\tag{8.58}$$

From (8.47) we see that the lhs of the above equation is effectively a sum over the pentagonal numbers less than $2m$, while the rhs is non-zero if and only if m is a pentagonal number. Furthermore, we can put ω equal to the complex number i in (8.57), which results in the lhs becoming the product of imaginary values of the $q(j,\omega)$, while the rhs gives the number of discrete partitions summing to m or $q(m,1)$. Obviously, since the rhs is real, the imaginary parts on the lhs cancel.

The foregoing analysis can also be extended by raising (8.48) to an arbitrary power ρ and applying Corollary 2 to Theorem 4.1. Then we obtain a similar power series expansion to (4.54), but now in powers of z with the coefficients depending upon ρ. Thus, we arrive at

$$\prod_{k=1}^{\infty}(1-\omega z^k)^{\rho}\equiv 1+\sum_{k=1}^{\infty}q(k,-\omega,\rho)\,z^k,\tag{8.59}$$

where the coefficients $q(k,\omega,\rho)$ can be determined from (4.55) by setting $D_i=q(i,\omega)$. Consequently, we find that these coefficients can be expressed as

$$q(k, \omega, \rho) = L_{P,k}\left[(-1)^{N_k}(-\rho)_{N_k}\prod_{i=1}^{k}\frac{q(i, \omega)^{\lambda_i}}{\lambda_i!}\right]. \tag{8.60}$$

It should be emphasized that in the above results ρ can be any value including a complex number. For integer values of ρ greater than zero the equivalence symbol in (8.59) can be replaced by an equals sign. As we shall see shortly, when $\rho = -1$ and $\omega = 1$ in the above result, the coefficients $q(k, -1, -1)$ become equal to the partition-number function, $p(k)$, while if ρ is equal to a positive integer, say j, then (8.60) simplifies drastically due to the fact that for $k > j$, the factor $(-\rho)_{N_k}$ vanishes for $N_k > j$. That is, the partitions with more than j parts do not contribute to the $q(k, \omega, \rho)$. Moreover, if $\omega = -1$, further redundancy occurs in (8.60) since the $q(i, \omega)$ become the discrete partition numbers or $q(i)$, which we have seen are only non-zero when i is a pentagonal number [66,67].

There is, however, another method of developing a power series expansion or generating function to the lhs of (8.59), which involves the partition operator, but it is more general because in this instance (8.25) is raised to an arbitrary power ρ. Therefore, we arrive at the following corollary.

Corollary to Theorem 8.2. The generalized version of the infinite product in (8.59), where the coefficients of z^k are equal to C_k, possesses the following power series expansion or generating function:

$$\prod_{k=1}^{\infty}(1 + C_k z^k)^{\rho_k} \equiv 1 + \sum_{k=1}^{\infty}B_k(\boldsymbol{\rho})\, z^k, \tag{8.61}$$

where the coefficients $B_k(\boldsymbol{\rho})$ are given by

$$B_k(\boldsymbol{\rho}) = L_{P,k}\left[(-1)^{N_k}\prod_{i=1}^{k}\frac{(-\rho_i)_{\lambda_i}}{\lambda_i!}\, C_i^{\lambda_i}\right]. \tag{8.62}$$

In these results $(\boldsymbol{\rho})$ denotes $(\rho_1, \rho_2, \ldots, \rho_k)$. If $S = \inf|C_k|^{-1/k}$, then for $|z| < S$ the equivalence symbol can be replaced by an equals sign. Furthermore, the coefficients $B_k(\boldsymbol{\rho})$ satisfy the following relations:

$$L_{P,k}\left[(-1)^{N_k}N_k!\prod_{i=1}^{k}\frac{B_i(\boldsymbol{\rho})^{\lambda_i}}{\lambda_i!}\right] = L_{P,k}\left[(-1)^{N_k}\prod_{i=1}^{k}\frac{(\rho_i)_{\lambda_i}}{\lambda_i!}\, C_i^{\lambda_i}\right], \tag{8.63}$$

and

$$B_k(\boldsymbol{\mu} + \boldsymbol{v}) = \sum_{j=0}^{k}B_j(\boldsymbol{\mu})\, B_{k-j}(\boldsymbol{v}). \tag{8.64}$$

Remark. The reader should observe that in previous cases involving a constant power of ρ the Pochhammer symbol appeared outside the product in the argument of the partition operator as in Corollary 2 to Theorem 4.1. This was because the ρ-dependence of the coefficients in the resulting power series was only affected by the total number of parts in each partition. In the above case each part i is assigned a value that is dependent upon ρ_i and thus, $(\rho_i)_{\lambda_i}$ now appears inside the product in the argument of the partition operator.

Proof. We begin by employing Lemma A.1 in Appendix A again. As a consequence, the infinite product on the lhs of (8.61) can be expressed as

$$\prod_{k=1}^{\infty}(1 + C_k z^k)^{\rho_k} \equiv \left(1 + \sum_{j=1}^{\infty}\frac{(-\rho_1)_j}{j!}(-C_1 z)^j\right)\left(1 + \sum_{j=1}^{\infty}\frac{(-\rho_2)_j}{j!}(-C_2 z^2)^j\right)$$

$$\times \left(1 + \sum_{j=1}^{\infty}\frac{(-\rho_3)_j}{j!}(-C_3 z^3)^j\right)\left(1 + \sum_{j=1}^{\infty}\frac{(-\rho_4)_j}{j!}(-C_4 z^4)^j\right)\cdots. \quad (8.65)$$

Also from Lemma A.1, each binomial series in the above product is absolutely convergent in the disk given by $|z| < |C_i|^{-1/i}$. In addition, each series is conditionally convergent in a specific region of the complex plane. From the proof of Theorem 8.1, which deals with the $\rho_k = -1$ and $C_k = -1$ case for all k, we found that these regions overlap each other when all the series appear together in an infinite product as on the lhs of (8.61). As a result, we were left with a region (the unit disk) in which all the series were only absolutely convergent. That is, there was a circle of singularities separating the divergent values from the absolutely convergent values. Therefore, if $S = \inf|C_k|^{-1/k}$ does not vanish, then all the series in (8.65) are absolutely convergent whenever $|z| < S$ and the equivalence symbol can be replaced by an equals sign. Furthermore, if the ρ_k all equal a positive integer, say ℓ, then the binomial series become polynomials of degree ℓ and we can also replace the equivalence symbol by an equals sign for any value of z.

Expanding the product of all the series in (8.65) in powers of z yields

$$\prod_{k=1}^{\infty}(1 + C_k z^k)^{\rho_k} \equiv 1 - (-\rho_1)_1\, C_1\, z + \left(\frac{(-\rho_1)_2}{2!}\, C_1^2 - (-\rho_2)_1\, C_2\right)z^2$$

$$+ \left(-\frac{(-\rho_1)_3}{3!}\, C_1^3 + (-\rho_1)_1\,(-\rho_2)_1\, C_1\, C_2 - (-\rho_3)_1\, C_3\right)z^3 + \cdots. \quad (8.66)$$

From this result we see that there is one term appearing in the first order coefficient in z, while there are two terms appearing in the second order coefficient. The third order coefficient is composed of three terms. Had the fourth order term been displayed, there would have been five terms in the coefficient. In fact, the number of terms in the k-th order coefficient is the number of partitions summing to k or $p(k)$. Therefore, we need to develop a means of coding the partitions as was done with the partition method for a power series expansion in Theorem 4.1.

Since the first order term corresponds to $\{1\}$, we assign a value of $-C_1$ to each occurrence of a one in a partition. The first term in the second order coefficient possesses a factor of C_1^2. Therefore, it must correspond to $\{1_2\}$, while the other term must correspond to the other partition summing to 2, viz. $\{2\}$. This term possesses a factor of $-C_2$. Hence each occurrence of a two in a partition yields a factor of $-C_2$. If we continue this process indefinitely for the higher orders, then we find that each occurrence of a part i yields a factor of $-C_i$.

This, however, is not the complete story. Accompanying the factor of $-C_1$ in the first order term is the factor of $(-\rho_1)_1$, while the first and second terms in the second order coefficient possess factors of $(-\rho_1)_2/2!$ and $(-\rho_2)_1$, respectively. That is, when there is one occurrence of a part i in a partition, its assigned value must be multiplied by $(-\rho_i)_1$, but if there are two occurrences of a part in a partition, then the assigned value must be multiplied by $(-\rho_i)_2/2!$. Therefore, if a part i occurs λ_i times in a partition, then it contributes a factor of $(-\rho_i)_{\lambda_i}(-C_i)^{\lambda_i}/\lambda_i!$. This can be checked with the various quantities comprising the third order term. Hence the total contribution made by a partition is given by the product over all parts, viz. $\prod_{i=1}^{k}(-\rho_i)_{\lambda_i}(-C_i)^{\lambda_i}/\lambda_i!$. Finally, the coefficient $B_k(\boldsymbol{\rho})$ is evaluated by summing over all partitions summing to k, which yields (8.62).

If we assume that $z < S$, which is actually not necessary, we can invert the equation form of (8.61). Hence we obtain

$$\prod_{k=1}^{\infty}(1 + C_k z^k)^{-\rho_k} = \frac{1}{1 + \sum_{k=1}^{\infty} B_k(\boldsymbol{\rho}) z^k} . \tag{8.67}$$

Because the rhs can be regarded as the regularized value of the geometric series, we can apply the method for a power series expansion where the coefficients of the inner series, viz. p_k in Theorem 4.1, are equal to $-B_k(\boldsymbol{\rho})$, while the coefficients of the outer series q_k are once again equal to $(-1)^k$. Then we find that

$$\prod_{k=1}^{\infty}(1 + C_k z^k)^{-\rho_k} \equiv 1 + \sum_{k=1}^{\infty} G_k z^k , \tag{8.68}$$

where according to (4.5), the coefficients of the above power series expansion are given by

$$G_k = L_{P,k}\left[(-1)^{N_k} N_k! \prod_{i=1}^{k} \frac{B_i(\boldsymbol{\rho})^{\lambda_i}}{\lambda_i!}\right].$$ (8.69)

From earlier in the proof we know that the above product can be expressed in terms of a power series of the form given on the rhs of (8.61) except that now the ρ_k are replaced by $-\rho_k$. Because the resulting power series expansions possess the same regularized value, they are equal to one another. Moreover, since z is arbitrary, we can equate like powers of z, thereby arriving at (8.63).

The final identity is easily proved by noting that

$$\prod_{i=1}^{\infty}\left(1 + C_k z^k\right)^{\mu_k + \nu_k} = \prod_{i=1}^{\infty}\left(1 + C_k z^k\right)^{\mu_k}\left(1 + C_k z^k\right)^{\nu_k}.$$ (8.70)

The lhs represents the regularized value for the series on the rhs of (8.61) with coefficients $B_k(\boldsymbol{\mu} + \boldsymbol{\nu})$, while the rhs represents the regularized value of the product of two series, one with coefficients $B_k(\boldsymbol{\mu})$ and the other with coefficients $B_k(\boldsymbol{\nu})$. Because the regularized value is the same for both cases, we have

$$1 + \sum_{k=1}^{\infty} B_k(\boldsymbol{\mu} + \boldsymbol{\nu}) z^k = \left(1 + \sum_{j=1}^{\infty} B_k(\boldsymbol{\mu}) z^k\right)\left(1 + \sum_{k=1}^{\infty} B_k(\boldsymbol{\nu}) z^k\right)$$

$$= 1 + \sum_{k=1}^{\infty} z^k \sum_{j=0}^{k} B_j(\boldsymbol{\mu}) B_{k-j}(\boldsymbol{\nu}).$$ (8.71)

Since z is arbitrary, we can equate like powers yet again. In doing so, we obtain (8.64), which completes the proof.

In order to make the material in the corollary clearer, let us consider a few examples. We have already generalized the discrete partition polynomials $q(k, \omega)$ by introducing the power of ρ into their associated product as demonstrated by (8.59). It was also found that the new polynomials $q(k, \omega, \rho)$ could be expressed in terms of the partition operator acting on the discrete partition polynomials given by (8.60). Now we apply the corollary to Theorem 8.2 to the product where the C_k and ρ_k are set equal to ω and ρ, respectively. Consequently, we arrive at

$$q(k, \omega, \rho) = L_{P,k}\left[(-\omega)^{N_k} \prod_{i=1}^{k} \frac{(-\rho)_{\lambda_i}}{\lambda_i!}\right].$$ (8.72)

So, let us evaluate $q(3, \omega, \rho)$ via (8.60) and (8.72). With regard to (8.60) we require $q(1, \omega)$, $q(2, \omega)$ and $q(3, \omega)$, which have already been evaluated and are displayed in the second column of Table 8.1. Hence we find that

$$q(3, \omega, \rho) = -\frac{(-\rho)_3}{3!}\omega^3 + (-\rho)_2\,\omega^2 - (-\rho)_1\,(\omega^2 + \omega)\,. \tag{8.73}$$

After a little algebra, the above result reduces to

$$q(3, \omega, \rho) = \left(\frac{\rho^3}{6} - \frac{\rho^2}{2} + \frac{\rho}{3}\right)\omega^3 + \rho^2\omega^2 + \rho\omega\,. \tag{8.74}$$

By setting $k = 3$ in (8.72), we obtain

$$q(3, \omega, \rho) = -\frac{(-\rho)_3}{3!}\omega^3 + (-\rho)_1\,(-\rho)_1\,\omega^2 - (-\rho)_1\,\omega\,. \tag{8.75}$$

Again after a little algebra, we end up with (8.74), although we can see that both (8.73) and (8.75) are composed of different quantities. This demonstrates that while (8.60) and (8.72) both employ the partition operator, they are in fact different representations of the $q(k, \omega, \rho)$.

The next example is a result attributed to Euler that appears on p. 56 of Ref. [13]. This is

$$\prod_{k=1}^{\infty}(1 - z^k)^3 = 1 - 3z + 5z^3 - 7z^6 + \cdots = \sum_{k=0}^{\infty}(-1)^k(2k+1)z^{\binom{k+1}{2}}\,. \tag{8.76}$$

From (8.59), we have

$$\prod_{k=1}^{\infty}(1 - z^k)^3 = 1 + \sum_{k=1}^{\infty}q(k, -1, 3)\,z^k\,. \tag{8.77}$$

Note that the equivalence symbol has been replaced by an equals sign because $\rho = 3$ in this instance. Euler's result means that the coefficients of the generating function in (8.77) are non-zero only when the power of z is equal to another type of figurate number known as a triangular number [72,73]. If we equate like powers of z in both power series expansions given above and use (8.60), then we arrive at

$$L_{P,k}\left[(-1)^{N_k}\,(-3)_{N_k}\prod_{i=1}^{k}\frac{q(i)^{\lambda_i}}{\lambda_i!}\right] = \begin{cases} (-1)^j(2j+1), & \text{if } k = \binom{j+1}{2}, \\ 0, & \text{otherwise}\,. \end{cases} \tag{8.78}$$

In deriving this result, $q_i(i, -1)$ has been substituted by the discrete partition numbers $q(i)$ according to (8.46). Alternatively, putting $C_i = -1$ and $\rho_i = 3$ in (8.72) yields

$$q(k, -1, 3) = L_{P,k}\left[\prod_{i=1}^{k}\frac{(-3)_{\lambda_i}}{\lambda_i!}\right]\,. \tag{8.79}$$

In the above result $(-3)_{\lambda_i}$ is only non-zero for λ_i equal to 1, 2 and 3, in which case it equals -3, 6 and -6, respectively. This is a different class of restricted partition from those we encountered in Chapter 7 since it means that partitions in which a part appears more than three times are excluded. To generate such partitions all that needs to be done is to scan each partition twice by introducing another for loop in **termgen** of **partgen** in Chapter 3. If in the first scan the multiplicity λ_i or variable *freq* is greater than 3, then a goto statement is required so that the program avoids the next for loop, which is responsible for printing out the specific partitions. When such a code is constructed, one will discover that out of the total of 627 partitions summing to 20, there are only 320 partitions in which all the parts occur at most three times. As a result of (8.79), we have

$$
L_{P,k}\left[\prod_{i=1}^{k} \frac{(-3)_{\lambda_i}}{\lambda_i!}\right] = \begin{cases} (-1)^j(2j+1), & \text{if } k = \binom{j+1}{2}, \\ 0, & \text{otherwise}. \end{cases} \tag{8.80}
$$

Moreover, it should be noted that if $\rho = 1$ such that $\lambda_i \le 1$ for all parts i in a partition, then the sum over all partitions in (8.80) reduces to the discrete partition operator, viz. $L_{DP,k}[\cdot]$, regardless of the values for the coefficients C_i.

If we put ω equal to unity in (8.59), take the cube power, i.e. set $\rho = 3$, and multiply out the three series on the rhs, then by equating like powers of the resulting series with the rhs of (8.77), we obtain

$$
q(k, -1, 3) = \sum_{j_1=0}^{k} \sum_{j_2=0}^{j_1} q(k-j_1)q(j_1-j_2)q(j_2), \tag{8.81}
$$

while multiplying the $\rho = 1$ and $\rho = 2$ versions of the series on the rhs of (8.59) gives

$$
q(k, -1, 3) = \sum_{j=0}^{k} q(j)q(k-j, -1, 2) = \begin{cases} (-1)^i(2i+1), & k = \binom{i+1}{2}, \\ 0, & \text{otherwise}. \end{cases} \tag{8.82}
$$

In the above result $q(k, \omega, 2)$ can be evaluated by putting $\rho = 2$ in (8.60). They can also be determined by equating like powers of z from taking the square of the $\rho = 1$ version on the rhs of (8.59) with those of the $\rho = 2$ version of the same equation. Then we find that

$$
q(k, \omega, 2) = \sum_{i=0}^{k} q(i, \omega)q(k-i, \omega). \tag{8.83}
$$

Table 8.2 Coefficients $q(k, \omega, 2)$ and $q(k, \omega, 3)$ in the power series expansions of the $\rho = 2$ and $\rho = 3$ cases of (8.59)

k	$q(k, \omega, 2)$	$q(k, \omega, 3)$
0	1	1
1	2ω	3ω
2	$\omega^2 + 2\omega$	$3\omega^2 + 3\omega$
3	$4\omega^2 + 2\omega$	$\omega^3 + 9\omega^2 + 3\omega$
4	$2\omega^3 + 5\omega^2 + 2\omega$	$9\omega^3 + 12\omega^2 + 3\omega$
5	$4\omega^3 + 8\omega^2 + 2\omega$	$3\omega^4 + 18\omega^3 + 18\omega^2 + 3\omega$
6	$\omega^4 + 10\omega^3 + 9\omega^2 + 2\omega$	$12\omega^4 + 37\omega^3 + 21\omega^2 + 3\omega$
7	$4\omega^4 + 14\omega^3 + 12\omega^2 + 2\omega$	$3\omega^5 + 33\omega^3 + 54\omega^3 + 27\omega^2 + 3\omega$
8	$9\omega^4 + 22\omega^3 + 13\omega^2 + 2\omega$	$12\omega^5 + 66\omega^4 + 81\omega^3 + 30\omega^2 + 3\omega$
9	$2\omega^5 + 16\omega^4 + 30\omega^3 + 16\omega^2$ $+ 2\omega$	$\omega^6 + 39\omega^5 + 114\omega^4 + 109\omega^3 + 36\omega^2$ $+ 3\omega$
10	$4\omega^5 + 30\omega^4 + 40\omega^3 + 17\omega^2$ $+ 2\omega$	$9\omega^6 + 81\omega^5 + 189\omega^4 + 144\omega^3 + 39\omega^2$ $+ 3\omega$
11	$12\omega^5 + 44\omega^4 + 50\omega^3 + 20\omega^2$ $+ 2\omega$	$27\omega^6 + 168\omega^5 + 279\omega^4 + 180\omega^3 + 45\omega^2$ $+ 3\omega$
12	$\omega^6 + 22\omega^5 + 66\omega^4 + 64\omega^3$ $+ 21\omega^2 + 2\omega$	$3\omega^7 + 73\omega^6 + 291\omega^5 + 402\omega^4 + 226\omega^3$ $+ 48\omega^2 + 3\omega$
13	$4\omega^6 + 40\omega^5 + 92\omega^4 + 76\omega^3$ $+ 24\omega^2 + 2\omega$	$12\omega^7 + 162\omega^6 + 483\omega^5 + 552\omega^4 + 270\omega^3$ $+ 54\omega^2 + 3\omega$
14	$9\omega^6 + 64\omega^5 + 127\omega^4 + 92\omega^3$ $+ 25\omega^2 + 2\omega$	$42\omega^7 + 315\omega^6 + 744\omega^5 + 741\omega^4 + 324\omega^3$ $+ 57\omega^2 + 3\omega$
15	$20\omega^6 + 104\omega^5 + 164\omega^4$ $+ 108\omega^3 + 28\omega^2 + 2\omega$	$3\omega^8 + 102\omega^7 + 569\omega^6 + 1119\omega^5 + 957\omega^4$ $+ 379\omega^3 + 63\omega^2 + 3\omega$

Table 8.2 displays both $q(k, \omega, 2)$ and $q(k, \omega, 3)$ up to $k = 15$. Interestingly, for $\omega = -1$, (8.83) reduces to

$$q(k, -1, 2) = \sum_{i=0}^{k} q(i)q(k - i) . \tag{8.84}$$

Thus, we see that $q(k, -1, 2)$ and $q(k, -1, 3)$ can be expressed as Cauchy products involving only the discrete partition numbers.

On p. 23 of Ref. [12] Andrews states that Gauss derived the following result:

$$\sum_{k=0}^{\infty} z^{(k^2+k)/2} = \prod_{k=1}^{\infty} \frac{(1 - z^{2k})}{(1 - z^{2k-1})} = \prod_{k=1}^{\infty} (1 - z^k)(1 + z^k)^2 . \tag{8.85}$$

Once again, the equals sign is only valid for $|z| < 1$. From this result we see that only the powers of z equal to $(k^2 + k)/2$, where k is any non-negative integer, possess a non-zero coefficient. For any power k we obtain contributions from all the partitions summing to k with only odd parts and from the discrete partitions summing to k consisting of only even parts. If the number of even parts in these discrete partitions is even, then the partition will yield a value of unity. Otherwise, it will yield a value of -1. In addition, the generating function possesses mixed partitions composed of discrete partitions with only even parts and standard partitions with only odd parts. The values contributed by the mixed partitions to the coefficients of the power series depend upon the number of parts in the discrete partitions. For example, if we consider z^{10} or $k = 4$ on the lhs of (8.85), then it will be composed of the contributions due to the partitions summing to 10 with only odd parts. There are 10 of these beginning with $\{1, 9\}$ and ending with $\{5_2\}$. Hence these partitions contribute a value of 10 to the coefficient. On the other hand, there are only 3 discrete partitions with even parts summing to 10. These are $\{10\}$, $\{2, 8\}$ and $\{4, 6\}$. Since the last two possess an even number of parts, they each contribute a value of unity, while the single part partition gives a value of -1. Overall, the discrete partitions summing to 10 contribute a value of unity to the coefficient, which now becomes 11 when the contribution from the standard partitions summing to 10 with only odd parts is included. However, when the discrete partition is $\{2\}$, we need to consider the partitions summing to 8 with odd parts. There are six of these, beginning with $\{1, 7\}$ and ending with $\{3, 5\}$. Because there is only one part in the discrete partition, these mixed partitions contribute a value of -6 to the coefficient, which now drops to 5. However, there are still more mixed partitions. We need to consider the discrete partition of $\{4\}$. In this case we need the partitions summing to 6 with odd parts. There are four of these, beginning with $\{1, 5\}$ and ending with $\{3_2\}$. In this instance the mixed partitions contribute a value of -4, yielding a value of 1 for the coefficient as indicated above. We do not need to consider the contributions where the discrete partitions sum to either 6 or 8 because in these cases there are only 2 partitions, one of which has two parts and the other with a single part. Hence they cancel each other yielding a value of 0. Thus, from this analysis we observe that the coefficient of z^{10} in (8.85) is equal to unity.

If we introduce (8.59) into (8.85), then we find that

$$\sum_{k=0}^{\infty} z^{k^2+k/2} = 1 + \sum_{k=0}^{\infty} z^k \sum_{j=0}^{k} q(j)q(k-j,1,2) . \tag{8.86}$$

By equating like powers of z, we obtain for i, a positive integer,

$$\sum_{j=0}^{k} q(j)q(k-j,1,2) = \begin{cases} 1, & k = (i^2 + i)/2 , \\ 0, & \text{otherwise} . \end{cases} \tag{8.87}$$

Therefore, for k not equal to a triangular number [72,73], combining the above result with (8.82) yields

$$\sum_{j=0}^{k} q(j)\Big(q(k-j,1,2) \pm q(k-j,-1,2)\Big) = 0 . \tag{8.88}$$

In this chapter we have been able to apply the partition method to the partition-number generating function, whose infinite product is $P(z)$. Inverting this product has enabled us to study discrete partitions in detail and demonstrate that the generating functions of more general infinite products of the form of (8.25) can be evaluated by using the discrete partition operator, which has a major advantage over the partition operator in that there are substantially less discrete or distinct partitions than integer partitions. As a consequence, combinatorial explosion will occur at much higher values of k than using the partition operator. E.g., for partitions summing to 100, there are only 44,205 discrete partitions compared with 190,569,292 integer partitions. Thus, there is a significant reduction in computational effort if the discrete partition operator can be applied instead of the partition operator, but, of course, this is dependent upon the form of the infinite product in the first place. Finally, since we have been able to generalize the inverted form of $P(z)$ by introducing a coefficient ω next to the power of z and considering arbitrary powers, we shall do likewise to $P(z)$ in the following chapter, thereby enabling us to consider problems where the discrete partition operator cannot be employed.

CHAPTER 9

Generalization of the Partition-Number Generating Function

As a result of the material presented in the previous chapter we are in a position to study the generating functions of more advanced infinite products. We begin by introducing the coefficient ω next to the power of z in the infinite product, $P(z)$ given by (6.2). Based on our examination of the inverted form of $P(z)$ in the previous chapter, we expect to obtain polynomial coefficients in the resulting generating function. Thus, the first generalization of $P(z)$ becomes

$$P(z, \omega) := \prod_{k=1}^{\infty} \frac{1}{(1 - \omega z^k)} . \tag{9.1}$$

According to p. 112 of Ref. [13], this infinite product can also be written as

$$P(z, \omega) = (1 - \omega) \sum_{k=0}^{\infty} \frac{\omega^k z^{k^2}}{(z; z)_k \, (\omega; z)_{k+1}} , \tag{9.2}$$

while inverting (8.48) yields

$$P(z, \omega) = \frac{1}{1 + \sum_{k=1}^{\infty} q(k, -\omega) z^k} . \tag{9.3}$$

We have seen that $P(z)$ or $\omega = 1$ in (9.1) yields a power series expansion whose coefficients are given by the partition-number function, $p(k)$. This expansion is obtained by expanding each term in the generating function into the geometric series for each value of i. It is this value in the generating function, which is responsible for yielding the specific parts in a partition, while the power of z^i in each geometric series gives the multiplicity of part i in the partition. For example, multiplying $(z^2)^3$ arising from the expansion of $1/(1 - z^2)$ by $(z^3)^4$ from the expansion of $1/(1 - z^3)$ means that the partition will have three twos and four threes. By introducing ω into the generating function as above, we see that the overall power of ω yields the total number of parts or length of a partition. In the example just mentioned, we would obtain $(\omega z^2)^3$ multiplied by $(\omega z^3)^4$ from the

The Partition Method for a Power Series Expansion.
DOI: http://dx.doi.org/10.1016/B978-0-12-804466-7.00009-7

expansion of (9.1), which yields $\omega^7 z^{18}$. The power of 7 for ω represents the total number of twos and threes in the partition. Therefore, we expect the coefficient of each power of ω in the coefficients of the resulting generating function to yield the total number of partitions where the number of parts equals the power of ω.

By regarding the rhs of (9.3) as the regularized value of the geometric series and the $q(k, -\omega)$ as coefficients of the inner series in Theorem 4.1, i.e. $p_k = q(k, -\omega)$, we can apply the theorem once more by setting $y = z$. If we express the generating function as a power series expansion with coefficients, $p(k, \omega)$, viz.

$$P(z, \omega) = \prod_{k=1}^{\infty} \frac{1}{(1 - \omega z^k)} \equiv 1 + \sum_{k=1}^{\infty} p(k, \omega) z^k, \qquad (9.4)$$

then from (4.5) the coefficients are given by

$$p(k, \omega) = L_{P,k} \left[(-1)^{N_k} N_k! \prod_{i=1}^{k} \frac{q(i, -\omega)^{\lambda_i}}{\lambda_i!} \right]. \qquad (9.5)$$

Furthermore, by putting $\rho = -1$ and $C_i = -\omega$ in the Corollary to Theorem 8.2, we find that the $p(k, \omega)$ become the coefficients, $B_k(-1)$ given by (8.62). Thus, we arrive at

$$p(k, \omega) = L_{P,k} \left[\omega^{N_k} \right]. \qquad (9.6)$$

This tells us that the coefficients in the $p(k, \omega)$ will be the number of partitions summing to k where each power of ω corresponds to the number of parts in the partitions. That is,

$$p(k, \omega) = \sum_{i=1}^{k} p_i(k) \omega^i, \qquad (9.7)$$

where from Chapter 3, $p_i(k) = \begin{vmatrix} k \\ i \end{vmatrix}$. It has already been stated that the sub-partition numbers obey the recurrence relation given by (3.1). In terms of the fixed number of parts operator defined by (6.16) we also obtain

$$p_i(k) = L_{P,k}^i \left[1 \right]. \qquad (9.8)$$

The first few partition function polynomials are found to be: $p(0, \omega) = 1$, $p(1, \omega) = \omega$, $p(2, \omega) = \omega^2 + \omega$, $p(3, \omega) = \omega^3 + \omega^2 + \omega$ and $p(4, \omega) = \omega^4 + \omega^3 + 2\omega^2 + \omega$. In fact, those up to $k = 15$ are displayed in the third column of

Table 8.1 next to the discrete partition polynomials, $q(k, \omega)$. The coefficients can also be obtained by running the program **numparts** discussed in Chapter 6. For example, by running the program with *tot* and *numparts* set equal to 13 and 5 respectively, we obtain 18 five-part partitions summing to 13, which represents the coefficient of ω^5 for $p(13, \omega)$ in the table. Moreover, the results have been checked by setting ω equal to unity, then summing the coefficients for each value of k and observing that these sums yield the same values as the PartitionsP routine in Mathematica [32].

Since the coefficients of the partition function polynomials represent the total number of partitions where the number of parts or length is equal to the power of ω, the highest order term of these polynomials is k, which is due to the k-part partition or $\{1_k\}$. The other partitions are unable to contribute an ω^k term because the highest order of all other $q(k, \omega)$ is less than k. The partition $\{1_{k-1}, 2\}$ only yields an ω^{k-1} term as its highest order term because the highest order term of $q(2, \omega)$ is 1. Hence $\deg p(k, \omega) = k$. Conversely, the lowest order term in the $p(k, \omega)$ is the lowest order term in $q(k, \omega)$ stemming from the single part partition, $\{k\}$. Its coefficient is unity. Therefore, we find that $p_k(k) = p_{k-1}(k) = p_1(k) = 1$, $p_{k-2}(k) = 2$ and $p(k, 1) = \sum_{i=1}^{k} p_i(k) = p(k)$. Furthermore, the total number of partitions with an even number of parts or even partitions is given by $\sum_{i=1}^{\lfloor k/2 \rfloor} p_{2i}(k)$, while the total number of partitions with an odd number of parts or odd partitions is equal to $\sum_{i=1}^{m} p_{2i-1}(k)$, where $m = k/2$ when k is even and $m = \lfloor k/2 \rfloor + 1$ when k is odd. On the other hand, the coefficient $p_2(k)$ can be evaluated by noting that it represents the number of all the 2-part partitions, namely $\{j, k - j\}$, where j ranges from 1 to $\lfloor k/2 \rfloor$. Therefore, the number of 2-part partitions summing to k is given by $p_2(k) = \lfloor k/2 \rfloor$, while according to Ref. [70], the number of 3-part partitions is given by $p_3(k) = \lfloor k^2/12 \rfloor$ for $k > 3$. In addition, a table of the subpartition numbers for k and i ranging from 0 to 11 is presented on p. 46 of Ref. [13]. All these results agree with those obtained from (9.5), confirming that the latter result does yield polynomials where the coefficient of the i-th power represents the number of partitions summing to k with i parts.

An interesting property of the $p_i(k)$ or $\begin{vmatrix} k \\ i \end{vmatrix}$ is that they appear to saturate or plateau once k is sufficiently large. For example, the coefficient of the third highest order term in the partition function polynomials in Table 8.1, viz. $p_{k+1-3}(k)$, is equal to one for $k = 3$, but for all values of $k > 3$ in the table it remains equal to two. Similarly, if we examine the coefficient of the sixth highest order terms of the $p(k, \omega)$ or $p_{k+1-6}(k)$, then for $k = 6, 7, 8$ and

9, the coefficients equal 1, 3, 5 and 6 respectively. However, for $k > 9$, the coefficient is always equal to 7. This property or behaviour applies to the other coefficients up to the eighth highest order terms in the table, while the coefficients of the ninth highest order term and beyond have not had the opportunity to reach the saturation point because k is not sufficiently large in the table.

Another interesting property of the partition function polynomials can be observed by setting $z = z^2$ and $\omega = \omega^2$ in (9.1). Then we obtain

$$P\left(z^2, \omega^2\right) = \prod_{k=1}^{\infty} \frac{1}{(1 - \omega^2 z^{2k})} \equiv 1 + \sum_{k=1}^{\infty} p\left(k, \omega^2\right) z^{2k}. \tag{9.9}$$

The quantity on the lhs of the above equivalence can also be expressed as

$$P\left(z^2, \omega^2\right) = P(z, \omega) P(z, -\omega). \tag{9.10}$$

Introducing the rhs of (9.1) into the above equation yields

$$P(z, \omega) P(z, -\omega) \equiv \left(1 + \sum_{k=1}^{\infty} p(k, \omega) z^k\right)\left(1 + \sum_{k=1}^{\infty} p(k, -\omega) z^k\right). \tag{9.11}$$

Since the series on the rhs's of the above equivalences possess the same regularized value, namely $P(z^2, \omega^2)$, they are equal to one another in accordance with the concept of regularization [4–6,18,19,43,44]. Moreover, because z is arbitrary, we can again equate like powers of z. Therefore, we arrive at

$$\sum_{j=0}^{2k+1} p(j, \omega) p(2k + 1 - j, -\omega) = 0, \tag{9.12}$$

and

$$\sum_{j=0}^{2k} p(j, \omega) p(2k - j, -\omega) = p\left(2k, \omega^2\right). \tag{9.13}$$

As in the previous chapter we can generalize the preceding analysis by replacing $-\omega$ in $P(z, \omega)$ by C_k, which effectively represents the inversion of (8.25). From Lemma A.1 in Appendix A we have

$$\begin{aligned} H(z)^{-1} &= \prod_{k=1}^{\infty} \frac{1}{(1 + C_k z^k)} \\ &\equiv \left(1 - C_1 z + C_1^2 z^2 + \cdots\right)\left(1 - C_2 z^2 + C_2^2 z^4 + \cdots\right) \\ &\quad \times \left(1 - C_3 z^3 + C_3^2 z^6 + \cdots\right)\left(1 - C_4 z^4 + C_4^2 z^8 + \cdots\right)\cdots, \end{aligned} \tag{9.14}$$

where the equivalence symbol can be replaced by an equals sign provided that $|C_i z^i| < 1$ for all i. Expanding the above yields

$$H(z)^{-1} \equiv 1 - C_1 z + \left(C_1^2 - C_2\right) z^2 + \left(-C_1^3 + C_2\, C_1 - C_3\right) z^3 + \cdots . \quad (9.15)$$

The coefficient of each power of z in the above result is composed of contributions that can be related to the partitions summing to k according to Theorem 4.1. The major difference between the above situation and that in the proof of Theorem 4.1 is that there is no multinomial factor associated with each contribution made by a partition as was the case in the proof of the corollary to Theorem 8.2. Furthermore, each part i in a partition is now assigned a value of $-C_i$. Hence (9.15) becomes

$$H(z)^{-1} \equiv \sum_{k=0}^{\infty} H_k\, z^k , \quad (9.16)$$

where $H_0 = 1$, and

$$H_k = L_{P,k}\left[(-1)^{N_k} \prod_{i=1}^{k} C_i^{\lambda_i} \right] . \quad (9.17)$$

This result represents the case of all the ρ_k being set equal to -1 in the corollary to Theorem 8.2.

If the C_i are set equal to unity in (9.14), then we find that

$$\prod_{i=1}^{\infty} \frac{1}{\left(1 + z^i\right)} \equiv 1 + \sum_{k=1}^{\infty} L_{P,k}\left[(-1)^{N_k} \right] z^k , \quad (9.18)$$

where the equivalence symbol can be replaced by an equals sign when $|z| < 1$. Therefore, the coefficients of the power series expansion on the rhs represent the difference between the number of even and odd partitions summing to k. Both classes of partitions are discussed at length in Chapter 6. In addition, the above equivalence is analogous to putting the C_i equal to unity in (8.27) and applying Theorem 4.1 to its inverted form. In this case the coefficients of the inner series are given by $p_k = L_{DP,k}[1]$, the number of discrete partitions, while the coefficients of the outer series are given by $q_k = (-1)^k$. Then the coefficients of the power series expansion on the rhs of (9.18) can be expressed in terms of the number of discrete partitions as

$$L_{P,k}\left[(-1)^{N_k} \right] = L_{P,k}\left[(-1)^{N_k} N_k! \prod_{i=1}^{k} L_{DP,i}[1]^{\lambda_i} / \lambda_i! \right] . \quad (9.19)$$

If we multiply the product on the lhs of (9.18) by $(1 - z^i)$ in both the numerator and denominator, then we find that

$$\prod_{i=1}^{\infty} \frac{1}{(1 + z^i)} = \prod_{i=0}^{\infty}(1 - z^{2i+1}) . \qquad (9.20)$$

When the product on the rhs of (9.20) appeared with powers of z^i only rather than z^{2i+1} at the beginning of Chapter 8, we saw that the resulting power series possessed coefficients which were equal to the number of discrete partitions summing to k. In the above result all the even powers are now missing. This means that the coefficients of the resulting power series will be the number of discrete partitions with only odd parts. That is,

$$\prod_{i=0}^{\infty}(1 - z^{2i+1}) = 1 + \sum_{k=0}^{\infty}(-1)^k L_{ODP,k}\Big[1\Big] z^k , \qquad (9.21)$$

where ODP denotes that only partitions with odd parts are to be considered in the sum over partitions. Therefore, $\lambda_{2i} = 0$ for all values of i. The phase factor of $(-1)^k$ in the series expansion arises from the fact that only an even number of odd discrete parts yields an even power of z, while only an odd number of discrete parts yields an odd power of z. That is,

$$(-1)^k L_{ODP,k}\Big[1\Big] = L_{DP,k}\Big[\prod_{i=1}^{\lfloor k/2 \rfloor}(-1)^{\lambda_{2i-1}}\Big]. \qquad (9.22)$$

In a similar fashion we arrive at

$$\prod_{i=0}^{\infty}(1 + z^{2i}) = 1 + \sum_{k=1}^{\infty} L_{EDP,2k}\Big[1\Big] z^{2k} , \qquad (9.23)$$

where only even parts are allowed in the even discrete partition operator, i.e. $\lambda_{2i+1} = 0$ for all i. Alternatively, we can replace the even discrete partition operator by the discrete partition operator since

$$L_{EDP,2k}\Big[1\Big] = L_{DP,k}\Big[1\Big]. \qquad (9.24)$$

Moreover, by equating like powers of z in the power series on both rhs's of (9.18) and (9.21), we arrive at

$$L_{P,k}\Big[(-1)^{N_k}\Big] = (-1)^k L_{ODP,k}\Big[1\Big]. \qquad (9.25)$$

From this result we see that when k is even, the number of even partitions is greater than the number of odd partitions, while for odd values of k, the opposite applies. Multiplying both sides by $(-1)^k$ results in taking the absolute value or modulus of the lhs. Thus, the above statement tells us that the absolute value of the difference between the number of even and odd partitions is equal to the number of discrete partitions with only odd parts in them or the number of partitions with distinct odd parts, a result first proved by Euler according to p. 14 of Ref. [12].

As a result of the previous chapter, it is a relatively simple exercise to produce a code that evaluates the difference between the number of partitions with an even number of parts and those with an odd number of parts. Two new global variables are required, one for evaluating the difference as each partition is scanned and another that is either equal to 1 or -1 depending on whether there is an even number of parts or an odd number of parts. Once the second value is determined, it needs to be added to the first global variable in **main**. The second global variable must be evaluated in **termgen** after the for loop has been altered to calculate the total number parts in the partition and is determined by summing all components of the integer array *part*. Therefore, one finds after running the code for several values of k that

$$\left|L_{P,k}\left[(-1)^{N_k}\right]\right| \geq \left|L_{P,j}\left[(-1)^{N_j}\right]\right|, \tag{9.26}$$

for $k \geq j$. Fig. 9.1 presents the graph of the ratio of the absolute value of the difference between odd and even partitions summing to k to the total number of partitions or $p(k)$ for $k \leq 50$. Whilst the absolute value of the $L_{P,k}[(-1)^{N_k}]$ increases with k, we see that in relation to the total number of partitions the ratio decreases monotonically, once k exceeds 15.

From (8.44) and (8.46) we notice that the infinite products, $\prod_{k=1}^{\infty}(1 + z^k)$ and $1/P(z)$ can be expressed as generating functions, whose the coefficients are equal to $L_{DP,k}[1]$ and $L_{DP,k}[(-1)^{N_k}]$, respectively. The first product can also be written as

$$\prod_{k=1}^{\infty}\left(1 + z^k\right) = \prod_{k=1}^{\infty}\frac{1}{\left(1 - z^{2k-1}\right)}. \tag{9.27}$$

This result is simply obtained by manipulating the rhs after multiplying the numerator and denominator by $\left(1 - z^{2k}\right)$. If we put $C_k = -1$ in (9.14), then using (9.16) and (9.17) we arrive at

$$\prod_{k=1}^{\infty}\left(1 + z^k\right) \equiv 1 + \sum_{k=1}^{\infty}L_{OP,k}\left[1\right]z^k, \tag{9.28}$$

where $L_{OP,k}[\cdot]$ represents the odd part partition operator, which we have seen has two different forms given by (7.8) and (7.9) depending upon whether k is an even or an odd number. Because the infinite product on

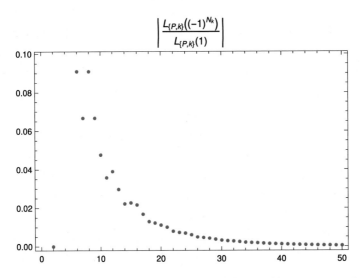

Figure 9.1 The ratio of the difference between even and odd partitions to the total number of partitions versus k.

the lhs also yields a generating function whose coefficients represent the number of discrete partitions summing to k, we see immediately that

$$L_{DP,k}\left[1\right] = L_{OP,k}\left[1\right]. \tag{9.29}$$

Hence the number of discrete partitions is equal to the number of partitions composed only of odd parts, which is another result obtained by Euler according to p. 5 of Ref. [12].

By putting $C_i = -1$ in (8.25), we ended up with (8.46). If we put $C_i = -1$ and $z = z^2$, then the product in (8.25) yields a generating function in powers of z^2 according to (8.26), but now the coefficients represent the number of distinct partitions summing to $2k$ with only even parts operating on $(-1)^{N_{2k}}$. Alternatively, this is equivalent to the number of distinct partitions summing to k operating on $(-1)^{N_k}$. Therefore, we can write

$$\prod_{k=1}^{\infty}\left(1 - z^{2k}\right) = 1 + \sum_{k=1}^{\infty} L_{EDP,2k}\left[(-1)^{N_{2k}}\right] z^{2k}$$

$$= 1 + \sum_{k=1}^{\infty} L_{DP,k}\left[(-1)^{N_k}\right] z^{2k}. \tag{9.30}$$

In (9.30) $L_{EDP,2k}[\cdot]$ denotes the even discrete partition operator, which acts on the number of discrete partitions summing to $2k$ where the parts are only even integers. This is opposed to the odd discrete partition operator, $L_{ODP,k}[\cdot]$, where the parts are odd numbers and discrete, but can sum to both even and odd integers. Since the infinite product of $(1 - z^{2k})$ is the product of two separate infinite products involving $(1 - z^k)$ and $(1 + z^k)$, we have

$$\prod_{k=1}^{\infty}\left(1 - z^{2k}\right) = \left(1 + \sum_{k=1}^{\infty} L_{DP,k}\left[(-1)^{N_k}\right] z^k\right)\left(1 + \sum_{k=1}^{\infty} L_{DP,k}\left[1\right] z^k\right).$$

(9.31)

Equating like powers of z in (9.31) with those of (9.30) yields

$$L_{DP,k}\left[(-1)^{N_k}\right] = \sum_{j=0}^{2k} L_{DP,j}\left[(-1)^{N_j}\right] L_{DP,2k-j}\left[1\right],$$

(9.32)

and

$$\sum_{j=0}^{2k+1} L_{DP,j}\left[(-1)^{N_j}\right] L_{DP,2k+1-j}\left[1\right] = 0.$$

(9.33)

The $j = k$ term in (9.32) can be separated and carried across to the lhs. Then (9.32) becomes

$$\left(1 - L_{DP,k}\left[1\right]\right) L_{DP,k}\left[(-1)^{N_k}\right] = \sum_{j=0}^{k-1}\left(L_{DP,j}\left[(-1)^{N_j}\right] L_{DP,2k-j}\left[1\right]\right.$$
$$\left. + L_{DP,2k-j}\left[(-1)^{N_j}\right] L_{DP,j}\left[1\right]\right).$$

(9.34)

Since $L_{DP,k}[(-1)^{N_k}]$ and $L_{DP,k}[1]$ are equal to $q(k)$ and $q(k,1)$ respectively, (9.32) and (9.33) can also be written as

$$q(k)\left(1 - q(k, 1)\right) = \sum_{j=0}^{k-1}\left(q(j)q(2k - j, 1) + q(2k - j)q(j, 1)\right),$$

(9.35)

and

$$\sum_{j=0}^{k}\left(q(j)q(2k + 1 - j, 1) + q(2k + 1 - j)q(j, 1)\right) = 0.$$

(9.36)

In these results it should be borne in mind that $q(0) = q(0, 1) = 1$. Isolating the $j = 0$ terms in the above equations yields

$$q(2k, 1) + q(2k) = q(k)\left(1 - q(k, 1)\right) - \sum_{j=1}^{k-1}\left(q(j)\, q(2k - j, 1)\right.$$
$$\left. + q(2k - j)\, q(j, 1)\right),$$

(9.37)

and

$$q(2k+1, 1) + q(2k+1) = \sum_{j=1}^{k} \Big(q(j)\, q(2k+1-j, 1)$$

$$+ q(2k+1-j)\, q(j, 1) \Big). \tag{9.38}$$

Both (9.37) and (9.38) represent recurrence relations for determining the number of discrete partitions or $q(k, 1)$. Like the Euler/MacMahon recurrence relation given by (8.10), they utilize the discrete partition numbers or $q(k)$ and consequently, most of the terms in the sums vanish when the summation index j is not equal to a pentagonal number.

It should also be noted that the analysis leading to (8.14) can be adapted to provide another representation for the partition function polynomials or $p(k, \omega)$. First, we re-write the generalized product in (9.4) as

$$P(z, \omega) = \exp\Big(\sum_{m=1}^{\infty} \sum_{j=1}^{\infty} \omega^j\, z^{mj}/j \Big), \tag{9.39}$$

where now it is assumed that $|\omega z| < 1$. Consequently, the modified version of (8.14) becomes

$$P(z, \omega) = 1 + \sum_{k=1}^{\infty} \frac{1}{k!} \Big(\omega z + (\omega + \omega^2/2)\, z^2 + (\omega + \omega^3/3)\, z^3$$

$$+ (\omega + \omega^2/2 + \omega^4/4)\, z^4 + \cdots + \gamma_j(\omega) z^j + \cdots \Big)^k, \tag{9.40}$$

where $\gamma_j(\omega) = \sum_{d|j} (d/j)\, \omega^{j/d}$ and d represents a divisor or factor of j as before. The first few of these polynomials are: $\gamma_0(\omega) = 1$, $\gamma_1(\omega) = \omega$, $\gamma_2(\omega) = \omega + \omega^2/2$, $\gamma_3(\omega) = \omega + \omega^3/3$, and $\gamma_4(\omega) = \omega + \omega^2/2 + \omega^4/4$. As expected, for $\omega = 1$ they reduce to the γ_j below (8.12). This means that (8.15) can be generalized to

$$q(k, -\omega) = L_{P,k}\Big[(-1)^{N_k} \prod_{i=1}^{k} \frac{\gamma_i(\omega)^{\lambda_i}}{\lambda_i!} \Big]. \tag{9.41}$$

Furthermore, the $\gamma_j(\omega)$ are polynomials in ω whose highest and lowest orders are respectively j and unity. That is, deg $\gamma_j(\omega) = j$. We shall refer to these unusual polynomials as the divisor polynomials. By applying Theorem 4.1 to (9.40) with the coefficients of the inner and outer series set equal to $\gamma_k(\omega)$ and $1/k!$ respectively, we arrive at

$$p(k, \omega) = L_{P,k}\Big[\prod_{i=1}^{k} \frac{\gamma_i(\omega)^{\lambda_i}}{\lambda_i!} \Big]. \tag{9.42}$$

Since $\deg \gamma_i(\omega) = i$ and $\prod_{i=1}^{k} \omega^{i\lambda_i} = \omega^k$, the highest order term in the $p(k, \omega)$ is k. On the other hand, since the lowest order term in the $\gamma_i(\omega)$ is ω, $\prod_{i=1}^{k} \omega^{\lambda_i} = \omega^{N_k}$ and N_k ranges from unity to k, the lowest order term in the $p(k, \omega)$ is unity, again confirming that the partition number polynomials are polynomials in ω with $\deg p(k, \omega) = k$.

The sixth and final program presented in Appendix B is called **dispfnpoly**. It prints out both $q(k, -\omega)$ and $p(k, \omega)$ in symbolic form so that they can be imported into Mathematica [32]. To run this program, the user must specify the order k of the polynomials. The program is different from the other programs in the appendix since it is not required to determine the factorial of the total number of distinct parts, i.e. $N_k!$. When the global variable *polytype* is equal to unity in the for loop in **main**, the program determines the discrete partition polynomial at the order specified by the user. When it equals 2 in the second iteration of the same for loop, the program proceeds to determine the corresponding partition function polynomial. E.g., for $k = 6$, the following output is generated:

Q[6,-w_]:= DP[6,w] (-1) + DP[1,w] DP[5,w] (-1)^(2) +
DP[1,w]^(2) DP[4,w] (-1)^(3)/2! + DP[1,w]^(3) DP[3,w] (-1)^(4)/3! +
DP[1,w]^(4) DP[2,w] (-1)^(5)/4! + DP[1,w]^(6) (-1)^(6)/6! +
DP[1,w]^(2) DP[2,w]^ (2) (-1)^(4)/(2! 2!) +
DP[1,w] DP[2,w] DP[3,w] (-1)^(3) + DP[2,w] DP[4,w] (-1)^(2) +
DP[2,w]^(3) (-1)^(3)/3! + DP[3,w]^(2) (-1)^(2)/2!

P[6,w_]:= DP[6,w] + DP[1,w] DP[5,w] + DP[1,w]^(2) DP[4,w]/2! +
DP[1,w]^(3) DP[3,w]/3! + DP[1,w]^(4) DP[2,w]/4! + DP[1,w]^(6)/6! +
DP[1,w]^(2) DP[2,w]^(2)/(2! 2!) + DP[1,w] DP[2,w] DP[3,w] +
DP[2,w] DP[4,w] + DP[2,w]^(3)/3! + DP[3,w]^(2) /2!

Time taken to compute the coefficients is 0.000000 seconds

The terms on the rhs denoted by DP[k,w] represent the divisor polynomials of order k. These can be obtained by typing in the following line in Mathematica:

DP[k_,w_]:= Sum[w^d/d,{d,Divisors[k]}].

From (9.41) and (9.42) we see that the discrete partition polynomials or rather the $q(k, -\omega)$ are almost identical to the partition function polynomials or $p(k, \omega)$ when they are both expressed in terms of the divisor polynomials, the only difference being the phase factor depending on N_k or the length of the partitions appearing in the former result. Since this factor is positive when the number of parts in a partition is even and negative for an odd number of parts, the only difference between the two sets of polynomials occurs when the partitions possess an odd number of parts. That is, they are symmetric (anti-symmetric) for an even (odd) number of parts.

If we multiply (8.48) with ω set equal to $-\omega$ by (9.4), then by equating like powers on both sides of the resulting equation we obtain

$$\sum_{j=0}^{k} q(j, -\omega)\, p(k-j, \omega) = 0 \,. \tag{9.43}$$

Alternatively, (9.43) can be expressed as

$$\sum_{j=0}^{k} L_{P,j}\left[(-1)^{N_j} \prod_{i=1}^{j} \frac{\gamma_i(\omega)^{\lambda_i}}{\lambda_i!}\right] L_{P,k-j}\left[\prod_{i=1}^{k-j} \frac{\gamma_i(\omega)^{n_i}}{n_i!}\right] = 0 \,. \tag{9.44}$$

These results represent the generalization of the Euler/MacMahon recurrence relation given by (8.10).

We can also generalize the infinite product in (9.4) by introducing the arbitrary power of ρ as we did for the discrete partition case in the corollary to Theorem 8.2. Therefore, we define $P_\rho(z, \omega)$ as

$$P_\rho(z, \omega) := \prod_{k=1}^{\infty} \frac{1}{(1-\omega z^k)^\rho} \equiv 1 + \sum_{k=1}^{\infty} p(k, \omega, \rho)\, z^k \,. \tag{9.45}$$

The lhs also represents the regularized value of the series on the rhs of (9.4) taken to the ρ-th power. That is,

$$\prod_{k=1}^{\infty} \frac{1}{(1-\omega z^k)^\rho} \equiv \left(1 + \sum_{k=1}^{\infty} p(k, \omega)\right)^\rho \,. \tag{9.46}$$

Consequently, we find that

$$1 + \sum_{k=1}^{\infty} p(k, \omega, \rho)\, z^k = \left(1 + \sum_{k=1}^{\infty} p(k, \omega)\, z^k\right)^\rho \,. \tag{9.47}$$

We now apply Theorem 4.1 to the rhs of (9.47) with the coefficients of the inner series p_k equal to $p(k, \omega)$ and the coefficients of the outer series q_k set equal to the binomial theorem coefficients of $(-1)^k(-\rho)_k/k!$. Then (4.5) yields

$$p(k, \omega, \rho) = L_{P,k}\left[(-1)^{N_k} (-\rho)_{N_k} \prod_{i=1}^{k} \frac{p(i, \omega)^{\lambda_i}}{\lambda_i!}\right]. \tag{9.48}$$

As a consequence of the preceding analysis, we are now in a position to study more advanced products. In particular, let us consider the following quotient:

$$P(z, \beta\omega, \alpha\omega) = Q(z, -\beta\omega)P(z, \alpha\omega) = \prod_{k=1}^{\infty} \frac{(1 - \beta\omega z^k)}{(1 - \alpha\omega z^k)} . \tag{9.49}$$

In deriving a power series expansion or generating function for the above product we expect the power of ω in the coefficients to yield the total number of parts or N_k in the partitions summing to the power of z. Furthermore, the power of β in the coefficients should represent the number of parts due to the discrete partitions, while the power of α should give the number of parts due to the standard integer partitions. By adopting the same approach as in the other infinite products presented in this chapter, we can express (9.49) as

$$P(z, \beta\omega, \alpha\omega) = \sum_{k=0}^{\infty} q(k, -\beta\omega)z^k \sum_{k=0}^{\infty} p(k, \alpha\omega)z^k$$

$$= \sum_{k=0}^{\infty} z^k \, QP_k(\omega, \beta, \alpha) \tag{9.50}$$

where

$$QP_k(\omega, \beta, \alpha) := \sum_{j=0}^{k} q(j, -\beta\omega) \, p(k - j, \alpha\omega) . \tag{9.51}$$

From (9.51) we see that $QP_0(\omega, \beta, \alpha) = 1$. In addition, since $P(z, \alpha\omega, \alpha\omega)$ equals unity, it follows that $QP_k(\omega, \alpha, \alpha) = 0$ for $k > 0$.

Table 9.1 presents the coefficients $QP_k(\omega, \beta, \alpha)$ up to $k = 15$. These results have been determined by introducing (9.51) and the results in Table 8.2 into Mathematica [32]. As can be seen from the table they are polynomials of degree k in ω. The power of ω in these polynomials gives the number of parts in the final partitions, which combine the parts from the standard integer partitions with those from the discrete partitions. As expected, the polynomials vanish when $\alpha = \beta$ since the product $P(z, \beta\omega, \alpha\omega)$ equals unity in this case. Furthermore, the highest power of α is k. This corresponds to the fact that the power of α represents the number of parts in the standard integer partitions. Therefore, the greatest number of parts in the final partitions appearing in the generating function will be due to the partition $\{1_k\}$ with no parts from a discrete partition. The highest power of β, however, is considerably lower since it is determined by the partition with the most number of discrete parts summing to k. In this instance

Table 9.1 Coefficients $QP_k(\omega, \beta, \alpha)$ in the generating function for the infinite product in (9.49)

k	$QP_k(\omega, \alpha, \beta)$
0	1
1	$(\alpha - \beta)\omega$
2	$(\alpha - \beta)(\omega + \alpha\omega^2)$
3	$(\alpha - \beta)(\omega + (\alpha - \beta)\omega^2 + \alpha^2\omega^3)$
4	$(\alpha - \beta)(\omega + (2\alpha - \beta)\omega^2 + \alpha(\alpha - \beta)\omega^3 + \alpha^3\omega^4)$
5	$(\alpha - \beta)(\omega + 2(\alpha - \beta)\omega^2 + 2\alpha(\alpha - \beta)\omega^3 + \alpha(\alpha - \beta)\omega^4 + \alpha^4\omega^5)$
6	$(\alpha - \beta)(\omega + (3\alpha - 2\beta)\omega^2 + (3\alpha^2 - 3\alpha\beta + \beta^2)\omega^3 + 2\alpha^2(\alpha - \beta)\omega^4$ $+ \alpha^3(\alpha - \beta)\omega^5 + \alpha^5\omega^6)$
7	$(\alpha - \beta)(\omega + 3(\alpha - \beta)\omega^2 + (4\alpha - \beta)(\alpha - \beta)\omega^3 + (3\alpha - \beta)(\alpha - \beta)\alpha\omega^4$ $+ 2\alpha^3(\alpha - \beta)\omega^5 + \alpha^4(\alpha - \beta)\omega^6 + \alpha^6\omega^7)$
8	$(\alpha - \beta)(\omega + (4\alpha - 3\beta)\omega^2 + (5\alpha - 2\beta)(\alpha - \beta)\omega^3 + (5\alpha^2 - 6\alpha\beta + 2\beta^2)\alpha\omega^4$ $+ \alpha^2(3\alpha - \beta)(\alpha - \beta)\omega^5 + 2\alpha^4(\alpha - \beta)\omega^6 + \alpha^5(\alpha - \beta)\omega^7 + \alpha^7\omega^8)$
9	$(\alpha - \beta)(\omega + 4(\alpha - \beta)\omega^2 + (7\alpha^2 - 9\alpha\beta + 3\beta^2)\omega^3 + 2\alpha(3\alpha - 2\beta)(\alpha - \beta)\omega^4$ $+ \alpha^2(5\alpha - 2\beta)(\alpha - \beta)\omega^5 + \alpha^3(3\alpha - \beta)(\alpha - \beta)\omega^6 + 2\alpha^5(\alpha - \beta)\omega^7$ $+ \alpha^6(\alpha - \beta)\omega^8 + \alpha^8\omega^9)$
10	$(\alpha - \beta)(\omega + (5\alpha - 4\beta)\omega^2 + 4(2\alpha - \beta)(\alpha - \beta)\omega^3 + (9\alpha^2 - 5\alpha\beta + \beta^2)(\alpha - \beta)\omega^4$ $+ \alpha^2(7\alpha^2 - 11\alpha\beta + 5\beta^2)\omega^5 + \alpha^3(5\alpha - 2\beta)(\alpha - \beta)\omega^6 + \alpha^4(3\alpha - \beta)(\alpha - \beta)\omega^7$ $+ 2\alpha^6(\alpha - \beta)\omega^8 + \alpha^7(\alpha - \beta)\omega^9 + \alpha^9\omega^{10})$
11	$(\alpha - \beta)(\omega + 5(\alpha - \beta)\omega^2 + 5(2\alpha - \beta)(\alpha - \beta)\omega^3 + (11\alpha^2 - 9\alpha\beta + \beta^2)(\alpha - \beta)\omega^4$ $+ \alpha(5\alpha - \beta)(2\alpha - \beta)(\alpha - \beta)\omega^5 + \alpha^3(7\alpha - 5\beta)(\alpha - \beta)\omega^6$ $+ 5\alpha^4(5\alpha - 2\beta)(\alpha - \beta)\omega^7 + \alpha^5(3\alpha - \beta)(\alpha - \beta)\omega^8 + 2\alpha^7(\alpha - \beta)\omega^9$ $+ \alpha^8(\alpha - \beta)\omega^{10} + \alpha^{10}\omega^{11})$
12	$(\alpha - \beta)(\omega + (6\alpha - 5\beta)\omega^2 + (12\alpha^2 - 18\alpha\beta + 7\beta^2)\omega^3 + (15\alpha^3 - 26\alpha^2\beta + 14\alpha\beta^2$ $- 2\beta^3)\omega^4 + \alpha(13\alpha^2 - 12\alpha\beta + 2\beta^2)(\alpha - \beta)\omega^5 + \alpha^2(11\alpha^3 - 18\alpha^2\beta + 9\alpha\beta^2$ $- \beta^3)\omega^6 + \alpha^4(7\alpha - 5\beta)(\alpha - \beta)\omega^7 + \alpha^5(5\alpha - 2\beta)(\alpha - \beta)\omega^8$ $+ \alpha^6(\alpha - \beta)(3\alpha - \beta)\omega^9 + \alpha^9(\alpha - \beta)\omega^{11} + \alpha^{11}\omega^{12})$
13	$(\alpha - \beta)(\omega + 6(\alpha - \beta)\omega^2 + 2(7\alpha - 4\beta)(\alpha - \beta)\omega^3 + (18\alpha^2 - 17\alpha\beta + 3\beta^2)(\alpha - \beta)\omega^4$ $+ \alpha(9\alpha - 4\beta)(2\alpha - \beta)(\alpha - \beta)\omega^5 + 2\alpha^2(7\alpha^2 - 7\alpha\beta + \beta^2)(\alpha - \beta)\omega^6$ $+ \alpha^3(11\alpha^2 - 8\alpha\beta + \beta^2)(\alpha - \beta)\omega^7 + \alpha^5(7\alpha - 5\beta)(\alpha - \beta)\omega^8 + \alpha^6(5\alpha - 2\beta)(\alpha - \beta)\omega^9$ $+ \alpha^7(3\alpha - \beta)(\alpha - \beta)\omega^{10} + 2\alpha^9(\alpha - \beta)\omega^{11} + \alpha^{10}(\alpha - \beta)\omega^{12} + \alpha^{12}\omega^{13})$
14	$(\alpha - \beta)(\omega + (7\alpha - 6\beta)\omega^2 + 2(8\alpha - 5\beta)(\alpha - \beta)\omega^3 + (23\alpha^2 - 21\alpha\beta + 5\beta^2)(\alpha - \beta)\omega^4$ $+ \alpha(23\alpha^2 - 25\alpha\beta + 7\beta^2)(\alpha - \beta)\omega^5 + 5\alpha^2(2\alpha - \beta)^2(\alpha - \beta)\omega^6$ $+ \alpha^3(15\alpha^3 - 29\alpha^2\beta + 17\alpha\beta^2 - 2\beta^3)\omega^7 + \alpha^4(11\alpha^2 - 8\alpha\beta + \beta^2)(\alpha - \beta)\omega^8$ $+ \alpha^6(7\alpha - 5\beta)(\alpha - \beta)\omega^9 + \alpha^7(5\alpha - 2\beta)(\alpha - \beta)\omega^{10} + \alpha^8(3\alpha - \beta)(\alpha - \beta)\omega^{11}$ $+ 2\alpha^{10}(\alpha - \beta)\omega^{12} + \alpha^{11}(\alpha - \beta)\omega^{13} + \alpha^{13}\omega^{14})$
15	$(\alpha - \beta)(\omega + 7(\alpha - \beta)\omega^2 + (19\alpha^2 - 30\alpha\beta + 12\beta^2)\omega^3 + (27\alpha^2 - 29\alpha\beta + 6\beta^2)(\alpha - \beta)\omega^4$ $+ (30\alpha^4 - 64\alpha^3\beta + 45\alpha^2\beta^2 - 11\alpha\beta^3 + \beta^4)\omega^5 + \alpha^2(13\alpha - 9\beta)(2\alpha - \beta](\alpha - \beta)\omega^6$ $+ \alpha^3(7\alpha - 5\beta)(3\alpha - \beta)(\alpha - \beta)\omega^7 + \alpha^4(15\alpha^2 - 15\alpha\beta + 2\beta^2)(\alpha - \beta)\omega^8$ $+ \alpha^5(11\alpha^2 - 8\alpha\beta + \beta^2)(\alpha - \beta)\omega^9 + \alpha^7(7\alpha - 5\beta)(\alpha - \beta)\omega^{10} + \alpha^8(5\alpha - 2\beta)(\alpha - \beta)\omega^{11}$ $+ \alpha^9(3\alpha - \beta)(\alpha - \beta)\omega^{12} + 2\alpha^{11}(\alpha - \beta)\omega^{13} + \alpha^{12}(\alpha - \beta)\omega^{14} + \alpha^{14}\omega^{15})$

the power of α will be zero. E.g., for $k = 8$, the highest power of β is three, which is in accordance with (8.51). When $\alpha = 0$, the polynomials reduce to the polynomials arising from the generating function for discrete partitions, namely $q(k, -\beta\omega)$, while for $\beta = 0$, they become the partition function polynomials or $p(k, \alpha\omega)$. In addition, for $\alpha = 0$ the coefficients in the resulting polynomials give the number of discrete partitions where the number of parts is equal to the power of β. For $\beta = 0$ the coefficients of the resulting polynomials become the subpartition numbers or $p_i(k)$, where i represents the power of α.

The interesting terms in the polynomials displayed in Table 9.1 are the cross-terms involving both α and β, which represent the mixture of the discrete partitions and standard integer partitions with the total number of the parts in such partitions equal to the power of ω. For example, in $QP_3(\omega, \alpha, \beta)$ the coefficient of ω^2 has a term equal to $-2\alpha\beta$ by expanding $(\alpha - \beta)^2$. This tells us that one part in the partition has come from the infinite product for standard integer partitions or the denominator of (9.49) and the other has come from the product for discrete partitions or the numerator of (9.49). There are two possibilities where this can occur: either the one has come from the $k = 1$ term in the numerator of (9.49) and the two has come from the $k = 2$ denominator of (9.49) or vice-versa. Hence we obtain a value of $-2\alpha\beta\omega^2$ for z^3. On the other hand, the coefficients of ω^2 in $QP_3(\omega, \alpha, \beta)$ involving only standard integer and discrete partitions are respectively α^2 and β^2. In this instance, the partition maintains its discreteness by accepting one term from $k = 1$ and $k = 2$ term in the numerator of (9.49). The standard integer partition of $\{1, 2\}$, which yields the α^2 term in the coefficient of ω^2 in $QP_3(\omega, \alpha, \beta)$ arises from the $k = 1$ and $k = 2$ terms in the denominator of (9.49). From the results in the table it is obvious that the power of α can be much higher than the power of β reflecting the fact that the greatest number of parts in a discrete partition summing to a particular value is significantly less than the greatest number of parts in a standard integer partition summing to the same value.

According to p. 23 of Ref. [12], Gauss derived the following result:

$$1 + 2\sum_{k=1}^{\infty}(-1)^k z^{k^2} = \prod_{k=1}^{\infty} \frac{(1 - z^k)}{(1 + z^k)}. \tag{9.52}$$

The rhs of the above result is a special case of $P(z, \beta\omega, \alpha\omega)$, namely $P(z, 1, -1)$. If the values for $\alpha\omega$ and $\beta\omega$ are introduced into the rhs of (9.49), then we can equate like powers of z with the lhs of (9.52). Consequently, for i equal to a positive integer we arrive at

$$QP_k(1, 1, -1) = \sum_{j=0}^{k} q(j)\, p(k-j, -1) = \begin{cases} 2(-1)^i, & k=i^2, \\ 0, & \text{otherwise}. \end{cases} \quad (9.53)$$

The above result is easily checked with the results appearing in Table 9.1 on setting $\alpha = -1/\omega$ and $\beta = 1/\omega$.

Although the partition method for a power series expansion was developed in order to solve the properties of the charged Bose gas [1,2], not much has been said about the method being applied to other disciplines outside mathematics. One discipline that is likely to benefit from the method, however, is statistical mechanics, particularly in the area of lattice models. For example, on p. 238 of Ref. [74] the spontaneous magnetization of the Ising model is given as

$$M_0 = \prod_{n=1}^{\infty} \left(\frac{1 - x^{4n-2}}{1 + x^{4n-2}} \right), \quad (9.54)$$

where $x^2 = \exp(-\pi K_l/K_l')$ and K and K' are the complete elliptic integrals of the first kind with moduli l and $l' = (1 - l^2)^{1/2}$. As stated on the preceding page of the same reference, Barber and Baxter [75] were only able to expand the above product to fourth order in x. However, we can do better by using the results in Table 9.1.

First, we note that to obtain the product in the spontaneous magnetization, we put $z = x^4$, $\omega = 1/x^2$, $\alpha = -1$ and $\beta = 1$ in (9.49). Then we introduce the results in Table 9.1 into a Mathematica notebook by typing the following statement:

```
Mzero[x_, N_] := Simplify[Sum[x^(4 k) QP[-1, 1, x^(-2), k], {k, 0, N}]].
```

In this line of code N is simply the truncation level for the series expansion. In fact, if we set it equal to 15, i.e., using all the results in Table 9.1, then we obtain

$$M_0 = 1 - 2x^2 + 2x^4 - 4x^6 + 6x^8 - 8x^{10} + 12x^{12} - 16x^{14} + 22x^{16} - 30x^{18}$$
$$+ 40x^{20} - 52x^{22} + 68x^{24} - 88x^{26} + 112x^{28} - 144x^{30} + \cdots. \quad (9.55)$$

These are not all the terms generated by setting N equal to 15. In actual fact, Mathematica prints out terms to x^{58}, but those beyond x^{30} are only partially correct. That is, to obtain the correct coefficients of these terms, we require more polynomials in Table 9.1. The above result has been determined by comparing it with the $N = 14$ result and observing which coefficients agree with each other. Thus, it is found that those up to x^{2N}

are always correct. From (9.55) we see that the spontaneous magnetization is in even powers of x, while the coefficients not only oscillate in sign, but are also even numbers. Furthermore, if we type the coefficients into the online encyclopedia of integer sequences, then we discover that the sequence represents the expansion of $f(-q)/f(q)$ [76] in powers of q, where $f(q) = f(q, -q^2)$ and $f(x, y)$ represents the Ramanujan two-variable theta function [77], which for $|xy| < 1$, is defined as

$$f(x, y) := \sum_{k=0}^{\infty} x^{k(k+1)/2} y^{k(k-1)/2} . \qquad (9.56)$$

On p. 319 of Ref. [74] Baxter presents the spontaneous magnetization of the triangular three-spin model, which is shown to be equivalent to the Kagomé lattice eight-vertex model for certain values of the interaction coefficients. Thus we have

$$M_0 = \prod_{n=1}^{\infty} \left(\frac{1 - p^{6n-3}}{1 + p^{6n-2}} \right) , \qquad (9.57)$$

where $0 < p < 1$ and p is itself expressed in terms of an infinite product given by

$$\exp\left(-\frac{J}{k_B T}\right) = \sqrt{p} \prod_{n=1}^{\infty} \left(\frac{1 - p^{8n-7}}{1 - p^{8n-5}} \right) \left(\frac{1 - p^{8n-1}}{1 - p^{8n-3}} \right) . \qquad (9.58)$$

In (9.58) J is the coupling constant between neighbouring sites on the lattice, T is the temperature and k_B is Boltzmann's constant. We shall study the infinite product in (9.58) after we introduce the Heine product. Returning to (9.57), we can still use the same code for Mzero as above except we now set $\alpha = -p$, $\beta = 1$, $\omega = p^{-3}$ and $z = p^6$. On this occasion, it is found that the coefficients between Mzero with N and $N + 1$ agree up to p^{4N+2}. Therefore, if we put $N = 15$, the expression for M_0 will be correct to p^{62}. Hence we find that the spontaneous magnetization of the triangular three-spin model has the following power series expansion:

$$\begin{aligned}
M_0 = &\ 1 - p^3 - p^4 + p^7 + p^8 - p^9 - p^{10} - p^{11} + 2p^{13} + p^{14} - p^{16} - 2p^{17} + 2p^{19} \\
&+ p^{20} + p^{21} - 2p^{22} - 3p^{23} + 2p^{25} + 3p^{26} + p^{27} - 3p^{28} - 4p^{29} - 2p^{30} \\
&+ 3p^{31} + 6p^{32} + 3p^{33} - 3p^{34} - 7p^{35} - 4p^{36} + 3p^{37} + 7p^{38} + 6p^{39} - 2p^{40} \\
&- 9p^{41} - 7p^{42} + p^{43} + 10p^{44} + 9p^{45} - 2p^{46} - 11p^{47} - 10p^{48} + 12p^{50} \\
&+ 13p^{51} - 13p^{53} - 16p^{54} - 2p^{55} + 17p^{56} + 19p^{57} + 3p^{58} - 17p^{59} - 23p^{60} \\
&- 6p^{61} + 18p^{62} + \cdots .
\end{aligned} \qquad (9.59)$$

Alternatively, the above result can be obtained by setting $\alpha = -p^{-2}$, $\beta = p^{-3}$, $\omega = 1$ and $z = p^6$, which represents a check.

We can also use the preceding analysis to derive a power series expansion or generating function for the product of two specific forms of $P(z, x, y)$ involving the three parameters, ω, x, and y, and the variable, z. This product was first studied by Heine. According to p. 55 of Ref. [13] he found that

$$\prod_{k=1}^{\infty} \frac{(1 - \omega x z^k)}{(1 - \omega z^k)} \frac{(1 - \omega y z^k)}{(1 - \omega x y z^k)} = \sum_{k=0}^{\infty} \frac{(1/x; z)_k}{(z; z)_k} \frac{(1/y; z)_k}{(\omega z; z)_{k+1}} (\omega x y z)^k. \quad (9.60)$$

If we set $a = 1/x$, $b = 1/y$, $c = \omega z$ and $q = z$ with $|c/ab| < 1$ and $|q| < 1$, which are the conditions for guaranteeing absolute convergence, then the above result can be expressed as a q-hypergeometric series. This is perhaps the more familiar form for the product, where it is written as

$$\sum_{k=0}^{\infty} \frac{(a; q)_k}{(q; q)_k} \frac{(b; q)_k}{(c; q)_k} \left(\frac{c}{ab}\right)^k = \prod_{k=0}^{\infty} \frac{\left(1 - (c/a)q^k\right)}{\left(1 - cq^k\right)} \frac{\left(1 - (c/b)q^k\right)}{\left(1 - (c/ab)q^k\right)}. \quad (9.61)$$

The above result appears as Corollary 2.4 on p. 20 of Ref. [12].

The lhs of (9.60) represents the product of $P(z, \omega x, \omega)$ and $P(z, \omega y, \omega x y)$. If we denote the product of $P(z, x, y)$ and $P(z, s, t)$ by $P_2(z, x, y, s, t)$ and introduce (9.50), then we find that

$$P_2(z, \omega x, \omega, \omega y, \omega x y) = \prod_{k=1}^{\infty} \frac{(1 - \omega x z^k)}{(1 - \omega z^k)} \frac{(1 - \omega y z^k)}{(1 - \omega x y z^k)}$$

$$= \sum_{k=0}^{\infty} z^k \, HP_k(\omega, x, y), \quad (9.62)$$

where

$$HP_k(\omega, x, y) = \sum_{j=0}^{k} QP_j(\omega, x, 1) \, QP_{k-j}(\omega, y, xy). \quad (9.63)$$

The equals sign appears in (9.62) only because of the conditions given below (9.60).

Table 9.2 presents the coefficients of the generating function for the Heine product denoted by $HP_k(\omega, x, y)$ up to $k = 10$. They have been obtained by implementing (9.63) in Mathematica [32] and using the results in Table 9.1. From the table it can be seen that the $HP_k(\omega, x, y)$ are polynomials in ω of degree k. The coefficient of the leading order term is

$$C_k^{HP} = \frac{(1 - x^k y^k)}{(1 - xy)} (1 - x)(1 - y), \quad (9.64)$$

Table 9.2 Coefficients of the polynomials $HP_k(\omega, \alpha, \beta)$ given in (9.63)

k	$HP_k(\omega, x, \gamma)$
0	1
1	$(1-x)(1-\gamma)\omega$
2	$(1-x)(1-\gamma)\omega\big(1+(1+x\gamma)\omega\big)$
3	$(1-x)(1-\gamma)\omega\big(1+(1-x)(1-\gamma)\omega+(1+x\gamma+x^2\gamma^2)\omega^2\big)$
4	$(1-x)(1-\gamma)\omega\big(1+(2-x-\gamma+2x\gamma)\omega+(1-x)(1-\gamma)(1+x\gamma)\omega^2$ $+(1+x\gamma)(1+x^2\gamma^2)\omega^3\big)$
5	$(1-x)(1-\gamma)\omega\big(1+2(1-x)(1-\gamma)\omega+2(1-x)(1-\gamma)(1+x\gamma)\omega^2$ $+(1-x)(1-\gamma)(1+x\gamma+x^2\gamma^2)\omega^3+(1+x\gamma+x^2\gamma^2+x^3\gamma^3+x^4\gamma^4)\omega^4\big)$
6	$(1-x)(1-\gamma)\omega\big(1+(3-2x-2\gamma+3x\gamma)\omega+(3(1-x)(1-\gamma)(1+x\gamma)$ $+x^2+x\gamma+\gamma^2)\omega^2+(1-x)(1-\gamma)(2+3x\gamma+2x^2\gamma^2)\omega^3$ $+(1-x)(1-\gamma)(1+x\gamma)(1+x^2\gamma^2)\omega^4+(1+x\gamma+x^2\gamma^2+x^3\gamma^3+x^4\gamma^4+x^5\gamma^5)\omega^5\big)$
7	$(1-x)(1-\gamma)\omega\big(1+3(1-x)(1-\gamma)\omega-(1-x)(1-\gamma)((1-x)(1-\gamma)+3(1+x\gamma))\omega^2$ $+(1-x)(1-\gamma)((1-x)(1-\gamma)(1+x\gamma)+2(1+x\gamma+x^2\gamma^2))\omega^3$ $+(1-x)(1-\gamma)(2+3x\gamma+3x^2\gamma^2+2x^3\gamma^3)\omega^4+(1-x)(1-\gamma)(1+x\gamma+x^2\gamma^2$ $+x^3\gamma^3+x^4\gamma^4)\omega^5+(1+x\gamma+x^2\gamma^2+x^3\gamma^3+x^4\gamma^4+x^5\gamma^5+x^6\gamma^6)\omega^6\big)$
8	$(1-x)(1-\gamma)\omega\big(1+(3(1-x)(1-\gamma)+1+x\gamma)\omega+(1-x)(1-\gamma)(5-2x-2\gamma+5x\gamma)\omega^2$ $+(5-6\gamma+2\gamma^2+x^3\gamma(2-6\gamma+5\gamma^2)+x^2(2-11\gamma+16\gamma^2-6\gamma^3)+x(-6+16\gamma$ $-11\gamma^2+2\gamma^3))\omega^3+(1-x)(1-\gamma)(3-\gamma+x^3\gamma^2(3\gamma-1)-x^2\gamma(\gamma^2-5\gamma+1)$ $-x(\gamma^2-5\gamma+1))\omega^4+(1-x)(1-\gamma)(2+3x\gamma+3x^2\gamma^2+3x^3\gamma^3+2x^4\gamma^4)\omega^5$ $+(1-x)(1-\gamma)(1+x\gamma+x^2\gamma^2+x^3\gamma^3+x^4\gamma^4+x^5\gamma^5)\omega^6+(1+x\gamma+x^2\gamma^2$ $+x^3\gamma^3+x^4\gamma^4+x^5\gamma^5+x^6\gamma^6+x^7\gamma^7)\omega^7\big)$
9	$(1-x)(1-\gamma)\omega\big(1+4(1-x)(1-\gamma)\omega+(x^2(7\gamma^2-9\gamma+3)-x(9\gamma^2-19\gamma+9)$ $+3\gamma^2-9\gamma+7)\omega^2+2(1-x)(1-\gamma)(x^2\gamma(3\gamma-2)-(2\gamma-3)-x(2\gamma^2-5\gamma+2))\omega^3$ $+(1-x)(1-\gamma)(x^3\gamma^2(5\gamma-2)-(2\gamma-5)-x^2\gamma(2\gamma^2-9\gamma+3)-x(3\gamma^2-9\gamma+2))\omega^4$ $+(1-x)(1-\gamma)(x^4\gamma^3(3\gamma-1)-(\gamma-3)-x^2\gamma(\gamma^2-6\gamma+1)-x^3\gamma^2(\gamma^2-5\gamma+1)$ $-x(\gamma^2-5\gamma+1))\omega^5+(1-x)(1-\gamma)(2x^5\gamma^5+3x^4\gamma^4+3x^3\gamma^3+3x^2\gamma^2+3x\gamma+2)\omega^6$ $+(1-x)(1-\gamma)(1+x\gamma+x^2\gamma^2+x^3\gamma^3+x^4\gamma^4+x^5\gamma^5+x^6\gamma^6)\omega^7+(1+x\gamma+x^2\gamma^2$ $+x^3\gamma^3+x^4\gamma^4+x^5\gamma^5+x^6\gamma^6+x^7\gamma^7+x^8\gamma^8)\omega^8\big)$
10	$(1-x)(1-\gamma)\omega\big(1+(5+5x\gamma-4x-4\gamma)\omega+4(1-x)(1-\gamma)(2+2x\gamma-x-\gamma)\omega^2$ $+(1-x)(1-\gamma)(x^2(9\gamma^2-5\gamma+1)-x(5\gamma^2-17\gamma+5)+\gamma^2-5\gamma+9)\omega^3$ $+(x^4\gamma^2(7\gamma^2-11\gamma+5)+5\gamma^2-11\gamma+7-x^3\gamma(11\gamma^3-30\gamma^2+25\gamma-7)+x(7\gamma^3$ $-25\gamma^2+30\gamma-11)+x^2(5\gamma^4-25\gamma^3+41\gamma^2-25\gamma+5))\omega^4+(1-x)(1-\gamma)(x^4\gamma^3(5\gamma$ $-2)-2\gamma+5-x^3\gamma^2(2\gamma^2-10\gamma+3)-x^2\gamma(3\gamma^2-11\gamma+3)-x(3\gamma^2-10\gamma+2))\omega^5$ $+(1-x)(1-\gamma)(x^5\gamma^4(3\gamma-1)-\gamma+3-x^2\gamma(1+x\gamma)(\gamma^2-6\gamma+1)-x(1+x^3\gamma^3))(\gamma^2$ $-5\gamma+1))\omega^6+(1-x)(1-\gamma)(2+3x\gamma+3x^2\gamma^2+3x^3\gamma^3+3x^4\gamma^4+3x^5\gamma^5+2x^6\gamma^6)\omega^7$ $+(1-x)(1-\gamma)(1+x\gamma+x^2\gamma^2+x^3\gamma^3+x^4\gamma^4+x^5\gamma^5+x^6\gamma^6+x^7\gamma^7)\omega^8$ $+(1+x\gamma+x^2\gamma^2+x^3\gamma^3+x^4\gamma^4+x^5\gamma^5+x^6\gamma^6+x^7\gamma^7+x^8\gamma^8+x^9\gamma^9)\omega^9\big)$

while that for the penultimate leading order term is found to be

$$C_{k-1}^{HP} = \frac{\left(1 - x^{k-2}y^{k-2}\right)}{(1 - xy)} (1 - x)^2 (1 - y)^2.$$

(9.65)

The above equation is only valid for $k > 2$. On the other hand, the lowest order term in ω for these polynomials is linear and its coefficient is given by

$$C_1^{HP} = (1 - x)(1 - y).$$

(9.66)

As expected, the polynomials are zero when either $x = 1$ or $y = 1$ since in these cases $P(z, \omega x, \omega, \omega y, \omega x y)$ is equal to unity. In addition, they are symmetrical in x and y in the sense that a fixed power of ω with a term of $\alpha x^i y^j$, where $i \neq j$, in its coefficient will also possess the term of $\alpha x^j y^i$. Consequently, $HP_k(\omega, x, y) = HP_k(\omega, y, x)$.

If we put $z = p^8$, $\omega = p^{-3}$, $x = p^2$, and $y = p^{-4}$, then the Heine product given by (9.62) reduces to

$$P_2\left(p^8, p^{-1}, p^{-3}, p^{-7}, p^{-5}\right) = \prod_{k=1}^{\infty} \frac{\left(1 - p^{8k-1}\right)\left(1 - p^{8k-7}\right)}{\left(1 - p^{8k-3}\right)\left(1 - p^{8k-5}\right)}$$

$$= \sum_{k=0}^{\infty} p^{8k} HP_k\left(p^{-3}, p^2, p^{-4}\right).$$

(9.67)

This is, of course, the same product appearing below the spontaneous magnetization of the triangular three-spin model discussed above. In this instance, it is found that when the truncation parameter N becomes the upper limit of the sum involving the $HP_k(\omega, x, y)$, the highest correct power in the resulting series expansion is p^{3N+3}. This is in spite of fact that the expansion yields terms up to p^{119} for $N = 15$. For $N = 15$, (9.58) becomes

$$\exp\left(-\frac{J}{k_B T}\right) = \sqrt{p}\left(1 - p + p^3 - p^4 + p^5 - 2p^7 + 2p^8 - p^9 + 2p^{11} - 3p^{12}\right.$$
$$+ 2p^{13} - 2p^{15} + 4p^{16} - 4p^{17} + 4p^{19} - 6p^{20} + 5p^{21} - 6p^{23}$$
$$+ 9p^{24} - 6p^{25} + 7p^{27} - 12p^{28} + 9p^{29} - 10p^{31} + 16p^{32}$$
$$- 13p^{33} + 15p^{35} - 22p^{36} + 17p^{37} - 20p^{39} + 29p^{40} - 21p^{41}$$
$$\left. + 25p^{43} - 38p^{44} + 28p^{45} + p^{46} - 32p^{47} + 50p^{48} + \cdots\right).$$

(9.68)

On a final note, Baxter also expresses the spontaneous polarization of the triangular three-spin model as an infinite product. From (11.10.30) of

Ref. [74], we have

$$P_0 = \prod_{n=1}^{\infty} \left(\frac{1 + p^{4n}}{1 - p^{4n}} \frac{1 - p^{3n}}{1 + p^{3n}} \right).$$ (9.69)

The above product is not in the appropriate form to be represented as a Heine product. However, we can express it as

$$P_0 = P\left(p^4, -1, 1\right) P\left(p^3, 1, -1\right),$$ (9.70)

where $P(z, \beta, \alpha)$ is given by (9.49) with the generating function appearing in (9.50). Introducing the generating function into (9.70) yields

$$P_0 = \sum_{k=0}^{\infty} p^{3k} \sum_{j=0}^{k} p^j \, QP_j(1, -1, 1) \, QP_{k-j}(1, 1, -1),$$ (9.71)

where the coefficients QP_j have already been defined by (9.51) and are displayed in Table 9.1. In this case because of the specific arguments inside the polynomials, the coefficients yield constants. Moreover, from (9.53) we know that many of the results in the table vanish for $\alpha = -1$ and $\beta = 1$, which is not the case for the other way around, i.e., $\alpha = 1$ and $\beta = -1$. If we replace the upper limit in (9.71) by the truncation parameter N, then it is found that the expressions for the spontaneous magnetization with $N = 13$ and $N = 14$ agree up to the p^{52}-term. For $N = 14$ and $N = 15$, the results agree up to the p^{56}-term. Therefore, we see that the terms generated by (9.71) with the upper limit replaced by N are correct up to the p^{4N}-term. For $N = 15$, the spontaneous polarization for the triangular three-spin model is given by

$$\begin{aligned}
P_0 = & 1 - 2p^3 + 2p^4 - 4p^7 + 4p^8 - 8p^{11} + 10p^{12} - 16p^{15} + 18p^{16} - 28p^{19} \\
& + 32p^{20} - 48p^{23} + 56p^{24} - 82p^{27} + 92p^{28} - 132p^{31} + 148p^{32} - 208p^{35} \\
& + 234p^{36} - 324p^{39} + 360p^{40} - 492p^{43} + 544p^{44} - 736p^{47} + 812p^{48} \\
& - 1088p^{51} + 1192p^{52} - 1456p^{55} + 1728p^{56} - 2080p^{59} + 1472p^{60} + \cdots.
\end{aligned}$$ (9.72)

The material presented in this chapter will be developed further and applied to other topics in statistical mechanics as discussed in the following chapter.

CHAPTER 10

Conclusion

Because of its success in solving important intractable problems in applied mathematics and theoretical physics [1–5], and its potential for solving others, it was deemed that the partition method for a power series expansion ought to be brought to the attention of mathematicians and physicists in the form of a book. The aim of this book was not only to present the theory behind the partition method for a power series expansion, but also with the aid of this theory, to devise a programming methodology whereby ultimately, the symbolic forms for the coefficients in these expansions could be introduced into a mathematical software package such as Mathematica [32]. Then the coefficients could be evaluated by employing: (1) the integer arithmetic routines, thereby avoiding round-off errors associated with scientific notation, or (2) the symbolic routines where the coefficients were expressed in terms of mathematical functions such as polynomials.

The first two chapters of this book are concerned with introducing the method for a power series expansion into various applications or problems. There we observe that the coefficients of the resulting power series for a function are obtained by summing the contributions made by each partition summing to the order k. These contributions are evaluated by: (1) assigning values p_i to each part i in a partition, (2) multiplying by a multinomial factor composed of the factorial of the total number of parts or length of the partition, $N_k!$, divided by the factorial of the multiplicity of each part, $\lambda_i!$ and (3) multiplying by the coefficient of an outer series involving the total number of parts, viz. q_{N_k}. Therefore, one needs to know the compositions of all the partitions summing to k in order to apply the partition method for a power series expansion, which means in turn that an algorithm that is capable of generating partitions in the multiplicity representation is required. Whilst Chapter 3 discusses various methods of generating partitions, it turns out that the novel bivariate recursive central partition or BRCP algorithm is the most suitable method for implementation in the partition method for a power series expansion, mainly because it utilizes modern computer programming techniques based on the graphical representation of the partitions in the form of a non-binary tree referred to here as a partition tree and is depicted in Fig. 1.2. As a consequence, the BRCP algorithm is able to generate partitions in the multiplicity representation more efficiently or

The Partition Method for a Power Series Expansion.
DOI: http://dx.doi.org/10.1016/B978-0-12-804466-7.00010-3

naturally than other algorithms. Programs that scan and generate partitions in standard lexicographic form are of limited value to the partition method for a power series expansion unless they can be modified to generate the partitions in the multiplicity representation. This, however, will result in an adverse effect on performance.

Initially, the partition method for a power series expansion was applied to problems where the original function was a composite function. Later, it was found that this condition could be relaxed to quotients of "pseudo-composite" functions. Nevertheless, the functions appearing in the quotients of the pseudo-composite functions must themselves be expressible in terms of power series expansions, which are referred to as the inner and outer series. Neither series, however, is required to be absolutely convergent. In fact, when divergent series are used, the resulting power series expansion can turn out to be convergent for all values of the variable.

Because divergent series can be used in the partition method for a power series expansion, the concept of regularization of a divergent series has been employed throughout this book. As described in Appendix A, this concept is defined as the removal of the infinity in the remainder so as to make the series summable or yield a finite limit. The finite limits obtained in this process are referred to as regularized values, whilst the statements in which they appear together with the series are no longer equations. Instead, they are referred to as equivalence statements or equivalences, for short. A necessary property of the regularized value is that it must be unique, particularly if it is identical to the value one obtains when the series is absolutely convergent within a finite radius of the complex plane such as the geometric series.

The general theory behind the partition method for a power series expansion is presented as Theorem 4.1, which shows how power series expansions for intractable mathematical functions can be derived by expressing them as a quotient of pseudo-composite functions. By using Lemma A.1 in Appendix A, a corollary to Theorem 4.1 is presented, whereby the partition method is adapted to the case of the quotient of the pseudo-composite functions being taken to an arbitrary power. As a result of Theorem 4.1 and the corollary, we observe that the process of evaluating the contributions made by each partition can be viewed as a discrete operation, giving rise to the partition operator denoted by $L_{P,k}[\cdot]$. While $L_{P,k}[1] = p(k)$ or the number of partitions summing to k, varying the argument inside the operator yields completely different identities as evidenced by the numerous examples derived in Chapter 6. Moreover, the partition operator can be modified so that it only applies to specific classes or types of partitions such as discrete or odd/even partitions, again producing new and fascinating identities when the arguments inside these alternative operators are altered.

Because the number of partitions increases exponentially, it becomes rather onerous to apply the partition method for a power series expansion when the order k is greater than 10. This problem is overcome by modifying the BRCP algorithm so that the contribution due to each partition can be expressed in symbolic form. Chapter 5 presents two programs with each applicable to different situations. The first **partmeth** calculates all the coefficients D_k and E_k in Theorem 4.1 up to and including the value of k specified by the user. Unfortunately, for $k > 20$ the output generated by this program becomes too large. Then we require a code that only evaluates the coefficients D_k and E_k for a particular value of k, which is accomplished by the second code **mathpm**. For much larger values of k, say for $k > 100$, where combinatorial explosion begins to become an issue, storing the coefficients is no longer a viable option. In these cases the output needs to be divided into smaller blocks. As one block is used to calculate the contribution to the coefficient, the result can be stored and the block removed. Then the next block can be imported into the software package. Once it is evaluated and the result stored or added to the result from the previous block, it too can be removed and the next block imported. Therefore, summing the stored results from all the blocks yields the value for the coefficient.

In the process of developing a programming methodology for the partition method for a power series expansion, it was evident that other programs using the BRCP algorithm could be created to determine various types or classes of integer partitions appearing in the theory of partitions. Thus, Chapter 7 was devoted to presenting codes that could determine: (1) partitions either with a fixed number of parts or with specific parts, (2) doubly-restricted partitions, where all the parts lie in a range of values, (3) discrete/distinct partitions, where each part occurs at most once, (4) perfect partitions, where the parts generate all integers up to their sum uniquely, and (5) conjugate partitions via Ferrers diagrams. Normally, different programming approaches are required to solve these problems, but here they have been solved by making minor modifications to the program **partgen** in Chapter 3. Consequently, new operators such as the discrete partition operator $L_{DP,k}[\cdot]$ and the odd- and even–part partition operators, $L_{OEP,k}[\cdot]$ and $L_{EEP,k}[\cdot]$ have been created. For example, the number of discrete partitions summing to k or $q(k, 1)$ can now be represented as $L_{DP,k}[1]$.

In Chapters 8 and 9 the partition method for a power series expansion is used to derive the generating functions from increasingly sophisticated extensions of the infinite product defined by $P(z) := \prod_{k=1}^{\infty}(1 - z^k)$ and its inverted form given by $1/P(z)$. This important product was found by Euler to yield a generating function whose coefficients are equal to the partition–number function, $p(k)$. Before proceeding with the study, it is shown that the generating function of $P(z)$ is absolutely convergent within the unit disk centred at the origin in the complex plane, but is divergent for all other values of z. In other words, $P(z)$ represents the regularized value of the generating function for $|z|$ not situated within the unit disk. This is often postulated in the literature, but no formal proof of this result has ever appeared. Then Theorem 4.1 is used to derive the generating function for $1/P(z)$ whose coefficients $q(k)$ are expressed in terms of the partition operator acting on $p(i)$, which is the value assigned to each part i in the partitions. In this case the coefficients $q(k)$, which are referred to as the discrete partition numbers, are only non-zero when k is a pentagonal number, again a result first obtained by Euler. By inverting the approach we obtain the partition–number function or $p(k)$ again, but in this case the partition operator acts on $q(i)$, since this is the value assigned to each part i.

Chapter 8 continues with the derivation of alternative representations for the generating functions of $P(z)$ and its inverted form. In these results the coefficients for the j-th power of the inner series are expressed in terms of a sum over the divisors or factors d of j divided by j and are denoted by γ_j. When Theorem 4.1 is applied to these new forms of the infinite products, the coefficients of the i-th power in the generating functions now have the partition operator acting on γ_i. The major difference between the resulting power series for $P(z)$ and its inverted form is the appearance of the phase factor $(-1)^{N_k}$ in the latter case.

The inverted form of $P(z)$ is then extended to the case where the coefficients of z^k in the infinite product are set equal to C_k rather than -1 in the case of the discrete partition numbers. From Theorem 8.2 the coefficients of the generating function for this product $H(z)$ are given by the discrete partition operator $L_{DP,k}[\cdot]$ acting on the C_i, as they are the assigned values of each part i. Conversely, this theorem implies that any power series can be expressed as an infinite product. If the C_k are set equal to the parameter ω, then the coefficients of the generating function for the infinite product denoted by $Q(z, \omega)$ not only become polynomials of degree k in ω and are thus referred to as the discrete partition polynomials $q(k, \omega)$, but also the powers of ω yield the number of parts in the discrete partitions.

Next identities involving the discrete partition polynomials are derived before a corollary to Theorem 8.2 appears, which deals with the case where each term in the infinite product $H(z)$ is taken to an arbitrary power ρ_k. Therefore, by adapting the partition method for a power series expansion one can derive generating functions for very complicated infinite products. The chapter concludes by examining the special case where the arbitrary powers ρ_k and coefficients C_k are both set equal to the constant values, ρ and ω, respectively. In this case the coefficients of the generating function become the polynomials $q(k, \omega, \rho)$, both of degree k in ρ and ω. In particular, the $\rho = 2$ and 3 cases are studied in detail since they feature in well-known products studied by Euler and Gauss.

Chapter 9 deals with the derivation of the generating functions for even more complicated products than those appearing in Chapter 8. As a result of the success of introducing the parameter ω in the inverted form of $P(z)$, the chapter begins by introducing ω next to the powers of z^k in $P(z)$. Consequently, the generating function for the infinite product denoted by $P(z, \omega)$ possesses polynomial coefficients $p(k, \omega)$ of degree k in ω. These partition function polynomials, which reduce to the partition-number function $p(k)$ when ω is set equal to unity, are expressed in terms of the partition operator acting where each part i is assigned the value, $q(i, -\omega)$. They also possess many interesting properties, while their coefficients yield the number of partitions in which the number of parts is given by the power of ω. As a result of this analysis, interesting recurrence relations are obtained for the number of discrete partitions or $q(i, 1)$.

As in the case of $P(k)$ and its inverted form, $P(k, \omega)$ and $Q(k, \omega)$ can also be expressed as an exponentiated double sum, both of which can be handled by Theorem 4.1. Thus, it is found that the partition function and discrete partition polynomials can be expressed in terms of the partition operator with the main difference being that in the case of the former polynomials the parts i are assigned the polynomial values, $\gamma_i(\omega)$, while for $Q(k, \omega)$ the parts are assigned the values of $-\gamma_i(\omega)$. These new polynomials called divisor polynomials represent the extension of the γ_i in Chapter 8. Their coefficients are equal to the divisors d of i divided by i, while each power of ω is equal to the reciprocal of the coefficient. Moreover, they reduce to the γ_i when $\omega = 1$. Because this is a somewhat unusual situation involving divisors, a special program is presented in Appendix B that evaluates the partition function and discrete partition polynomials in symbolic form so that they can be imported into Mathematica. Consequently, the final forms for the polynomials, $p(k, \omega)$ and $q(k, -\omega)$, can be determined by evaluating the divisor polynomials using the Divisors[k] routine in Mathematica.

Chapter 9 continues with the introduction of an arbitrary power ρ into the product $P(z, \omega)$ and determining the coefficients $p(k, \omega, \rho)$ of the generating function. These coefficients are given in terms of the partition operator acting where the parts i are assigned to negative values of the partition function polynomials, viz. $-p(i, \omega)$, multiplied by the Pochhammer symbol of $(-\rho)_{N_k}$. Next the generating function of the product of $Q(z, -\beta\omega)$ and $P(z, \alpha\omega)$ is studied in detail. This infinite product denoted by $P(z, \beta\omega, \alpha\omega)$ combines the properties of discrete partitions with standard partitions. The coefficients of the resulting generating function, which are denoted by $QP_k(\omega, \alpha, \beta)$, are polynomials of degree k in ω, while the powers of α and β indicate the number of parts in the standard and discrete partitions, respectively. The chapter concludes with the derivation of the generating function for Heine's product, which is given by the product of $P(z, \omega x, \omega)$ and $P(z, \omega y, \omega x y)$. The coefficients of the generating function for this infinite product, which arises in q-hypergeometric function theory, are denoted by $HP_k(\omega, x, y)$ and are obtained by summing the product of $QP_j(\omega, x, 1)$ with $QP_{k-j}(\omega, y, xy)$ for j ranging from 0 to k. Several coefficients in the last two examples are tabulated in order to display their complicated nature.

The material presented in Chapters 8 and 9 represents an introduction to the subject of how generating functions or series expansions can be obtained from infinite products such as $P(z, \omega)$ and Heine's product. Infinite products arise frequently in the lattice models of statistical mechanics as demonstrated by some of the examples presented in Chapter 9. Many of the solutions to these models involve the theory of elliptic functions including the theta functions and Jacobian elliptic functions, which according to Chapter 15 of Ref. [74] can be expressed in terms of infinite products. Moreover, quantities such as the half-period magnitudes K and K', the modulus k and its conjugate can be expressed as infinite products whose generating functions can be determined from combinations of (8.59) and (9.45). In addition, the famous Ising model can be re-worked by applying the partition method for a power series expansion to the fairly intractable integrals in Onsager's derivation of the free energy and internal energy as discussed in Chapter 5 of Ref. [78]. In fact, it is often stated that Onsager's formula for the spontaneous magnetization of the Ising model is simple, but still to this day no one has ever produced a power series expansion for this quantity in powers of $J/k_B T$, where J is the coupling constant. As the isotropic form of the spontaneous magnetization is given by $M_0 = (1 - 1/\sinh(2J/k_B T)^4)^{1/8}$, we know from Chapter 2 that such an expansion will involve the generalized cosecant numbers. Furthermore, as an integer power is involved, these generalized cosecant numbers will be related to the cosecant numbers, which from (1.16) are related to $\zeta(2k)/\pi^{2k}$. Thus, an expansion in powers of $J/k_B T$ for the spontaneous magnetization of the Ising model will involve even integer values of the Riemann zeta function. The issues raised in this paragraph are to be investigated in the near future.

APPENDIX A

Regularization

Throughout this book we have used the concept of regularization to express many of the results in terms of equivalence statements rather than as equations. In this appendix we present a summary of this important concept, while a more detailed exposition can be found in Chapter 4 of [19] and Refs. [6], [43] and [44].

Broadly speaking, regularization is defined as the removal of the infinity in the remainder of a divergent series so as to make it summable. Typically, this means that when a power series representation for a function is divergent, as in the case of an asymptotic expansion or even a Taylor series expansion, it must be regularized in order that meaningful values can be obtained. The remaining value after the infinity is removed is referred to as the regularized value. The regularized value does not equal the series, but is equivalent to it. Moreover, it is often unknown when a power series expansion is convergent or divergent. Therefore, in both instances one should not use the equals sign in a mathematical statement. Instead, the less stringent equivalence symbol should replace the equals sign with the resulting expression becoming an equivalence statement or equivalence for short. Only for those values where the series is convergent can one replace the equivalence symbol by an equals sign.

Let us begin this description of regularization with perhaps one of the most elementary series that is known to be divergent, the geometric series. Although this series is one of the simplest of all divergent series since its coefficients are all equal to unity, it serves as the basis for regularizing more complicated asymptotic series. That is, if a more complicated asymptotic series can be expressed in terms of the geometric series, then it will also be regularized once the regularized value of the geometric series is introduced into it. We shall express the geometric series as

$$
{}_1\mathcal{F}_0(1; z) = \sum_{k=0}^{\infty} z^k , \tag{A.1}
$$

where

$$
{}_p\mathcal{F}_q(\alpha_1, \ldots, \alpha_p; \beta_1, \ldots \beta_q; z) = \sum_{k=0}^{\infty} \frac{\Gamma(k + \alpha_1) \cdots \Gamma(k + \alpha_p)}{\Gamma(\alpha_1) \cdots \Gamma(\alpha_p)}
$$

The Partition Method for a Power Series Expansion.
DOI: http://dx.doi.org/10.1016/B978-0-12-804466-7.00016-4

$$\times \frac{\Gamma(\beta_1)\cdots\Gamma(\beta_q)}{\Gamma(k+\beta_1)\cdots\Gamma(k+\beta_q)} \frac{z^k}{k!} . \qquad (A.2)$$

The reader should note that the standard notation for a hypergeometric function, viz. $_pF_q$, has been modified in the above definition by using the slightly different notation of $_p\mathcal{F}_q$. The standard notation for relating a hypergeometric function to the series on the rhs is only valid when it is absolutely convergent, i.e., when $q \geq p$ or when $|z| < 1$ for $q = p - 1$. Since we are interested in the behaviour of divergent series, we have introduced the above notation as a shorthand means of expressing all series of the form given by (A.2). That is, the lhs of (A.2) is merely an alternative means of expressing any series of the form on the rhs. It does not mean that the series is convergent as in the case of the standard notation for hypergeometric functions. On the other hand, the reader should realize that the function $_pF_q(\alpha_1, \ldots, \alpha_p; \beta_1, \ldots \beta_q; z)$ still exists and represents the regularized value of the series on the rhs of (A.2) when the series is divergent.

On p. 19 of Ref. [17], it is stated that the geometric series is absolutely convergent inside the unit disk of $|z| < 1$ and divergent for all other values of z. However, we shall now show that the series is actually conditionally convergent for $\Re z < 1$ and divergent for $\Re z > 1$ when $|z| > 1$. For the latter case, the series will need to be regularized to yield a finite value, but in doing so, we still have to maintain the connection to the limit when the series is absolutely convergent. This is what is meant by removing the infinity so as to make a divergent series summable in the definition of regularization. To observe this more clearly, we express the geometric series as

$$\sum_{k=0}^{\infty} z^k = \sum_{k=0}^{\infty} \Gamma(k+1) \frac{z^k}{k!} = \lim_{p\to\infty} \sum_{k=0}^{\infty} \frac{z^k}{k!} \int_0^p e^{-t} t^k \, dt . \qquad (A.3)$$

As the integral in (A.3) is finite, technically, the order of the summation and integration can be interchanged. In actual fact, an impropriety occurs when this is done, as discussed in more detail shortly. The method consisting of: (1) introducing the gamma function, (2) replacing it by its integral representation in the numerator and (3) interchanging the order of the summation and integration, is known as Borel summation. Therefore, Borel summation of the geometric series yields

$$\sum_{k=0}^{\infty} z^k = \lim_{p\to\infty} \int_0^p e^{-t} \sum_{k=0}^{\infty} \frac{(zt)^k}{k!} \, dt = \lim_{p\to\infty} \int_0^p e^{-t(1-z)} \, dt$$

$$= \lim_{p\to\infty} \left[\frac{e^{-p(1-z)}}{z-1} + \frac{1}{1-z} \right] . \qquad (A.4)$$

For $\Re z < 1$, the first term of the last member of the above result vanishes and thus, the series yields a finite value of $1/(1-z)$. In addition, we see that the same value is obtained for the series when $\Re z < 1$ as for when z lies in the disk of absolute convergence given by $|z| < 1$. According to p. 18 of Ref. [17], this means the series is conditionally convergent for $\Re z < 1$ and $|z| > 1$. For $\Re z > 1$, however, the first term in the last member of (A.4) is infinite. Since we have defined regularization as the process of removing the infinity that arises from employing an improper mathematical method, we remove or neglect the first term in the last member of (A.4). Thus, we are left with a finite part that equals $1/(1-z)$. This represents the regularized value. Hence for all complex values of z except $\Re z = 1$, we arrive at

$$\sum_{k=0}^{\infty} z^k \begin{cases} \equiv \frac{1}{1-z}, & \Re z > 1, \\ = \frac{1}{1-z}, & \Re z < 1. \end{cases} \tag{A.5}$$

Often, it is not known for which values of z an asymptotic series is convergent and for which it is divergent. In these cases we replace the equals sign by the less stringent equivalence symbol on the understanding that we may be dealing with a series that is absolutely or conditionally convergent for some values of z. As a result, we adopt the shorthand notation of

$$\sum_{k=N}^{\infty} z^k = z^N \sum_{k=0}^{\infty} z^k \equiv \frac{z^N}{1-z}. \tag{A.6}$$

We refer to such results as equivalence statements or simply equivalences since they cannot be regarded strictly as equations. In fact, it is simply incorrect to refer to the above as an equation because we have seen for $\Re z > 1$ that the lhs equals infinity. Furthermore, the above notation is only applicable when the form for the regularized value of the divergent series is identical to the form of the limiting value of the convergent series. This is not always the case as described in Chapter 4 of Ref. [3]. In cases where they are different, we need to specify the two different values in a similar form to (A.5).

At the barrier of $\Re z = 1$, the situation appears to be unclear. For $z = 1$ the last member of (A.4) vanishes, which is consistent with removing the infinity due to $1/(1-z)$. For other values of $\Re z = 1$, the last member of (A.4) is clearly undefined. This is to be expected as this line forms the border between the domains of convergence and divergence for the series. Because the finite value remains the same to the right and to the left of the barrier at $\Re z = 1$ and in keeping with the fact that regularization is effectively the removal of the first term on the rhs of (A.4), we take $1/(1-z)$ to be the finite or regularized value when $\Re z = 1$. Hence (A.5) becomes

$$\sum_{k=0}^{\infty} z^k \begin{cases} \equiv \frac{1}{1-z}, & \Re z \geq 1, \\ = \frac{1}{1-z}, & \Re z < 1. \end{cases} \tag{A.7}$$

It is also interesting to note that at the point where absolute convergence meets divergence, i.e. $z = 1$, there is a singularity, while the series is indeterminate along the line $\Re z = 1$, representing the border between conditional convergence and divergence. In Theorem 8.1 it was found that the generating function of the infinite product $P(z)$ does not possess a region of conditional convergence. Instead it has a unit disk of absolute convergence which is separated from the divergent values outside by a ring of singularities as opposed to one singularity in the case of the geometric series.

As described in Refs. [3,18], the regularized value of a divergent series is analogous to the Hadamard finite part that occurs in the regularization of divergent integrals in the theory of generalized functions [21,22]. For example, consider the divergent integral of

$$I = \int_0^{\infty} e^{ax}\, dx = \lim_{p \to \infty} \int_0^p e^{ax}\, dx = \lim_{p \to \infty} \left[\frac{e^{ap} - 1}{a} \right]. \tag{A.8}$$

For $\Re a > 0$, this integral is divergent, but removing the first term in the last member yields a finite part of $-1/a$, which is the same result one obtains when $\Re a < 0$. To show the connection with regularization of a divergent series, the above integral can be written in terms of an arbitrary positive real parameter, say b, as

$$I = \int_0^{\infty} e^{-bx} e^{(a+b)x}\, dx = \int_0^{\infty} e^{-bx} \sum_{k=0}^{\infty} \frac{(a+b)^k x^k}{k!}\, dx. \tag{A.9}$$

In obtaining this result we have used the asymptotic method of expanding most of the exponential discussed on p. 113 of Ref. [79]. Now the order of the summation and integration is interchanged. Because most of the exponential has been expanded, an impropriety has occurred, which means that evaluating the integral can result in a divergent series depending on the values of a and b. As long as the series is divergent when I is divergent, this does not pose a problem and we can retain the equals sign. Then the integral I becomes

$$I = \sum_{k=0}^{\infty} \frac{(a+b)^k}{k!} \int_0^{\infty} e^{-bx} x^k\, dx = \frac{1}{a+b} \sum_{k=1}^{\infty} \left(1 + \frac{a}{b}\right)^k. \tag{A.10}$$

Now that the geometric series has appeared, we know that the rhs is divergent for $\Re(a/b) > 0$. Furthermore, because b is an arbitrary positive real parameter, the rhs is divergent for $\Re a > 0$, which is when I is divergent. So the above result is consistent. This would not have been the case had we applied the method of expanding most of the exponential to a convergent integral as is usually done. Now if we introduce the regularized value of the geometric series, viz. (A.5), into the above equation, then we obtain a value of $-1/a$ for the finite part of I. That is, by regularizing the series, we find that $I \equiv -1/a$, which is identical to evaluating the divergent integral directly and removing the infinity or the first term in the last member of (A.8). Therefore, evaluating the Hadamard finite part of a divergent integral is equivalent to regularization of a divergent series.

Farassat [80] discusses the issue of whether the appearance of divergent integrals in applications constitutes a breakdown in physics or mathematics. He concludes that divergent integrals arise as a result of incorrect mathematics because an ordinary derivative has been wrongly evaluated inside an improper integral. Consequently, he regards regularization of a divergent integral or taking the finite part as a necessary corrective measure. The same applies to divergent series in asymptotic expansions. As an entity, a divergent series yields infinity, but when regularized, one obtains a finite value or part. When an asymptotic method is used to obtain a power series expansion, there is an impropriety or flaw associated with the method. For example, the iterative method that is used to derive power series expansions from differential equations produces an infinity in the solution as described in Chapter 2 of Ref. [3]. Moreover, deriving asymptotic expansions from integral representations invariably involves integrating over a range that is outside the circle of absolute convergence of the expanded function. This observation was first made in Ref. [18] and is discussed extensively in Ref. [81]. Therefore, while regularization represents a mathematical abstraction for obtaining the finite value of a divergent series, it is necessary in asymptotics for correcting the impropriety in the method used to derive the asymptotic expansion in the first place. After all, the original function or integral from which an asymptotic expansion is derived is finite, even though its asymptotic expansion is not for specific values of the power variable.

We can use the regularization of the geometric series as the basis for deriving the regularized value of other more complicated divergent series as described in Ref. [3]. However, here we shall only consider the regularization of the binomial series in the following lemma since it is necessary for the studying the generating functions in Chapters 8 and 9.

Lemma A.1. Regularization of the binomial series yields

$$
{}_1\mathcal{F}_0(\rho; z) = \sum_{k=0}^{\infty} \frac{\Gamma(k+\rho)}{\Gamma(\rho)\, k!} z^k \begin{cases} = 1/(1-z)^\rho, & \Re z < 1, \\ \equiv 1/(1-z)^\rho, & \Re z \geq 1. \end{cases} \tag{A.11}
$$

Remark. As expected, putting $\rho = 1$ yields the case for the geometric series.

Proof. For $|z| < 1$, the lemma becomes the standard form of the binomial theorem, which is discussed on p. 95 of Ref. [17]. So, we do not need to consider these values of z. We now show that the series is conditionally convergent for $\Re z < 1$ and $|z| > 1$. Without loss of generality, we assume that $\Re \rho < 0$. Then there is a positive integer N, where $\rho^* = \rho + N$ and $\Re \rho^* > 0$. Therefore, the binomial series can be expressed as

$$
{}_1\mathcal{F}_0(\rho; z) = \sum_{k=0}^{N-1} \frac{\Gamma(k+\rho)}{\Gamma(\rho)\, k!} z^k + L_N(z) \sum_{k=0}^{\infty} \frac{\Gamma(k+\rho^*)}{\Gamma(\rho)\, k!} z_N^k. \tag{A.12}
$$

In the above result the operator $L_N(z)$ is defined as

$$
L_N(z) = \int_0^z dz_1 \int_0^{z_1} dz_2 \cdots \int_0^{z_{N-1}} dz_N. \tag{A.13}
$$

By introducing the integral representation for the gamma function into the numerator of the infinite sum and interchanging the order of the summation and integration, we find that

$$
{}_1\mathcal{F}_0(\rho; z) = \sum_{k=0}^{N-1} \frac{\Gamma(k+\rho)}{\Gamma(\rho)\, k!} z^k + \frac{1}{\Gamma(\rho)} L_N(z) \lim_{p \to \infty} \int_0^p e^{-t(1-z_N)} t^{\rho^*-1}\, dt. \tag{A.14}
$$

As for the geometric series, an impropriety has occurred, which means that the series must be regularized in order to yield a finite limit for certain values of z. Introducing No. 3.381(1) from Ref. [14], which is

$$
\int_0^p x^{\nu-1} e^{-\mu x}\, dx = \mu^{-\nu} \left(\Gamma(\nu) - \Gamma(\nu, p\mu) \right), \tag{A.15}
$$

allows us to express (A.14) in terms of the incomplete gamma $\Gamma(\rho^*, p(1-z))$. Therefore, (A.14) becomes

$$
{}_1\mathcal{F}_0(\rho; z) = \sum_{k=0}^{N-1} \frac{\Gamma(k+\rho)}{\Gamma(\rho)\, k!} z^k + \frac{1}{\Gamma(\rho)} L_N(z) \left((1-z_N)^{\rho^*} \Gamma(\rho^*) \right.
$$

$$
\left. - \lim_{p \to \infty} (1-z_N)^{\rho^*} \Gamma(\rho^*, (1-z_N)p) \right), \tag{A.16}
$$

where $\Re \rho^* > 0$. The last term in (A.16) can be evaluated by taking the leading order term of the large $|z|$-asymptotic expansion for the incomplete gamma function, which appears as No. 8.357 in Ref. [14]. Thus, we arrive at

$$_1\mathcal{F}_0(\rho; z) = \sum_{k=0}^{N-1} \frac{\Gamma(k+\rho)}{\Gamma(\rho)\,k!}\, z^k + \frac{1}{\Gamma(\rho)}\, L_N(z)\left((1-z_N)^{\rho^*}\Gamma(\rho^*)\right.$$
$$\left. - (1-z_N)^{2\rho^*-1}\lim_{p\to\infty} p^{\rho^*-1}\, e^{-(1-z_N)p}\right). \tag{A.17}$$

Next we apply the operator $L_N(z)$ to the last term in (A.17). This yields

$$-\frac{1}{\Gamma(\rho)}\, L_N(z)\left((1-z_N)^{2\rho^*-1}\lim_{p\to\infty} p^{\rho^*-1}\, e^{-(1-z_N)p}\right) = -\frac{1}{\Gamma(\rho)}$$
$$\times \lim_{p\to\infty} p^{\rho^*-1} L_{N-1}\int_0^{z_{N-1}} (1-z_N)^{2\rho^*-1} e^{-(1-z_N)p}\, dz_N\,. \tag{A.18}$$

In evaluating the integral over z_N, there will be a term amongst others that will possess the exponential factor of $\exp(-(1-z_{N-1})p)$ due to the upper limit of integration. To observe this more clearly, let $\rho^* = 1/2$, which, in turn, yields a term of $\exp(-(1-z_{N-1})p)/p$ in the limit as $p \to \infty$. Therefore, by carrying out the other integrals in $L_{N-1}(z)$, there will ultimately be one term that consists of powers of p multiplied by the exponential factor of $\exp(-(1-z)p)$. It does not matter if the powers of p are positive since it is the exponential factor that determines whether the series is convergent or divergent. For $\Re z < 1$, this term vanishes and we are left with a finite limit to the binomial series. On the other hand, if $\Re z > 1$, then the exponential factor yields infinity, which has to be removed in order to obtain a finite value for the binomial series. Thus we see that the binomial series is conditionally convergent for $\Re z < 1$ and $|z| > 1$ and divergent for $\Re z > 1$.

We now show that the regularized value of the series $_1\mathcal{F}_0(\rho; z)$ is the same value irrespective of whether $\Re z$ is greater or less than unity and is also equal to the value in the lemma. Initially, it is assumed that $0 < \Re \rho < 1$. Hence the series can be written as

$$\sum_{k=0}^{\infty} \frac{\Gamma(k+\rho)}{\Gamma(\rho)\,k!}\, z^k = \frac{1}{B(\rho, 1-\rho)} \sum_{k=0}^{\infty} \int_0^1 z^k t^{k+\rho-1}(1-t)^{-\rho}\, dt\,, \tag{A.19}$$

where $B(x, y)$ represents the beta function and we have introduced its integral representation into the numerator. Consequently, we see the connection with the geometric series when the order of the summation and integration is interchanged. If we replace the geometric series by its regularized value, then we obtain the regularized value of the binomial series.

This gives

$$\sum_{k=0}^{\infty} \frac{\Gamma(k+\rho)}{\Gamma(\rho)\,k!} z^k \equiv \frac{1}{B(\rho,1-\rho)} \int_0^1 \frac{t^{\rho-1}(1-t)^{-\rho}}{1-zt}\,dt. \tag{A.20}$$

From No. 9.111 of Ref. [14], the integral on the rhs of the above equivalence represents the integral representation for $_2F_1(1,\rho;1;z)$. It may not look as if much has been gained, especially since the Gauss hypergeometric function reduces to $_1F_0(\rho;z)$. However, the major difference is that the Gauss hypergeometric function has been analytically continued beyond its series representation, the latter being only convergent for $|z| < 1$. Moreover, from No. 7.3.1.1 in Ref. [82], we find that $_1F_0(\rho;z) = (1-z)^{-\rho}$ for all values of z.

So far, it has been shown that the regularized value for the binomial series is equal to the value in the lemma for $0 < \Re\rho < 1$. We now consider $\Re\rho > 1$. Then there exists a positive integer N such that $\rho = \rho^* + N$ and $0 < \Re\rho^* < 1$. Consequently, the binomial series can be expressed as

$$\sum_{k=0}^{\infty} \frac{\Gamma(k+\rho)}{\Gamma(\rho)\,k!} z^k = \sum_{k=0}^{\infty} \frac{\Gamma(k+\rho^*+N)}{\Gamma(\rho^*+N)\,k!} z^k = \frac{z^{1-\rho^*}\,\Gamma(\rho^*)}{\Gamma(\rho^*+N)}$$

$$\times \frac{d^N}{dz^N} z^{\rho^*+N-1} \sum_{k=0}^{\infty} \frac{\Gamma(k+\rho^*)}{\Gamma(\rho^*)\,k!} z^k. \tag{A.21}$$

The final series in (A.21) represents the binomial series for ρ^*, whose regularized value has already been obtained in (A.20). Then the regularized value of the above series becomes

$$\sum_{k=0}^{\infty} \frac{\Gamma(k+\rho)}{\Gamma(\rho)\,k!} z^k \equiv \frac{z^{1-\rho^*}\,\Gamma(\rho^*)}{\Gamma(\rho^*+N)} \frac{d^N}{dz^N} \frac{z^{\rho^*+N-1}}{(1-z)^{\rho^*}}. \tag{A.22}$$

To show that the rhs of (A.22) yields the value as given in the lemma, we use induction. Putting $N = 1$ into the rhs of the above result gives

$$\frac{z^{1-\rho^*}\,\Gamma(\rho^*)}{\Gamma(\rho^*+1)} \frac{d}{dz} \frac{z^{\rho^*}}{(1-z)^{\rho^*}} = \frac{z^{1-\rho^*}}{\rho^*} \left(\frac{\rho^* z^{\rho^*-1}}{(1-z)^{\rho^*}} - \frac{z^{\rho^*}(-\rho^*)}{(1-z)^{\rho^*+1}} \right)$$

$$= \frac{1}{(1-z)^{\rho^*+1}} = \frac{1}{(1-z)^{\rho}}. \tag{A.23}$$

Next we assume that the above result holds for $N = n$ and put $N = n+1$. Thus, the rhs of (A.22) becomes

$$\frac{z^{1-\rho^*}\,\Gamma(\rho^*)}{\Gamma(\rho^*+n+1)} \frac{d^{n+1}}{dz^{n+1}} \frac{z^{n+\rho^*}}{(1-z)^{\rho^*}} = \frac{z^{1-\rho^*}\,\Gamma(\rho^*)}{\Gamma(n+\rho^*+1)}$$

$$\times \frac{d^n}{dz^n} \left(\frac{(n+\rho^*)z^{n+\rho^*-1}}{(1-z)^{\rho^*}} - \frac{z^{n+\rho^*}(-\rho^*)}{(1-z)^{\rho^*+1}} \right). \tag{A.24}$$

Since the $N = n$ case has been assumed to be valid, this means that

$$\frac{\Gamma(\rho^*)\, z^{1-\rho^*}}{\Gamma(n+\rho^*)}\, \frac{d^n}{dz^n}\, \frac{z^{n+\rho^*-1}}{(1-z)^{\rho^*}} = \frac{1}{(1-z)^{\rho^*+n}}. \tag{A.25}$$

By introducing (A.25) into (A.24), one obtains

$$\frac{z^{1-\rho^*}\,\Gamma(\rho^*)}{\Gamma(\rho^*+n+1)}\, \frac{d^{n+1}}{dz^{n+1}}\, \frac{z^{n+\rho^*}}{(1-z)^{\rho^*}} = \frac{1}{(1-z)^{n+\rho^*}}$$
$$-\frac{z^{1-\rho^*}\,\Gamma(\rho^*+1)}{\Gamma(n+\rho^*+1)}\, \frac{d^n}{dz^n}\, \frac{z^{n+\rho^*}}{(1-z)^{\rho^*+1}}. \tag{A.26}$$

The last derivative can be viewed as the ρ^*+1 case of the assumption given by (A.25). Consequently, (A.26) can be written as

$$\frac{z^{1-\rho^*}\,\Gamma(\rho^*)}{\Gamma(\rho^*+n+1)}\, \frac{d^{n+1}}{dz^{n+1}}\, \frac{z^{n+\rho^*}}{(1-z)^{\rho^*}} = \frac{1}{(1-z)^{n+\rho^*}} - \frac{z^{1-\rho^*}\,\Gamma(\rho^*+1)}{\Gamma(n+\rho^*+1)}$$
$$\times\, \frac{\Gamma(n+\rho^*+1)\, z^{\rho^*}}{\Gamma(\rho^*+1)(1-z)^{n+\rho^*+1}} = \frac{1}{(1-z)^{n+\rho^*+1}} = \frac{1}{(1-z)^{\rho}}. \tag{A.27}$$

We have already seen that the rhs of (A.22) yields the regularized value as given in the lemma for $N = 1$. From (A.27), we see that it also holds for $N = 2, 3, \ldots$, i.e., for all N. Thus, the regularized value of the binomial series as given by (A.22) reduces to the value given in the lemma for $\Re\rho > 1$.

In order to complete the proof, we need to show that the regularized value in the lemma also applies when $\Re\rho < 0$. In this case we let $\rho = \rho^* - N$, where $0 < \Re\rho^* < 1$ and N is again a positive integer. Consequently, the binomial series can be expressed as

$$\sum_{k=0}^{\infty} \frac{\Gamma(k+\rho)}{\Gamma(\rho)\, k!}\, z^k = \sum_{k=0}^{N-1} \frac{\Gamma(k+\rho)}{\Gamma(\rho)\, k!}\, z^k + \sum_{k=N}^{\infty} \frac{\Gamma(k+\rho^*-N)}{\Gamma(\rho)\, k!}\, z^k. \tag{A.28}$$

By replacing $k-N$ by k in the second sum on the rhs of (A.28) and carrying out a little algebra, we arrive at

$$\sum_{k=0}^{\infty} \frac{\Gamma(k+\rho)}{\Gamma(\rho)\, k!}\, z^k = \sum_{k=0}^{N-1} \frac{\Gamma(k+\rho)}{\Gamma(\rho)\, k!}\, z^k + \frac{z^N\,\Gamma(\rho^*)}{\Gamma(\rho)} \sum_{k=0}^{\infty} \frac{\Gamma(k+\rho^*)}{\Gamma(\rho^*)\, k!}$$
$$\times\, \frac{\Gamma(k+1)}{\Gamma(k+N+1)}\, z^k. \tag{A.29}$$

We now prove by induction that the regularized value of the rhs of (A.29) yields the value given in the lemma. We begin by examining what happens when $N = 1$ in (A.29). Then we obtain

$$
\sum_{k=0}^{\infty} \frac{\Gamma(k+\rho)\, z^k}{\Gamma(\rho)\, k!} = 1 + \frac{z\,\Gamma(\rho^*)}{\Gamma(\rho)} \sum_{k=0}^{\infty} \frac{\Gamma(k+\rho^*)}{\Gamma(\rho^*)\, k!\,(k+1)}\, z^k
$$

$$
= 1 + (\rho^* - 1) \int_0^z \sum_{k=0}^{\infty} \frac{\Gamma(k+\rho^*)}{\Gamma(\rho^*)\, k!}\, t^k \, dt . \tag{A.30}
$$

The final sum represents the binomial series for ρ^*, whose regularized value has been already been determined in (A.20). By introducing the regularized value for the series into (A.30), we find that

$$
\sum_{k=0}^{\infty} \frac{\Gamma(k+\rho)}{\Gamma(\rho)\, k!}\, z^k \equiv 1 + (\rho^* - 1) \int_0^z (1-t)^{-\rho^*}\, dt = \frac{1}{(1-z)^{\rho^*-1}} . \tag{A.31}
$$

We now assume that for $N = n$ the rhs of (A.29) yields the regularized value of $(1-z)^{-\rho}$, while for $N = n+1$, (A.29) becomes

$$
\sum_{k=0}^{\infty} \frac{\Gamma(k+\rho)}{\Gamma(\rho)\, k!}\, z^k = \sum_{k=0}^{n} \frac{\Gamma(k+\rho)}{\Gamma(\rho)\, k!}\, z^k + \frac{z^{n+1}\,\Gamma(\rho^*)}{\Gamma(\rho)} \sum_{k=0}^{\infty} \frac{\Gamma(k+\rho^*)}{\Gamma(\rho^*)\, k!}
$$

$$
\times\, \frac{\Gamma(k+1)}{\Gamma(k+n+2)}\, z^k . \tag{A.32}
$$

Alternatively, the above equation can be expressed as

$$
\sum_{k=0}^{\infty} \frac{\Gamma(k+\rho)}{\Gamma(\rho)\, k!}\, z^k = 1 + \sum_{k=0}^{n-1} \frac{\Gamma(k+1+\rho)}{\Gamma(\rho)\,(k+1)!}\, z^{k+1} + \frac{\Gamma(\rho^*)}{\Gamma(\rho^* - n - 1)}
$$

$$
\times \sum_{k=0}^{\infty} \frac{\Gamma(k+\rho^*)}{\Gamma(\rho^*\, k!)} \frac{\Gamma(k+1)}{\Gamma(k+n+1)} \int_0^z t^{k+n}\, dt . \tag{A.33}
$$

In the truncated sum on the rhs of (A.33), we replace the factor of $z^{k+1}/(k+1)$ by $\int_0^z dt\, t^k$. After a little algebra we obtain

$$
\sum_{k=0}^{\infty} \frac{\Gamma(k+\rho)}{\Gamma(\rho)\, k!}\, z^k = 1 + \frac{1}{\Gamma(\rho^* - n - 1)} \int_0^z \Bigl[\sum_{k=0}^{N-1} \Gamma(k+\rho+1) \frac{t^k}{k!} \Bigr.
$$

$$
\Bigl. + \Gamma(\rho^*) \sum_{k=0}^{\infty} \frac{\Gamma(k+\rho^*)}{\Gamma(\rho^*)\, k!} \frac{\Gamma(k+1)}{\Gamma(k+n+1)}\, t^{k+N} \Bigr] dt . \tag{A.34}
$$

Both the truncated series and the infinite series on the rhs can be combined into one series, which becomes the binomial series for $\rho^* - n$. However, it has been assumed that this series has a regularized value of $(1-z)^{-\rho^*+n}$.

Therefore, we arrive at

$$\sum_{k=0}^{\infty} \frac{\Gamma(k+\rho)}{\Gamma(\rho)\,k!}\, z^k \equiv 1 + \frac{\Gamma(\rho^*-n)}{\Gamma(\rho^*-n-1)} \int_0^z \frac{1}{(1-t)^{\rho^*-n}}\, dt$$

$$= \frac{1}{(1-z)^{\rho^*-n-1}} = \frac{1}{(1-z)^{\rho}}\,. \tag{A.35}$$

We have already shown that the result in the lemma is valid for $N=1$ or $\rho = \rho^* - 1$ by deriving (A.31). The above implies that the result in the lemma is valid for $N=2$ or $\rho^* - 2$, and then for $N=3$ and so on for all positive integer values of N.

To complete the proof, we consider the cases when (1) ρ is an integer and (2) $\Re z = 1$. For $\rho = 1$, the result in the lemma reduces to the geometric series, whose regularized value has already been determined. For $\rho = -N$, where $N \geq 0$, the series on the lhs becomes finite. Then we find that

$$\sum_{k=0}^{\infty} \frac{\Gamma(k+\rho)}{\Gamma(\rho)}\, \frac{z^k}{k!} = \sum_{k=0}^{N} \frac{\Gamma(k-N)}{\Gamma(-N)}\, \frac{z^k}{k!}\,. \tag{A.36}$$

By introducing the reflection formula for the gamma function, viz. (6.55), we find that the series becomes $\sum_{k=0}^{N} \binom{N}{k}(-z)^k$. This, of course, equals $(1-z)^N$. For $\rho = N$, where $N > 1$, the series can be expressed as

$$\sum_{k=0}^{\infty} \frac{\Gamma(k+N)}{\Gamma(N)}\, \frac{z^k}{k!} = \frac{1}{\Gamma(N)} \frac{d^{N-1}}{dz^{N-1}} \sum_{k=0}^{\infty} z^k \equiv \frac{1}{(N-1)!} \frac{d^{N-1}}{dz^{N-1}} \frac{z^{N-1}}{1-z}\,. \tag{A.37}$$

The rhs of the above result can be shown to equal $1/(1-z)^N$ by induction.

For the second case the line $\Re z = 1$ is the border between convergence and divergence as we observed for the geometric series. Because the regularized value of the binomial series has the same value as when the series is convergent for all values of ρ, it is also set equal to $1/(1-z)^{\rho}$ when $\Re z = 1$.

One final remark needs to be made about this proof. The bulk of the proof has been concerned with obtaining the regularized value of the series for $\Re \rho > 1$ and $\Re \rho < 1$ by avoiding the introduction of the integral representation for the beta function in (A.20) because it is divergent for these values of ρ. However, if we had stated that for all values of ρ the integral for the beta function was given by

$$\int_0^{\infty} t^{\rho-1}(1-t)^{-\rho}\, dt \equiv \frac{\Gamma(\rho)\,\Gamma(1-\rho)}{\Gamma(k+1)}\,, \tag{A.38}$$

then it would have simplified the proof of the lemma drastically. That is, the rhs of the above result gives the finite or Hadamard part of a divergent integral [22]. As an aside, Ninham [83] has confirmed this result by applying his general formula for the finite of part of a divergent integral to the integral representation for $B(-1/2, -1/2)$. There, he demonstrates that the finite part for these values of the beta function integral vanishes, which is consistent with the value of $\Gamma(-1/2)\Gamma(-1/2)/\Gamma(-1)$ obtained from the rhs of the above result. In proving the lemma we chose not to extend the integral representation of the beta function to the divergent values of ρ because the reader could have claimed that the divergence in the binomial series had been masked or hidden by the divergence in the integral representation for the beta function. This is actually not the case since the process of regularization occurs in the step immediately after the integral representation has been introduced. So, to avoid any objection in using (A.38), we opted for the much longer and less controversial method of proving the lemma. This completes the proof of the lemma.

APPENDIX B

Computer Programs

The first code presented in this appendix is the program **partmeth** discussed in detail at the beginning of Chapter 5. It employs the BRCP algorithm of Chapter 3 to generate the coefficients arising from the partition method for a power series expansion according to Theorem 4.1 in symbolic form, thereby enabling the values of the coefficients to be evaluated either in integer form or as algebraic expressions in Mathematica.

Program No. 1 partmeth

```
/* This code is concerned with the application of the
   partition method for a power series expansion to
   the pseudo-composite function g(af(z)). Here it is
   assumed that g(z)= h(z)(1+ q_1 z + q_2 z^2+ ...+
   q_k z^k +...), where h(z) can be any function, but
   is often equal to unity or some factor multiplied
   by a non-integer power of z. In addition, f(z) is
   assumed to be a power series expansion of the form
   (p_0 + p_1 y+ p_2 y^2 + ...+ p_k y^k +...) with
   y=z^alpha. The code, however, is valid only for the
   case of p_0=0. The coefficients DS[k,n] and ES[k,n]
   are those presented in Theorem 4.1 and are computed
   in a format suitable for processing in Mathematica. */

#include <stdio.h>
#include <memory.h>
#include <stdlib.h>
#include <math.h>
#include <time.h>

int dim,sum,*part,inv_case;
time_t init_time, end_time;

/* If inv_case equals unity, then the code will evaluate
   the coefficients of the inverse power series, i.e.
   the power series for h(af(z))/g(af(z)). */

void termgen(int p)
{
int freq,i,num_parts=0,l,spacing=0,num_dis_parts=0,
dis_parts=0;
```

The Partition Method for a Power Series Expansion.
DOI: http://dx.doi.org/10.1016/B978-0-12-804466-7.00017-6

```
double  sign , dnum_parts ;

/*  num_parts  is  the  total  number  of  parts  in  the  partition ,
    while  num_dis_parts  is  the  number  of  distinct  parts
    with  greater  than  unity  frequency  and  is  required
    for  the  multinomial  factor .        */

if (p==sum )
     printf (( inv_case  >0)?"ES[%i , n_]:=  ":"DS[%i , n_]:=  ",p);
if ( inv_case ==0   && p!=sum)  printf ("+");
for ( i =1;  i <=dim;  i ++){
             freq=part [ i ];
             if ( freq ){
                     dis_parts ++;
                     if ( inv_case ==0){
                             printf ("p[%i , n]", i );
                             if ( freq >1)  printf ("^(% i)  ", freq );
                             else  printf (" ");
                                     }
                     num_dis_parts  += ( freq >1);
                     num_parts=num_parts+freq ;
                     }
                     }
if ( inv_case >0){
             dnum_parts =(double)  num_parts ;
             sign =pow ( -1.0 , dnum_parts );
             printf (( sign >0.0)?"+":" -");
                     }

if ( inv_case >0){
             printf ("DS[0,0]^(-% i)  ", num_parts +1);
                     }
else {
             printf ("q[%i]  a", num_parts );
             printf (( num_parts >1)?  "^(% i)  "  :  "  ", num_parts );
     }

if ( inv_case >0){
             for ( i =1;i <=dim ; i ++){
                 freq=part [ i ];
                 if ( freq ==1)  printf ("DS[%i , n]  ", i );
                 else  if ( freq >1)
                             printf ("DS[%i , n]^(% i)  ", i , freq );
                             }
                     }
if ( num_parts >1 && dis_parts >1){
             printf ("%i!", num_parts );
```

```
                    if(num_dis_parts){
                            printf((num_dis_parts>1)? "/(" : "/");
                            for(i=1; i<=dim; i++){
                                    freq=part[i];
                                    if(freq>1){
                                            if(spacing++) printf(" ");
                                            printf("%i!",freq);
                                            }
                                    }
                            if (num_dis_parts>1) printf(")");
                                    }
                            printf(" ");
                                    }
printf("\n");
}

void brcp(int p,int q)
{
part[p]++;
termgen(p);
part[p]--;
p -= q;
while(p >= q){
        part[q]++;
        brcp(p--,q);
        part[q++]--;
        }
}

int main(int argc, char *argv[])
{
int i;
double delta_t;

time(&init_time);
if(argc != 2) printf("usage: ./partmeth <#partitions>\n");
else{
        dim=atoi(argv[1]);
        for(sum=1; sum<=dim; sum++){
                part=(int *) malloc((dim+1)*sizeof(int));
                inv_case=0;
                for(i=1; i<=dim; i++) part[i]=0;
                brcp(sum,1);
                free(part);
                printf("\n");
                inv_case=1;
                part=(int *) malloc((dim+1)*sizeof(int));
```

```
                    for (i=1;i<=dim; i++) part[i]=0;
                    brcp (sum,1);
                    free (part);
                    printf("\n");
                                                }
        }
printf("\n");
time(&end_time);
delta_t= difftime (end_time, init_time);
printf("Computation time is %f seconds\n", delta_t);
return (0);
}
```

Once the order of coefficients becomes sufficiently large, viz. for $k \geq 20$, the code needs to be adapted so that only specific values for one of the two types of coefficient are evaluated in symbolic form. This requires: (1) separating the inverse case or the E_k from the D_k, and (2) removing the first for loop in **main** so that only $i = dim$ is computed. The modified code called **mathpm**, which determines only the D_k in symbolic form, is presented below.

Program No. 2 mathpm

```
/* The code mathpm determines the coefficients of the
   power series for the pseudo-composite function
   g(af(z)), where g(z)= h(z) (1+ q_1 z + q_2 z^2+ ...
   +q_k z^k +...) and h(z) can be any function. The
   function f(z) must be expressed as (p_0 + p_1 y+
   p_2 y^2 + ... + p_k y^k+...) where y=z^alpha and
   p_0 =0.   */


#include <stdio.h>
#include <memory.h>
#include <stdlib.h>
#include <time.h>

int dim,*part;
long unsigned int term=1;
time_t init_time, end_time;

void termgen(int p)
{
int f,i,num_parts=0,dis_part_cnt=0,1,num_dis_parts=0;
/* num_parts is the total number of parts in the partition,
   while num_dis_parts is the number of distinct parts.
   The latter is required for the multinomial factor. */
```

```
if (p==dim)  printf ("DS[%i ,n_]:=  ",p);
else {
     printf("+  ");
     if (term %3 == 0)  printf("\n");
     term++;
     }
for (i =1;  i <=dim;  i ++){
        f=part [i ];
        if (f){
                printf("p[%i]",i);
                if (f >1)  printf("^(%i)  ",  f);
                else  printf("  ");
                num_parts += 1 = f;
                num_dis_parts += (f >1);
             }
                }
   printf("q[%i]  a",num_parts);
   printf((num_parts >1)? "^(%i)  " : "  ",num_parts);
if ( num_parts > 1 ){
        printf("%i!",num_parts);
        if (num_dis_parts){
                printf((num_dis_parts >1)? "/(" : "/");
                for (i =1;  i <=dim;  i ++){
                        f=part [i ];
                        if (f >1){
                                if (dis_part_cnt++) printf(" ");
/* dis_part_cnt counts the number of distinct parts with
   greater than unity frequency. It is required to
   insert blank spaces in the denominator of the
   multinomial factor.  */
                                printf("%i!",  f);
                             }
                        }
                if (num_dis_parts >1)  printf (")");
                }
        printf("  ");
                }
}

void brcp (int p, int q)
{
part [p]++;
termgen (p);
part [p]--;
p -= q;
while (p >= q){
```

```
                part [q]++;
                brcp (p--,q);
                part [q++]--;
                    }
    }

int  main ( int  argc ,  char  *argv [])
{
int  i;
double  delta_t;
FILE  *ptr;
char  filename [10]=" times ";

time (& init_time );
if ( argc  != 2)  printf (" usage :  ./ mathpm  <#partitions >\n"  );
else {
    dim=atoi ( argv [1]);
    part =(int  *)  malloc (( dim +1)* sizeof ( int ));
    if ( part==NULL)  printf (" unable  to  allocate  array \n\n");
    else {
            free ( part );
            }
        }
printf ("\n");
time (& end_time );
delta_t=  difftime ( end_time , init_time );
ptr=fopen ( filename ," a ");
fprintf ( ptr ," Time  to  compute  p[%i ,n]  is  %f  seconds \n",
    dim , delta_t );
fclose ( ptr );
return (0);
}
```

Chapter 7 discusses the problem of generating discrete or distinct partitions where each part appears at most only once. Because of their importance in the theory of partitions, the entire code called **dispart.cpp** is presented below.

Program No. 3 dispart

```
#include <stdio.h>
#include <memory.h>
#include <stdlib.h>
#include <time.h>

int tot , *part;
long unsigned int term =1;
```

```
      time_t init_time , end_time ;

void termgen ()
{
      int freq , i ;

      for ( i =0; i <tot ; i ++){
                    freq=part [ i ];
                    if ( freq >1) goto end ;
                    }
      printf ("%ld :   ", term ++);
      for ( i =0; i <tot ; i ++){
                    freq=part [ i ];
                    if ( freq ) printf ("%i (%i )  ", freq , i +1);
                    }
      printf ("\n");
end : ;
}


void brcp ( int p , int q )
{
part [ p −1]++;
termgen () ;
part [ p −1]−−;
p −= q ;
while ( p >= q ){
        part [ q −1]++;
        brcp ( p −−,q ) ;
        part [ q ++ − 1]−−;
           }
}


int main ( int argc , char *argv [] )
{
int i ;
double delta_t ;
FILE *ptr ;
char filename [10]=" times ";

time (& init_time ) ;
if ( argc != 2) printf (" usage : ./ dispart <#partitions >\n");
else {
  tot=atoi ( argv [1]) ;
  part= ( int * ) malloc ( tot *sizeof ( int )) ;
  if ( part == NULL) printf (" unable to allocate array \n\n");
  else {
      for ( i =0; i <tot ; i ++) part [ i ]=0;
```

```
        brcp(tot,1);
        free(part);
        }
    }
printf("\n");
time(&end_time);
delta_t=difftime(end_time, init_time);
ptr=fopen(filename ,"a");
fprintf(ptr ,"Time taken to compute discrete partitions
    summing to %i is %f secs.\n", tot, delta_t);
fclose(ptr);
return(0);
}
```

The code presented below determines the conjugate partition as a partition is generated. This program called **transp** casts the original partition in the form of a Ferrers diagram, but, instead of the diagram being composed of dots, the rows are composed of ones. Hence the conjugate partition is determined by summing the ones in each column of the Ferrers diagram.

Program No. 4 transp

```
/* This program evaluates the partitions and their
conjugates for summation values or tot greater than
or equal to 2.*/

#include <stdio.h>
#include <memory.h>
#include <stdlib.h>

int tot, *part;
long unsigned int term=1;

void termgen()
{
int freq ,i ,j ,k, next_el , index , prev_el ,* part2 ,
        *sum_col , row_cnt=0;
int *ferrers ,** rptr;
/* The Ferrers array and its array of pointers */

printf(" Partition %ld is: ", term++);
ferrers =(int *) malloc(tot*tot*sizeof(int));
rptr =(int **) malloc(tot*sizeof(int *));

/* Get the pointers to the rows of ferrers */
```

```
for (i=0;i<tot;i++) rptr[i]= ferrers+(i*tot);
/* Here tot refers to the number of columns */

for(j=0;j<tot;j++){
    for(i=0;i<tot;i++){
        rptr[j][i]=0;
    }
}
/* Creation of the Ferrers diagram. Rather than
   being composed of dots the Ferrers diagram is
   composed of unit values since these will be
   used to determine the conjugate partition. */

for(j=0;j<tot;j++){
    freq=part[j];
    if(freq){
        for(i=row_cnt;i<row_cnt+freq;i++){
            for(k=0;k<=j;k++){
                rptr[i][k]=1;
            }
        }
    }
    row_cnt=row_cnt+freq;
}
/* Summation of the columns in the Ferrers
   diagram yielding a new array called sum_col. */

sum_col=(int *) malloc(tot*sizeof(int));
/* Initializing the array elements to zero */

for(i=0;i<tot;i++) sum_col[i]=0;
/* Summation of the columns now occurs */

for (j=0;j<tot;j++){
    for(i=0;i<tot;i++) sum_col[j]=sum_col[j]+rptr[i][j];
}

/* The array sum_col is reduced to the conjugate
   of the original partition through another array
   called part2 */

part2=(int *) malloc(tot*sizeof(int));
/* Initialization of the array elements to zero */

for(i=0;i<tot;i++) part2[i]=0;

/* Now the conjugate partition is arranged as
```

```
in  the  order  of  part  */

prev_el=sum_col[0];
/* the  highest  part  is  the  value  of  sum_col[0]  */
index=prev_el −1;
/* the  index  in  the  partition  array  is  one  less  */
part2[index]=1;
for (i=1;i<=tot;i++){
       next_el=sum_col[i];
       if(next_el==0) goto  out;
       if(next_el==prev_el)  part2[index]=part2[index]+1;
       else{
              index=next_el −1;
              part2[index]=1;
              prev_el=next_el;
              }
              }
out:  for(i=0;i<tot;i++){
       freq=part[i];
       if(freq) printf("%i(%i)  ",freq,i+1);
              }
/* The  conjugate  is  actually  the  reverse  order
   of part2 */
printf(" and  its  conjugate  is:  ");
for(i=0;i<tot;i++){
       freq=part2[tot−i −1];
       if(freq) printf("%i(%i)  ",freq,tot−i);
              }
free(sum_col);
free(part2);
free(ferrers);
free(rptr);
printf("\n");
}

void brcp(int p,int q)
{
part[p−1]++;
termgen();
part[p−1]−−;
p −= q;
while(p >= q){
           part[q−1]++;
           brcp(p−−,q);
           part[q++ −1]−−;
              }
}
```

```
int main(int argc, char *argv [])
{
int i;
if(argc != 2) printf("usage:./transp <#partitions >\n");
else {
        tot=atoi(argv [1]);
        part=(int *) malloc(tot*sizeof(int));
        if(part == NULL) printf("unable to allocate
            array\n\n");
        else {
            for(i=0;i<tot;i++) part[i]=0;
            brcp(tot ,1);
            free(part);
            }
    }
printf("\n");
return (0);
}
```

In Chapter 8 the partition method for a power series expansion was applied to the exponential version of the generating function for the partition-number function $p(k)$, which yielded (8.18). Although this result represents a sum over partitions involving the special numbers $q(i)$ known as the discrete partition numbers, it incorporates much redundancy because these numbers are often zero except when i represents a pentagonal number or $(3j^2 \pm j)/2$, where j is any integer integer. Then they are equal to $(-1)^j$. Appearing below is the program **pfn** which generates the partition-number function in symbolic form expressed in terms of the discrete partition numbers, which are given by Q[i]. The actual values of the partition-number function are evaluated by importing the output generated by **pfn** into Mathematica [32] and applying the latters integer arithmetic routines.

Program No. 5 pfn

```
#include <stdio.h>
#include <memory.h>
#include <stdlib.h>
#include <time.h>
#include <math.h>

int dim ,* part , limit , freq , first_term =0;
long unsigned int term =1;
time_t init_time , end_time ;
```

```
void termgen(int p)
{
int f,i,num_parts=0,j,jval,dis_part_cnt=0,l,
    num_dis_parts=0;
/* num_parts is the total number of parts in the partition
   while num_dis_parts is the number of distinct parts.
   The latter is required for the multinomial factor. */
/* jval=0; */

if (p==dim){
        printf("p[%i]:= ",p);
        for (i=1;i<=dim;i++){
                    jval=0;
                    freq=part[i];
                    if (freq >0){
                            for (j=1;j<=limit;j++){
                                    if(i == (3*j-1)*j/2) jval=j;
                                    if(i == (3*j+1)*j/2) jval=j;
                                                }
                            if ((jval==0) && (freq >0)) goto end;
                                first_term++;
                                }
                        }
                }
        }
else {
        for (i=1;i<=dim;i++){
                    jval=0;
                    freq=part[i];
                    if (freq >0){
                            for (j=1;j<=limit;j++){
                                    if(i == (3*j-1)*j/2) jval=j;
                                    if(i == (3*j+1)*j/2) jval=j;
                                                }
                            if ((jval==0) && (freq >0)) goto end;
                                }
                        }
        if (first_term !=0) printf("+ ");
        if (first_term ==0) first_term++;

        if (term %3 == 0) printf("\n");
        term++;
        }


for (i=1; i<=dim; i++){
        f=part[i];
```

```
           if(f ) {
/*                     printf("Q[%i]",i);     */
                for  (j=1;j<=limit;j++){
                         if  (i  ==  (3*j−1)*j/2)  jval=j;
                         if  (i  ==  (3*j+1)*j/2)  jval=j;
                                   }
                printf("((−1)^(%i))",jval);
                if(f>1) printf("^(%i)  ", f);
                else  printf("  ");
                num_parts += 1 = f;
                num_dis_parts += (f>1);
                   }
                     }
printf("(−1)");
printf((num_parts>1)? "^(%i)  " : "  ",num_parts);
if( num_parts > 1 ){
        printf("%i!",num_parts);
        if(num_dis_parts){
             printf((num_dis_parts>1)? "/(" : "/");
             for(i=1; i<=dim; i++){
                  f=part[i];
                  if(f>1){
                       if(dis_part_cnt++) printf(" ");
/* dis_part_cnt counts the number of distinct parts
with greater than unity frequency. It is required
for inserting blank spaces in the denominator of
the multinomial factor */
                       printf("%i!", f);
                       }
                     }
             if (num_dis_parts>1) printf(")");
                  }
        printf("  ");
                 }
end:  ;
}
void brcp(int p,int q)
{
part[p]++;
termgen(p);
part[p]−−;
p −= q;
while(p >= q){
        part[q]++;
        brcp(p−−,q);
        part[q++]−−;
        }
```

```
}

int main ( int argc , char *argv [] )
{
int i;
double delta_t;

time (& init_time );
if (argc != 2) printf (" usage :  ./ pfn <#partitions >\n" );
else {
          dim=atoi (argv [1]);
          limit= floor (1+ sqrt ((1+ 24 * dim ))/6);

          part =(int *) malloc (( dim +1)* sizeof (int ));
          if ( part ==NULL)
              printf (" unable to allocate array \n\n");
          else {
/*                for (sum=1; sum <=dim; sum++){       */
                      for ( i =1; i<=dim; i++) part [i] = 0;
                      brcp (dim ,1);
/*                        printf ("\n");
                      }       */
                  free (part );
                  }

              }
printf ("\n");
time (& end_time );
delta_t= difftime (end_time , init_time );
printf ("Time taken to compute the coefficient is
    %f seconds \n", delta_t );
return (0);
}
```

In Chapter 9 the coefficients of the generating function for the prod-
uct $Q(z, \omega) = \prod_{k=1}^{\infty} 1/(1 + \omega z^k)$ were referred to as the discrete partition
polynomials $q(k, \omega)$, which are given by (8.48). Soon after it was found
that these polynomials of degree i in ω can be expressed in terms of the
partition operator acting with the parts i equal to special polynomials $\gamma_i(\omega)$.
The latter were called divisor polynomials since their coefficients are divisors or factors of i. The relationship between both types of polynomials is
given by (9.41), while (9.42) shows how the divisor polynomials are related
to the partition function polynomials $p(k, \omega)$. In this instance the partition operator does not act on the phase dependence due to the number
of parts N_k or length of the partitions. Appearing below is the program
called **dispfnpoly**. This program prints out both the discrete partition and

partition function polynomials at a specified order in terms of the divisor polynomials in symbolic form so that they can be imported into Mathematica [32].

Program No. 6 dispfnpoly

```c
#include <stdio.h>
#include <memory.h>
#include <stdlib.h>
#include <time.h>

int dim,*part,polytype;
long unsigned int term=1;
time_t init_time, end_time;

void termgen(int p)
{
int f,i,num_parts=0,dis_part_cnt=0,1,num_dis_parts=0;
/* num_parts is the total number of parts in the
   partition while num_dis_parts is the number of
   distinct parts. The latter is required for the
   multinomial factor */

if(p==dim) printf((polytype==1)?"Q[%i,-w_]:=  ":
   "P[%i,w_]:=  ",p);

for(i=1; i<=dim; i++){
        f=part[i];
        if(f){
                num_parts += 1 = f;
                num_dis_parts += (f>1);
                }

        }
    if(term !=1) printf(" + ");
    if(term % 3 ==0) printf("\n");
    term++;
    for(i=1; i<=dim; i++){
        f=part[i];
        if(f){
                printf("DP[%i,w]",i);
                if(f>1) printf("^(%i) ",f);
                else printf(" ");
                }

        }
    printf((polytype==1)?"(-1)":"");
    if (num_parts>1) printf((polytype==1)?"^(%i)":"",
            num_parts);
```

```
          if (num_parts > 1  ){
               if (num_dis_parts){
                    printf ((num_dis_parts >1)?  "/("  :  "/");
                    for (i =1;  i <=dim;  i ++){
                         f=part [i ];
                         if (f >1){
                              if (dis_part_cnt ++)  printf (" ");
/*  dis_part_cnt  counts  the  number  of  distinct  parts  with
    greater  than  unity  frequency.  It  is  required  for
    inserting  blank  spaces  in  the  denominator  of  the
    multinomial  factor */
                              printf ("%i !" , f );
                         }
                    }
                    if (num_dis_parts >1)  printf (")");
               }
          }
}

void brcp (int p , int q)
{
part [p]++;
termgen (p );
part [p]−−;
p −= q;
while (p >= q){
          part [q]++;
          brcp (p−−,q );
          part [q++]−−;
          }
}

int main (int argc ,  char *argv [])
{
int i;
double delta_t ;

time (&init_time );
if (argc != 2)  printf (" usage :  ./ dispfnpoly
          <#partitions >\n");
else {
          dim=atoi (argv [1]);
          part =(int *)  malloc ((dim +1)* sizeof (int ));
          if (part ==NULL)  printf (" unable  to  allocate
               array \n\n");
          else {
                    for (polytype =1;  polytype <=2;polytype ++){
```

```
                              for (i=1; i<=dim; i++) part[i] = 0;
                              brcp (dim, 1);
                              if (polytype==1) printf("\n\n");
                              term=1;
                                                                    }

                     free (part);
               }
         }
printf("\n");
time(&end_time);
delta_t= difftime (end_time, init_time);
printf("Time taken to compute the coefficients is
     %f seconds\n", delta_t);
return (0);
}
```

REFERENCES

1. V. Kowalenko, N.E. Frankel, Asymptotics of the Bose Kummer function, J. Math. Phys. 35 (1994) 6179–7009.
2. V. Kowalenko, The non-relativistic charged Bose gas in a magnetic field II. quantum properties, Ann. Phys. (NY) 274 (1999) 165–250.
3. V. Kowalenko, Properties and applications of the reciprocal logarithm numbers, Acta Appl. Math. 109 (2009) 413–437, http://dx.doi.org/10.1007/s10440-008-9325-0.
4. V. Kowalenko, Generalizing the reciprocal logarithm numbers by adapting the partition method for a power series expansion, Acta Appl. Math. 106 (2009) 369–420, http://dx.doi.org/10.1007/s104440-008-9304-5.
5. V. Kowalenko, Applications of the cosecant and related numbers, Acta Appl. Math. 114 (2011) 15–134, http://dx.doi.org/10.1007/s10440-011-9604-z.
6. V. Kowalenko, On a finite sum involving inverse powers of cosines, Acta Appl. Math. 115 (2011) 139–151, http://dx.doi.org/10.1007/s10440-011-9609.
7. C.M. da Fonseca, M.L. Glasser, V. Kowalenko, An integral approach to the Gardner-Fisher and untwisted Dowker sums, submitted for publication; see also arXiv:1603.03700 [math.NT].
8. F. Bornemann, E.W. Weisstein, Power Series, From MathWorld—A Wolfram Web Resource, http://mathworld.wolfram.com/PowerSeries.html, Jun. 19, 2016.
9. E.W. Weisstein, Generating Function, From MathWorld—A Wolfram Web Resource, http://mathworld.wolfram.com/GeneratingFunction.html, Jun. 19, 2016.
10. Wikipedia, the free encyclopedia, Hardy-Littlewood circle method, http://en.wikipedia.org/wiki/HardyLittlewood_circle_method, March 15, 2015.
11. J.K.S. McKay, Partitions in natural order, Commun. ACM 13 (1970) 52.
12. G.E. Andrews, The Theory of Partitions, Cambridge University Press, Cambridge, 2003.
13. D.E. Knuth, The Art of Computer Programming, vol. 4, Fascicle 3: Generating All Combinations and Partitions, Addison-Wesley, Upper Saddle River, NJ, 2005; see also: The Art of Computer Programming, vol. 4A: Combinatorial Algorithms, Part I, Addison-Wesley, NY, 2011.
14. I.S. Gradshteyn, I.M. Ryzhik, Table of Integrals, Series and Products, fifth ed., Academic Press, London, 1994, edited by A. Jeffrey.
15. Z. Cèsaro, Elementary Class Book of Algebraic Analysis and the Calculation of Infinite Limits, first ed., ONTI, Moscow and Leningrad, 1936.
16. Wikipedia, the free encyclopedia, Sine, https://en.wikipedia.org/wiki/Sine, July 23, 2015.
17. E.T. Whittaker, G.N. Watson, A Course in Modern Analysis, Cambridge University Press, Cambridge, 1973.
18. V. Kowalenko, Towards a theory of divergent series and its importance to asymptotics, in: Recent Research Developments in Physics, vol. 2, Transworld Research Network, Trivandrum, India, 2001, pp. 17–68.
19. V. Kowalenko, The Stokes Phenomenon, Borel Summation and Mellin-Barnes Regularisation, Bentham ebooks, http://www.bentham.org, 2009.
20. Wikipedia, the free encyclopedia, Hadamard regularization, https://en.wikipedia.org/wiki/Hadamard_regularization, Jan. 11, 2015.

The Partition Method for a Power Series Expansion.
DOI: http://dx.doi.org/10.1016/B978-0-12-804466-7.00018-8

21. M.J. Lighthill, Introduction to Fourier Analysis and Generalised Functions, Cambridge University Press, Cambridge, 1975, Ch. 3.
22. I.M. Gel'fand, G.E. Shilov, Generalized Functions, vol. 1: Properties and Operations, Academic Press, New York, 1964.
23. M. Abramowitz, I.A. Stegun, Handbook of Mathematical Functions, Dover, New York, 1964.
24. Wikipedia, the free encyclopedia, Composition (combinatorics), http://en.wikipedia.org/wiki/Composition_(combinatorics), Oct. 13, 2015.
25. A. Apelblat, Volterra Functions, Nova Science Publishers, New York, 2008.
26. Ia.V. Blagouchine, Two series expansions for the logarithm of the gamma function involving Stirling numbers and containing only rational coefficients for certain arguments related to π^{-1}, J. Math. Anal. Appl. 442 (2016) 404–434.
27. E.W. Weisstein, Logarithmic Number, From MathWorld—A Wolfram Web Resource, http://mathworld.wolfram.com/LogarithmicNumber.html, Aug. 31, 2015.
28. N.J.A. Sloane, The On-Line Encyclopedia of Integer Sequences, http://www.research.att.com/~njas/seequences.
29. J. Spanier, K.B. Oldham, An Atlas of Functions, Hemisphere Publishing, New York, 1987.
30. A. Apelblat, N. Kravitsky, Improper integrals associated with the inverse Laplace transforms of $[\ln(s^2 + a^2)]^{-n}, n = 1, 2$, Z. Angew. Math. Mech. 78 (1998) 565–571.
31. E.T. Copson, An Introduction to the Theory of Functions of a Complex Variable, Clarendon Press, Oxford, 1976.
32. S. Wolfram, Mathematica—A System for Doing Mathematics by Computer, Addison-Wesley, Reading, 1992.
33. E.W. Weisstein, Bernoulli Polynomial, From MathWorld—A Wolfram Web Resource, http://mathworld.wolfram.com/BernoulliPolynomial.html, Sept. 5, 2015.
34. E.W. Weisstein, Euler Polynomial, From MathWorld—A Wolfram Web Resource, http://mathworld.wolfram.com/EulerPolynomial.html, Sept. 11, 2015.
35. E.W. Weisstein, Bell Polynomial, From MathWorld—A Wolfram Web Resource, http://mathworld.wolfram.com/BellPolynomial.html, Sept. 17, 2015.
36. J. Riordan, Combinatorial Identities, Wiley & Sons, New York, 1968, Chapter 5.
37. C.A. Charalambides, Enumerative Combinatorics, Chapman & Hall/CRC, Boca Raton, 2002.
38. J. Quaintance, H.W. Gould, Combinatorial Identities for Stirling Numbers—The Unpublished Notes of H.W. Gould, World Scientific Publishing Co., Singapore, 2016.
39. E.W. Weisstein, Stirling Number of the Second Kind, From MathWorld—A Wolfram Web Resource, http://mathworld.wolfram.com/StirlingNumberoftheSecondKind.html, Feb. 15, 2016.
40. R.P. Stanley, Enumerative Combinatorics, vol. 1, Cambridge University Press, NY, 2007.
41. I.G. Macdonald, Symmetric Functions and Hall Polynomials, Clarendon Press, Oxford, 1995.
42. A.P. Prudnikov, Yu.A. Brychkov, O.I. Marichev, Integrals and Series, vol. 1: Elementary Functions, Gordon and Breach, NY, 1986.
43. V. Kowalenko, Euler and divergent series, Eur. J. Pure Appl. Math. 4 (2011) 370–423.
44. V. Kowalenko, Euler and Divergent Mathematics, BestThinking Science, http://www.bestthinking.com/article/permalink/1255?tab=article&title=euler-and-divergent-mathematics, 2011.

45. J. Sondow, E.W. Weisstein, Harmonic Number, From MathWorld—A Wolfram Web Resource, http://mathworld.wolfram.com/HarmonicNumber.html, August 10, 2016.

46. Ia.V. Blagouchine, Two series expansions for the logarithm of the gamma function involving Stirling numbers and containing only rational coefficients for certain arguments related to π^{-1}, arXiv:1408.3902v9 [math.NT], May 30, 2016.

47. A. Zoghbi, I. Stojmenovic, Fast algorithms for generating integer partitions, Int. J. Comput. Math. 70 (1998) 319–332.

48. J. Kelleher, Encoding Partitions as Ascending Compositions, PhD Thesis, University College Cork, Ireland, 2005.

49. J. Kelleher, B. O'Sullivan, Generating all partitions: a comparison of two encodings, arXiv:0909.2331v1 [cs.Ds], 12 Sep. 2009.

50. K. Yamanaka, S. Kawano, Y. Kikuchi, S. Nakano, Constant time generation of integer partitions, IEICE Trans. Fundam. E90-A (2007) 888–895, Gordon and Breach, NY, 1986.

51. S. Skiena, Implementing Discrete Mathematics—Combinatorics and Graph Theory with Mathematica, Addison-Wesley, Redwood City, CA, 1990.

52. J. Kelleher, private communication, e-mail dated May 23, 2011.

53. L. Comtet, Advanced Combinatorics, D. Reidel, Dordrecht, 1974.

54. T.I. Fenner, G. Loizou, A binary tree representation and related algorithms for generating integer partitions, The Comput. J. 23 (1980) 332–337.

55. Wikipedia, the free encyclopedia, Partition (number theory), http://en.wikipedia.org/wiki/Partition_(number_theory), Dec. 22, 2015.

56. Wikipedia, the free encyclopedia, Rank of a partition, https://en.wikipedia.org/wiki/Rank_of_a_partition, Dec. 22, 2015.

57. Wikipedia, the free encyclopedia, Durfee square, https://en.wikipedia.org/wiki/Durfee_square, June 5, 2014.

58. P. Flajolet, R. Sedgewick, Analytic Combinatorics, Cambridge University Press, Cambridge, 2009, p. 188.

59. Faà di Bruno, Sullo sviluppo delle funzioni, Ann. Sci. Mat. Fis. Tortolini 6 (1855) 479–480.

60. Faà di Bruno, Note sur un nouvelle formule de calcul différentiel, Q. J. Math. 1 (1857) 359–360.

61. J. Riordan, Introduction to Combinatorial Analysis, Wiley & Son, NY, 1967, Ch. 4, http://mathworld.wolfram.com/PentagonalNumber.html, June 19, 2016.

62. E.W. Weisstein, Perfect Partition, From MathWorld—A Wolfram Web Resource, http://mathworld.wolfram.com/PerfectPartition.html, June 19, 2016.

63. G.H. Hardy, S. Ramanujan, Asymptotic formulae in combinatory analysis, Proc. London Math. Soc. (2) 17 (1918) 75–115.

64. H. Rademacher, On the partition function $p(n)$, Proc. London Math. Soc. 43 (1937) 241–254.

65. Wikipedia, the free encyclopedia, q-Pochhammer symbol, http://en.wikipedia.org/wiki/Q_Pochhammer_symbol, Nov. 15, 2015.

66. E.W. Weisstein, Pentagonal Number, MathWorld—A Wolfram Web Resource, http://mathworld.wolfram.com/PentagonalNumber.html, August 18, 2016.

67. N.J.A. Sloane, The On-Line Encyclopedia of Integer Sequences, oeis.org/A000326, May 10, 2015, A000326 Pentagonal Numbers.

68. E.W. Weisstein, Figurate Number, From MathWorld—A Wolfram Web Resource, http://mathworld.wolfram.com/FigurateNumber.html, August 18, 2016.

69. Wikipedia, the free encyclopedia, Figurate number, http://en.wikipedia.org/wiki/Figurate_Number, March 9, 2016.

70. E.W. Weisstein, Partition Function P, From MathWorld—A Wolfram Web Resource, http://mathworld.wolfram.com/PartitionFunctionP.html, June 19, 2016.

71. E.W. Weisstein, Divisor, From MathWorld—A Wolfram Web Resource, http://mathworld.wolfram.com/Divisor.html, June 19, 2016.

72. Wikipedia, the free encyclopedia, Triangular number, http://en.wikipedia.org/wiki/Triangular_Number, July 19, 2016.

73. N.J.A. Sloane, The On-Line Encyclopedia of Integer Sequences, oeis.org/A000217, May 5, 2015, Triangular Numbers.

74. R.J. Baxter, Exactly Solved Models in Statistical Mechanics, Dover, NY, 2007.

75. M.N. Barber, R.J. Baxter, On the spontaneous order of the eight-vertex model, J. Phys. C: Solid State Phys. 6 (1973) 2913–2921.

76. M. Somos, The On-Line Encyclopedia of Integer Sequences, oeis.org/A108494, June 6, 2005, Expansion of f(-q)/f(q) in powers of q where f() is a Ramanujan theta function.

77. E.W. Weisstein, Ramanujan's Theta Functions, MathWorld—A Wolfram Web Resource, http://mathworld.wolfram.com/RamanujanThetaFunctions.html, August 18, 2016.

78. C.J. Thompson, Mathematical Statistical Mechanics, Princeton University Press, Princeton, 1979.

79. R.B. Dingle, Asymptotic Expansions: Their Derivation and Interpretation, Academic Press, London, 1973.

80. F. Farassat, Introduction to Generalized Functions with Applications in Aerodynamics and Aeroacoustics, NASA Technical Paper 3428, Langley Research Center, Hampton, VI, 1994.

81. J.P. Boyd, Hyperasymptotics and the linear boundary layer problem: why asymptotic series diverge, SIAM Rev. 47 (2005) 553–575.

82. A.P. Prudnikov, Yu.A. Brychkov, O.I. Marichev, Integrals and Series, vol. 3: More Special Functions, Gordon and Breach, New York, 1990.

83. B.W. Ninham, Generalised functions and divergent integrals, Numer. Math. 8 (1966) 444–457.

INDEX

The Partition Method for a Power Series Expansion.
DOI: http://dx.doi.org/10.1016/B978-0-12-804466-7.00019-X

Printed in the United States
By Bookmasters